CROSSINGS

"Beyond the staggering data and the constructive ideas, *Crossings* is an important book because it is timely: Road ecology is bleeding into the public consciousness at a moment when we can still act on its lessons."

—Jonathan C. Slaght, *Atlantic*

"A badly needed corrective. . . . [D]eserves to make the reading lists of policymakers around the world."

—Marina Bolotnikova, *Vox*

"Deeply researched and compelling. . . . [O]ffers readers a look behind the scenes of a rich but underappreciated field of study that has the potential to affect our everyday lives."

—Sarah Boon, *Science*

"Engrossing. . . . [Ben] Goldfarb invites us to contemplate a future of roads that could be much brighter, if we would just adopt an ethic, he says, in which roads embrace the land instead of conquer it."

—*Smithsonian*

"Delves into the burgeoning field of road ecology and introduces the impassioned, sometimes eccentric scientists who invite us to perceive our roads as animals do to better understand the ecological impacts."

—Amanda Heidt, *Science News*

"Goldfarb's absorbing, highly intelligent book gently shakes us awake from our ethical torpor and helps us confront the conservation problem we perpetrate each time we get behind the wheel, accept a package, or use public transportation."

—M. R. O'Connor, *Undark*

"An elegant—at times startling—account of how our built environment has become an environmental crisis. . . . A manifesto against unnecessary death."

—Jimmy Tobias, *Nation*

"The book is teeming with horrifying statistics. . . . While that may sound bleak, *Crossings* is at times surprisingly funny."

—Jackie Flynn Mogensen, *Mother Jones*

"Chronicles the enormous ecological damage caused by roadbuilding. . . . Goldfarb guides the reader through an array of often heartbreaking stories."

—David Zipper, *Bloomberg*

"[*Crossings* has] so many cool stories . . . frogs and turtles being ushered across roads by volunteer hands, a wildlife crossing for cougars in California, citizen roadkill reporting networks. In many ways, it's a book about the people trying to correct our mistakes."

—Colleen Stinchcombe, *Seattle Times*

"Goldfarb traveled across the country, and the globe, to learn more about how roads have shaped not just our communities but the natural world around us. . . . [R]oads may be nearly invisible to the modern human, just another necessary part of everyday infrastructure. But to the other species on this planet, roads have fundamentally changed their existence."

—Emily Baron Cadloff, *Modern Farmer*

"Through expert interviews, compelling research and analysis, and dogged experiential reporting, Goldfarb brings to life some of the core impacts our 40 million miles of roads have had, and are having, on the natural world."

—Brett Berk, *Car and Driver*

"Eye-opening. . . . This is a rare, beautifully written book, which tells us hard truths about roads, cars and life on Earth, but still manages to make us feel positive about the road ahead."

—Vijaysree Venkatraman, *New Scientist*

"*Crossings* is science writing at its best. . . . [A] hopeful reminder of our responsibilities in the Anthropocene." —Miranda Weiss, *American Scholar*

"Written elegantly and convincingly, *Crossings* acknowledges that most of us can't make do without automobiles but urges individual responsibility . . . as well as public works initiatives of global proportions."

—David Luhrssen, *Shepherd Express*

"A fresh and startling history. . . . An astute, funny, and imaginative writer, Goldfarb pairs horror with hope." —*Booklist*, starred review

"Illuminating, witty. . . . [*Crossings* is] an astonishingly deep pool of wonders."

—*Kirkus Reviews*, starred review

"Captivating. . . . This one's a winner." —*Publishers Weekly*, starred review

"Roads aren't going away anytime soon, but *Crossings* will spark conversation around the future of motorized vehicles and transportation."

—*Bookpage*, starred review

"Goldfarb writes with such grace, flair, and wit, and he reveals just how thoroughly roads have reshaped the animal world. This is one of the very best science books that I read last year."

—Ed Yong, best-selling author of *An Immense World*

"Ben Goldfarb is the kind of gonzo environmental journalist Hunter S. Thompson would have loved. *Crossings*, his meditation on the ecological devastation roads and highways inflict—and on the very clever responses from humans and other creatures that road life demands—is an absolute shining star of a book. Modernity and the mobility all we Earth animals require is never going to look the same again."

—Dan Flores, best-selling author of *Coyote America* and *Wild New World*

"A brilliantly panoptic look at our planet's sprawling network of roads: what's wrong with them, how they got that way, and how they could be set right. Precise in detail but vast in scale, Goldfarb's storytelling carries echoes of Michael Pollan and John McPhee, but with a wry humor that is uniquely his own." —Robert Moor, best-selling author of *On Trails*

"Ben Goldfarb approaches our fellow animals with delighted curiosity and rare perception. A deeply researched, wonderfully vivid, and genuinely hopeful book." —Michelle Nijhuis, author of *Beloved Beasts*

"Like some David Attenborough of the asphalt, Ben Goldfarb has written a fascinating guide to understanding the wilder side of roads, both symbols of freedom and harbingers of unnatural selection."

—Tom Vanderbilt, best-selling author of *Traffic*

"A truly important and landmark book on a subject whose full impacts continue to be disregarded or underestimated in considering conservation efforts. *Crossings* is a moving, compassionate, and indispensable guide to navigating the issue of wildlife survival—and our own."

—Jeff VanderMeer, best-selling author of the *Southern Reach Trilogy*

"*Crossings*, Ben Goldfarb's impassioned quest to understand the ecology of roads and its impact on the natural world, is a marvel. The reader learns something new on every page, disturbed and amazed in equal measure. Goldfarb moves us briskly along the manipulated ecosystem of the highway, with vivid, evocative pitstops for environmental history, ecology, and the built environment. With 15 million additional miles of road scheduled to be built over the globe in the near future, the time for this book is now. *Crossings* adds a new perspective to conversations on how humans have reshaped life on earth." —The Whiting Award Judges' citation

ALSO BY BEN GOLDFARB

Eager:
The Surprising, Secret Life of Beavers and Why They Matter

CROSSINGS

HOW ROAD ECOLOGY
IS SHAPING
THE FUTURE OF OUR PLANET

Ben Goldfarb

W. W. NORTON & COMPANY
Independent Publishers Since 1923

For information about permission to reproduce selections from this book, write to Permissions,
W. W. Norton & Company, Inc., 500 Fifth Avenue, New York, NY 10110

For information about special discounts for bulk purchases, please contact
W. W. Norton Special Sales at specialsales@wwnorton.com or 800-233-4830

Manufacturing by Lakeside Book Company
Book design by Chris Welch
Production manager: Louise Mattarelliano

Library of Congress Control Number: 2023290736

ISBN 978-1-324-08631-4 pbk.

W. W. Norton & Company, Inc.
500 Fifth Avenue, New York, N.Y. 10110
www.wwnorton.com

W. W. Norton & Company Ltd.
15 Carlisle Street, London W1D 3BS

1 2 3 4 5 6 7 8 9 0

To Elise, my road-trip companion, who built
a bower of pine boughs for a red squirrel.

CONTENTS

PART III: THE ROADS AHEAD

CROSSINGS

Introduction

THE WING OF THE SWALLOW

If you've ever driven across the United States of America, you have passed beneath the wings of a plucky songbird—smaller than your palm, light as your pocket change, feathered in jaunty blue and umber—called the cliff swallow. Where other animals flee the human footprint, cliff swallows shelter in its tread. *Petrochelidon pyrrhonota* should properly be called the bridge swallow, for our steel spans have furnished it with more nesting sites than bluffs and canyons ever did. Once a bird of the western mountains, in the last century cliff swallows have spread onto the Great Plains and across them, plastering their gourd-shaped mud nests to girders and trusses, feats of avian engineering no less impressive than our own viaducts.

"Once the environment is ruined," a biologist named Charles Brown told me, "all we'll have left is rats, cockroaches, and cliff swallows."

Cliff swallows are gregarious birds whose colonies can number in the thousands. Like most civilizations, theirs are messy: they steal nests, bully others into mating, fight so viciously that they sometimes tumble into rivers and drown. For the last four decades Brown has paid annual visits to more than two hundred nesting sites in Nebraska, trying to figure out what makes swallow societies thrive or fail. He has studied how well they catch insects, how they spread diseases, how they fend off snakes. He has captured more than four hundred thousand swallows in nets and encircled more than two hundred thousand slender legs with coded metal bracelets. Mostly what he has done is drive, more miles than he can hazard, from bridge to bridge, colony to colony. "Ninety-eight percent of cliff swallows in western Nebraska," Brown said, "are going to be within fifty feet of a road."

Near a road, of course, is the most dangerous place an animal can live, and swallows, for all their agility, occasionally fall victim to passing cars and trucks. When Brown and his longtime collaborator, the ornithologist Mary Bomberger Brown, began studying swallows in the 1980s, they picked up these casualties, wings broken and heads crushed, and brought them to their lab. They dressed the bodies—replacing eyes and viscera with cotton, lacing up feathered breasts like shoes—and tucked them in a drawer. They didn't have any plan for the birds; it just seemed proper. The waste of death diminished by the salvage of data.

Years passed. The number of publications mounted; the swallows flourished. In 2012 a new assistant asked Brown to teach him the art of dressing birds. Brown promised there would be plenty of roadkill on which to practice. When summer arrived, though, there was virtually no roadkill to be found—just empty asphalt and swallows, gloriously alive, taunting the vehicles below.

The epiphany hit Brown like an eighteen-wheeler: this was not a single-summer fluke. Swallow roadkill had been dwindling for years. He'd collected twenty dead birds in 1984, when the project began, and twenty more in 1985 and 1986. Then the trend line slanted downward, straight as a ski slope: fifteen in 1989, thirteen in 1991, eight in 2002. By 2011 the toll had dropped to four.

Brown considered various explanations, then dismissed them. There weren't fewer swallows to be whacked or more vultures bearing away carcasses, nor was there less traveling on his part. No, Brown thought: somehow swallows had become harder to kill.

He found his answer in the corpses themselves. When he stretched a tape measure from the birds' shoulders to their outermost feathers, he found that car-struck swallows had longer wings than the average bird he snared in his nets. The difference was slight, no more than a few millimeters, but the gap had unmistakably grown over the years. Brown immediately understood the significance. Long wings were good for straight, lengthy flights: between nests and feeding grounds, for example. Shorter wings were better for maneuverability, for performing the tight pivots and rolls with which a swallow would evade a falcon—or a flatbed hauling a load of lumber to Omaha. Traffic was weeding clumsier, long-winged swallows from the population and favoring their nimbler, short-winged flockmates. It was Darwinian selection in action, so clean and rapid it belonged in a textbook.

"Morphologically, these are not the same birds anymore," Brown said. But it seemed to me they were also different in some deeper, more metaphysical sense. Centuries ago, before we paved North America, cliff swallows had existed largely beyond human influence; now they were so enmeshed in our world that our infrastructure had infiltrated their DNA. Cliff swallows were a success story, rare beneficiaries of concrete and steel. Yet their triumph had come at a cost—to the long-winged martyrs culled from the population and to the birds' altered genes themselves. Cliff swallows had survived, but as a changed thing. They had been shaped, subtly but intimately, by the road.

When alien archaeologists exhume the rubble of human civilization, they may conclude that our raison d'être was building roads. Some forty million miles of roadways encircle the earth, from the continent-spanning Pan-American Highway to the hundred thousand miles of illegal logging routes that filigree the Amazon. Our planet is burdened by three thousand tons of infrastructure for every human, nearly a third of an Eiffel Tower per person. Roads predate the wheel: Mesopotamian builders began laying mud-brick paths in 4000 BCE, centuries before anyone thought to drop a chariot onto a couple of potter's disks. Today it's impossible to imagine life without the asphalt arteries that connect goods with markets, employees with jobs, families with each other. "Everything in life is somewhere else," wrote E. B. White, "and you get there in a car."

Roads are both logistical essentials and cultural artifacts. They epitomize freedom—the "architecture of our restlessness," per Rebecca Solnit, the "two lanes [that] take us anywhere," per Bruce Springsteen. To us, roads signify connection and escape; to other life-forms, they spell death and division. Sometime during the twentieth century, scientists have written, roadkill surpassed hunting as "the leading direct human cause of vertebrate mortality on land." Name your environmental ill—dams, poaching, megafires—and consider that roads kill more creatures with less fanfare than any of them. (More birds die on American roads every *week* than were slain by the *Deepwater Horizon* oil spill, with the road deaths accompanied by a fraction of the hand-wringing.) And it's only getting worse as traffic swells. A half-century ago, just 3 percent of land-dwelling mammals met their end on a road; by 2017 the toll had quadrupled. It has never been more dangerous to set paw, hoof, or scaly belly on the highway.

Roads distort the planet in other, more insidious ways. No sooner was Rome's Via Cassia completed around 100 BCE than its surface began to shed sediment into Lago di Monterosi, spawning algal blooms that permanently distorted the lake's ecosystem. *Phytophthora lateralis*, an invasive fungus that attacks cedar trees, hitchhikes in the patterns of truck tires. The little red fire ant, a merciless insect notorious for stinging the eyes of elephants, has exploited logging tracks to spread through Gabon sixty times faster than it would have otherwise. Pavement itself blankets less than 1 percent of the United States, yet its influence—the "road-effect zone," to use ecological jargon—covers a full 20 percent. Park your car on the shoulder and bushwhack half a mile into the woods, and you'll still see fewer birds than you would in an unroaded wilderness. Hike two miles more, and you'll still see fewer mammals. If you're a Kerouac reader, you grew up steeped in the dogma that highways represent freedom. If you're a grizzly bear, they might as well be prison walls.

The repercussions of roads are so complex that it's hard to pinpoint where they end. British Columbia's caribou herds have dwindled to furtive bands, in part because logging and mining roads have permitted the ingress of wolves—a human-caused disaster disguised as natural predation. Nearly a fifth of America's greenhouse gas emissions are coughed out by cars and trucks, and the transportation sector is the fastest-growing contributor to climate change; meanwhile, the rise of electric vehicles, whose batteries depend on lithium and other metals, has catalyzed a mining boom that threatens to disfigure landscapes in places as disparate as Chile, Zimbabwe, and Nevada. Even habitat loss, the most thorough eraser of wildlife, is a road problem. Before you can log Alaska's rainforests or convert Bornean jungles into oil-palm monocultures, you need roads to transport the machinery in and the product out. Roads are, you might say, the routes of all evil.

Yet roads select winners as well as losers. Arizona's highways funnel rainfall into ditches and thus soften desert soils for pocket gophers, whose tunnels parallel the shoulder like subway lines. Vultures, ravens, and other cunning scavengers are ascendant, their diets subsidized by roadkill. Butterflies whose prairies have been devoured by cornfields find succor in unkempt strips of roadside milkweed. In Britain such habitat is called the *soft estate*—a suggestion that roads are capable of creating new ecosystems, even as they shatter existing ones. A biologist once led me beneath a highway bridge to show me hundreds of little brown bats roosting in its crevices, seemingly unbothered by the traffic thumping overhead.

Considering the outsized effects of roads, it's perhaps surprising that they didn't truly receive their scientific due until the late twentieth century. One afternoon in 1993 a landscape ecologist named Richard Forman was standing in his Harvard office with a few students, admiring a satellite photograph of a forest. Forman was expounding on the forest's features—where the water flowed, why people had put houses where they had, how the animals moved through it—when he paused. "I noticed the long slice going diagonally across the image," he recalled to me. "It was a two-lane road through the forest. I said, gee, we know a lot about the ecology of everything else in this image, but we don't know much about the ecology of *that*." Inspired by inattention, Forman soon coined an English term: *road* *ecology*, defined loosely as the study of how "life change[s] for plants and animals with a road and traffic nearby."*

He did not immediately attract disciples. When a major government committee invited Forman to present his new field to transportation higher-ups the next year, he was met with polite laughter. "You're not here to make us stop running over animals, are you?" one engineer asked, cocking an eyebrow. As the 1990s wore on, though, road ecology gained steam. Forman and other pioneers published papers, wrote textbooks, held conferences that lured curious officials. "All of a sudden," Forman said, "it became mainstream."

- - - - - - - - -

My own introduction to road ecology came in 2013, the year I embarked on a trip across the continent to write about an extraordinary scheme called the Yellowstone to Yukon Conservation Initiative. The goal of Yellowstone to Yukon, or Y2Y, is boggling: its advocates envision a network of connected habitats that would permit animals to wander unhindered along the spine of the Rockies, a region that spans five American states and four Canadian provinces and territories. Such a corridor would preserve migration routes for elk and caribou, permit far-ranging creatures such as wolves to mingle and mate, and help sensitive animals like wolverines flee northward as climate change nips at their heels. The initiative's emblem is the grizzly, whose expansive requirements make it a useful proxy for other forms of life. An ecosystem that can support bears is probably healthy enough for everyone else.

To the uninitiated, it sounded far-fetched. Soon after Y2Y's inception, *The West Wing* parodied it as the "Wolves Only Roadway," the vanity project of humorless tree huggers who get laughed out of the White House. But the show's writers, like most of Y2Y's critics, misunderstood the concept. Y2Y

---

* As European road ecologists are quick to remind their American colleagues, "road ecology" was actually a translation of *straßenökologie*, a neologism coined by a German scientist named Heinz Ellenberg in 1981.

wasn't a discrete pathway; it was a continental jigsaw riddled with missing pieces, most of them at the fragile margins where wildlands and settlements collided. The mission of Y2Y and its many partners was to plug those holes, to help bears and other animals safely navigate the Rockies without running afoul of humans. In British Columbia, I toured protected grainfields that grizzlies used to commute between mountain ranges at night. In Montana, I sniffed offal in an electric-fenced paddock where ranchers were composting their dead cows rather than permitting them to fester in bear-enticing boneyards. (Few travelers, human or ursine, can resist fast food.)

Yet Y2Y's deepest cuts remained mostly unhealed. The region was riven by enough numbered roads to fill a sudoku puzzle: I-90 and Highway 3 and Highway 20, routes 95 and 40 and 12 and 212, spiderwebbed otherwise wild lands. I drove highways that ended lives—I lost track of how many elk littered the shoulder on Crowsnest Pass—and others that cleaved grizzly populations into lonely clusters. Roads, I began to realize, were not merely a symptom of civilization but a distinct disease.

Among the roads within the Y2Y corridor's ambit was U.S. 93, which traverses Montana on its 1,300-mile jaunt from Arizona to the Canadian border. Like so many highways, U.S. 93 had been built heedlessly in the 1950s, plowing through wetlands, elk meadows, and a vast reservation belonging to the Confederated Salish and Kootenai Tribes. When, in the 1990s, state and federal agencies sought to expand U.S. 93 from two lanes to four, tribal officials demanded the chance to provide input on the reconstruction. A wider, faster road might be safer for drivers, but it would also slaughter more deer, elk, bears, and other animals foundational to the tribes' culture. "The road is a visitor," the tribes insisted, that should "respond to and be respectful of the land and the Spirit of Place."

The Salish and Kootenai flexed their legal and moral muscles, and, when U.S. 93 was finally reconstructed, engineers included around forty wildlife crossings—a network of underpasses, tunnels, and culverts that allowed animals to slink beneath the highway unimpeded. Roadside fencing kept creatures off the highway and guided them toward the passages. The project's flagship structure was an elegant bridge designed principally for that avatar of wildness, the grizzly bear. In aerial photos, the overpass looked at once futuristic and anachronistic, a green parabola that vaulted over the highway with Middle Earthish grace. If roads were a disease, wildlife crossings seemed like a treatment.

That October, I drove U.S. 93 in the company of Marcel Huijser, a lean, grizzled road ecologist who had begun studying the highway back when the crossings were still in their planning stages. I was then more or less ignorant of road ecology, let alone how one became a road ecologist, so, as we headed north from Missoula, I asked Huijser to tell me about his past. He'd grown up in the Netherlands, a country that packs one of the world's densest road networks into a landmass one-ninth the size of Montana. Bears and wolves had long since fled the overbuilt Dutch landscape, so Huijser had studied hedgehogs, which popped in and out of gardens like cheerful neighbors. "Everybody thinks they're cute and wonderful," Huijser said as we rolled past golden cottonwood galleries. "They're a very sympathetic animal. People want to make things pleasant for hedgehogs."

Alas, hedgehogs—small, plodding, nocturnal—were practically designed to be roadkill, and Huijser's calculations suggested that hundreds of thousands were being crushed each year. This was a familiar story in the Netherlands. Roads, dikes, canals, and towns had broken the country's landscape into pieces, leaving little space for hedgehogs and other fauna. In 1990 the country, with typical Dutch ingenuity, had launched a national defragmentation plan that ultimately led to the construction of more than eight hundred wildlife crossings on national highways, from badger pipes to deer bridges. Huijser's research had shown that hedgehogs prefer the ecotones where forests and grasslands meet, helping planners situate new passages. "One of my statements during my PhD defense was that they're really *edge*hogs," he said, sheepish at the pun.

In 1998, Huijser met his wife, an American conservationist named Bethanie Walder, at—what else?—a road conference. He eventually relocated to Montana to work for the Western Transportation Institute, the research team tasked with studying the U.S. 93 crossings. In the years that followed, Huijser and his colleagues scooped up deer pellets, pored over photos snapped by motion-activated cameras, and crouched to inspect hoof- and pawprints deposited in sandy soil. By the time I visited, animal collisions had fallen by around three-quarters, and Huijser's team had documented tens of thousands of successful traversals through the crossings: coyotes, foxes, bobcats, elk, otters, porcupines, moose, grizzlies. Highway 93, Huijser told me, "compares favorably with anywhere else in the world in the number and density of structures"—even the Netherlands.

As Huijser parked beside the overpass, dusk descended. He unlocked a gate in the roadside fence, and we slipped through as though into a portal, passing from the world of the highway to a wild parallel dimension. We climbed the gentle slope, autumn grass crispy beneath our boots. I reached the crossing's apex and peered down at the muffled stream of Missoula-bound traffic. The horizon glowed orange. The evening was sharp with oncoming winter. Though we were just three stories above the earth, I felt buoyant. The air shimmered with supernatural possibility; at any moment, I thought, a grizzly would emerge from the pines and trudge onto the bridge.

Huijser didn't share my muted awe. He paced the crossing and pointed out design flaws. He was displeased with the bridge's spartan landscaping: a few carefully positioned brush piles, he suspected, would aid mice and voles. And its sightlines were too exposed. His cameras had recently captured a black bear fleeing the headlights of an approaching car. "A visual screen would be helpful," he said. Shrubbery or an earthen berm might do the trick. "It could just be a wooden fence."

Huijser, it occurred to me, was attempting to inhabit other beings' *Umwelt*, their subjective lived experience. Road ecology was an act of interspecies imagination, a field whose radical premise asserted that it was possible to perceive our built world through nonhuman eyes. How does a moose comprehend traffic? What sort of tunnel appeals to a mink? Why do grizzly bears prefer crossing over highways while black bears go under? These questions had empirical answers, but they also required ecologists to think like wild animals—empathy manifested as science.

To us, roads are so mundane they're practically invisible; to wildlife, they're utterly alien. Other species perceive the world through senses we

cannot fathom and experience stressors and enticements we hardly register. Bats are lured astray by streetlights, snails desiccate as they slog across
deserts of asphalt, and seabirds crash-land on the shiny tarmac they mistake for the ocean. Consider the sensory experience of, say, a fox approaching a highway: the eerie linear clearing that gashes the landscape, the acrid
stench of tar and blood, the blinding lights staring from the faces of thun-
dering predators. When Hazel, the rabbit protagonist of *Watership Down*,
encounters his first road, he confuses it for a river, "black, smooth and
straight between its banks." A passing car "fill[s] the whole world with
noise and fear." "Now that I've learnt about it," he adds, "I want to get away
from it as soon as I can."

Road ecology inverts our oldest joke about animals and transportation:
Why did the chicken cross the road? Embedded in that chestnut is an assump-
tion—that the road is inviolable and eternal, as fixed in its course as a river.
The road is a given; it's the fowl whose actions demand explanation. But
the riddle's logic is backward. It's the animals who have always moved, the
road that's the upstart. A better question might be, *Why did the road cross
the land?*

    This framing isn't always comfortable. When we don't ignore roads,
we dismiss their toll as the inevitable cost of modernity. Other forms of
human-caused animal death are deliberate: we pull the trigger, set the trap,
order the cheeseburger. But few among us ever flatten an animal on purpose. Like most people, I at once cherish animals and think nothing of
piloting a three-thousand-pound death machine. The allure of the car is so
strong that it has persuaded Americans to treat forty thousand *human* lives
as expendable each year; what chance does wildlife have? One summer, in
Alaska, I hit a feisty songbird called a yellow-rumped warbler—a death I
didn't discover until I found the delicate splash of feathers wedged in the
grille the next day. I'd killed not with malice but with mobility. "We treat
the attrition of lives on the road like the attrition of lives in war," the writer
Barry Lopez lamented. "Horrifying, unavoidable, justified."

This is particularly true in the United States, home to the world's longest road network, at four million miles. Our mid-century automotive revolution spawned not only highways but also parking lots, driveways, suburbs, pipelines, gas stations, car washes, drive-throughs, tire shops, and strip malls—a totalizing ecosystem engineered for its dominant organism, the car. For all its grandeur, though, America's highway network is relatively static. Although we spend almost $200 billion on our roads annually, most goes toward repair rather than new construction. Granted, American wildlands are hardly safe from ill-conceived development: Florida, for one, has been scheming up new toll roads in panther habitat, and even routine highway maintenance projects have an uncanny knack for adding lanes and worsening traffic. Even so, our country's asphalt limbs have mostly ceased to elongate, petrified into something like their eternal shape.

Instead, we're exporting our autocentric lifestyle. More than fifteen million miles of new paved road lanes will be built worldwide by 2050, many through the world's remaining intact habitats, a concrete wave that the ecologist William Laurance has described as an "infrastructure tsunami." Astoundingly, as of 2016, three-quarters of the infrastructure that will exist by the middle of this century had yet to be built. Although it's easy to denounce the tsunami, I benefit from roads as much as anyone: I eat avocados trucked from California; I get pizza delivered to my doorstep; I rely on America's marvel of a highway system to reach friends and hospitals and airports. (And I confess to feeling what one Volkswagen ad campaign called *Fahrvergnügen*, the pleasure of driving.) Roads pose the same queasy conundrum as climate change: having profited wildly from growth, can wealthy nations deny less-developed countries the benefits of connectivity?

Road ecology offers one path through this thicket. North America and Europe constructed their road networks with little regard for how they would affect nature and even less comprehension of how to blunt those effects. Today, in theory, we know better. Road ecology has revealed the perils of reckless development and pointed us toward solutions. Over the last several decades, its practitioners have constructed bridges for bears, tunnels for turtles, rope webs that allow howler monkeys to swing over highways without descending to the forest floor. On Christmas Island, red crabs clamber over a steel span during their beachward migrations; in Kenya, elephants lumber beneath highways and railroads via passages as tall as two-story houses. And road ecology has yielded more than crossings: we've also learned to map and protect the migrations of cryptic animals, to design roadsides that nourish bees and butterflies, and to deconstruct the derelict logging tracks that lace our forests—proof that old mistakes need not be permanent.

Quoth the late-night sage John Oliver, "Infrastructure isn't sexy." Clearly he hasn't talked to a road ecologist. Roads have become one of conservation's most urgent topics, the focus of hundreds of scientists operating in dozens of countries. Over several years I traveled the world meeting some of them: the biologists tracking anteaters across Brazilian highways, the conservationists building bridges for California's mountain lions, the animal rehabbers caring for Tasmania's car-orphaned wallabies. While *Crossings* is rife with other species—mule deer and capybaras, wombats and monarch butterflies—it also considers how our own lives have been captured by pavement and how we can reclaim them. Wild animals, the naturalist Henry Beston wrote, are neither our brethren nor our underlings; instead, they are "other nations, caught with ourselves in the net of life and time." And the road ensnares us both.

This book is about how we escape.

PART I

KILLER ON THE ROAD

1

AND NOW THE DEVIL-WAGON!

*How the rise of cars made animal life more perilous,
sabotaged evolution, and birthed a new science.*

I f road ecology has a birth date, it's June 13, 1924—the morning a biologist named Dayton Stoner and his ornithologist wife, Lillian, left their
home in Iowa City, bound for a research station three hundred miles
away. The couple planned to spend the month capturing and banding birds,
as they did most summers; that year their tally would include kingfishers,
house wrens, and brown thrashers. Posterity doesn't record their vehicle's
make, but it's a decent bet they drove a Model T.

For the Stoners, who'd met at the University of Iowa, joint research trips
were a chief pleasure of married life. In 1918, six years after they wed, they
had traveled together on an insect-collecting expedition to Barbados. It was
an eventful voyage: locals concluded that the quiet American with the dark
mustache and the butterfly net was a German spy, hurled rocks at him, and
threatened to tie him up. The couple, reported a companion, "reaped a rich
entomological harvest at the peril of their lives." They returned with a haul
of tarantulas, locusts, and stink bugs—"Mr. Stoner's particular pets"—
and their partnership reinforced. Dayton and Lillian would publish several
papers together, including one on bank-swallow chicks and another on an
owl sick with bumblefoot.

Little wonder, then, that the Stoners didn't get far that June day in 1924 before they found something to investigate. Within minutes of departing Iowa City, they noticed the roadside was strewn with "a considerable number of dead animals, apparently casualties from passing motor cars." The first commercial car radio was still six years off; a couple had to entertain themselves somehow on a road trip across the Midwest. Dayton and Lillian invented a morbid game to pass the time: "an enumeration and actual count" of the dead.

Over the next two days the Stoners jounced through farmland and woodlots, rarely exceeding twenty-five miles per hour, noting each victim they passed. It was a perfect activity for Dayton, who, as a colleague noted, was "ruled rather strictly by two compulsions—one emanating from his personal reserve, the other from his scientific discipline." I imagine Dayton behind the wheel, Lillian beside him with pencil and pad, the light golden on the corn as they gently squabble. Was that a woodchuck or a rabbit, a catbird or a grackle? All told, they counted 84 animals on their way to the field station and 141 on the drive home, among them 19 flickers, 18 ground squirrels, 14 garter snakes, 2 weasels, and a shrike.

The Stoners weren't the first to recognize the environmental hazards of transportation. Roadkill predates the automobile. The novelist and poet Thomas Hardy eulogized the animals trampled by horse-drawn war wagons at Waterloo: "The mole's tunnelled chambers are crushed by wheels / The lark's eggs scattered, their owners fled." (And you thought Napoleon had a bad time.) Yet roadkill remained absent from the scientific literature until 1925, when Dayton Stoner published his and Lillian's surveys in *Science*. Their study, a road-ecology urtext, diagnosed a malady with no name; the word *roadkill* was still two decades from coinage.* Call it what you will: America's burgeoning need for speed, Stoner claimed, had become "one of the important checks upon the natural increase of many forms of life."

---

* *Roadkill* appears to have entered the lexicon in 1943, courtesy of Robert McCabe, an ecologist who observed juvenile partridges being killed by cars. The *Oxford English Dictionary*'s first citation occurs in 1972, in a highway guide that referenced magpies feeding on "carrion or road-kills."

Stoner's angst didn't arise in a societal vacuum. Like many Americans, he judged the automobile a disruptive and fearsome technology. By the 1920s cars had claimed tens of thousands of human lives, frayed social contracts, and demoted pedestrians to second-class citizens. "Even the widely heralded 'dirt roads' of Iowa are tainted with human blood," Stoner wrote. His roadkill concerns emerged from this broader automotive anxiety. "Not only is the mortality among human beings high," he carped, "but the death-dealing qualities of the motor car are making serious inroads on our native mammals, birds and other forms of animal life." Where some perceived progress, the Stoners saw a threat.

That roads threaten animals is ironic, for it was animals who inscribed the first etchings on the palimpsest that is America's road network. Many of our best-used roads began as wildlife trails—often bulldozed by bison, renowned for their "sagacious selection of the most sure and direct courses"—that solidified into Native American footpaths. Indigenous roads paralleled the Columbia, Colorado, and Mississippi Rivers; tracked the Pacific coast from Mexico to Alaska; and breached the Rocky Mountains. Roger Williams, the minister who founded Rhode Island, traveled the Northeast in 1636 via a Pequot and Narragansett trade route and marveled that Native feet had beaten a track "in the most stony and rockie places." English travelers on foot and horseback adopted this Pequot Path, the most expedient route from Boston to New York, and mail carriers incorporated it in the Boston Post Road, which today comprises portions of U.S. 1, 5, 20, and other traffic-choked highways. When Nathaniel Hawthorne wrote that the Native architects of New England's roads were perhaps "saddened by a flitting presentiment that [the white man's] heavy tread will find its way over all the land," he hadn't even anticipated Dunkin drive-throughs.

As colonial roads superimposed themselves atop Native trails, they became instruments of empire, the westward-grasping tentacles of a rapacious nation. In 1816, when South Carolina representative John C. Calhoun lobbied Congress to fund new wagon routes, he spared no grandiosity.

"Let us then bind the Republic together with a perfect system of roads and canals," Calhoun thundered. "Let us conquer space." His bill was vetoed, but his cause triumphed. Between 1807 and 1880 the U.S. Army carved twenty-one thousand miles of roads through prairies, forests, and mountains, cracking open America's interior to mail delivery and agriculture. Roads became tools of subjugation: when the government strong-armed the Potowatomi into ceding its territory in 1826, federal negotiators insisted the tribe include a hundred-foot-wide belt of land—a strip that became the Michigan Road, the main route to Indianapolis. The state of Indiana funded the road's construction by selling off the Potowatomi's dispossessed lands, the ill-gotten gains of imperialism converted to infrastructure.

America's early roads were shoddy, only rarely surfaced with gravel or stone, and at constant risk of being reclaimed by nature. In 1755 a British general named Edward Braddock hacked a twelve-foot-wide road through the Maryland scrub, a chore so dangerous that three wagon teams tumbled off a cliff in the process. Just three years later Braddock's hard-won road "had reverted to a trace through the forest." In the early nineteenth century, legions of turnpike companies paved toll roads with macadam, stone blocks that locked together beneath the pressure of wheels and hooves. By the mid-1800s, though, railroads had usurped horses, and neglected turnpikes devolved into muddy troughs, an era known as the "Dark Age of the Rural Road." Frustrated farmers often abandoned wagons in the mire, a loss they knew as the "mud tax." Desperate roadbuilders resorted to whatever materials they had on hand. Engineers in coastal Texas razed oyster beds to pave roads with shells, while officials in Washington promoted an annual "straw day," during which wheat growers carpeted boggy tracks with a layer of, yes, straw.

In the 1890s, America's deplorable roads were salvaged by a newfangled invention: the modern bicycle. The nation went wild for cycling, which promised freedom and good health. (Earlier bikes, with their ludicrously outsized front wheels, had been too dangerous to provide reliable transport.) Women pedaled to unchaperoned dates; ministers deemed cycling "the very highest, fullest and completest physical, mental and spiritual culture." As cycle-mania swept the country, the condition of its roads became an embarrassment. *Good Roads*, a literally muckraking magazine founded by the League of American Wheelmen, called out lousy roads and lauded ones that had been drained, graded, and paved. Farmers initially distrusted the "eastern bicycle fellers" meddling in rural affairs but soon made common cause. The U.S. government waged a national public-relations campaign, building high-quality "object lesson roads" that showed locals how pleasant their byways could be with a little investment. In 1904 one official crowed that on the roadbuilding front "more has been done in [the last decade] than during the one hundred preceding years."

Yet American roads were no match for the next transportation revolution. As the twentieth century dawned, automobiles exploded like bacteria in a petri dish: 78,000 registered vehicles in 1905, 2.3 million in 1915, more than 5 million three years after that. The only impediments to the horseless carriage's growth were the abysmal surfaces on which they rolled. Slop-spattered French drivers withdrew from American races in disgust; German imports broke down on roads "of a kind wholly undreamed of in Europe." By 1909, despite the efforts of the Good Roads movement, only 8 percent of America's 2.2 million road miles qualified as "improved," a designation that included everything from gravel surfacing to sand-and-clay mixtures to rudimentary rut-filling. "The American who buys an automobile finds himself with this great difficulty," sighed one manufacturer. "He has nowhere to use it."

Into the muddy breach at last stepped an engineer named Thomas MacDonald. Like Dayton Stoner, MacDonald was an Iowan, raised in a town whose rich, damp soils at once produced bountiful crops and prevented farmers from hauling them to market. In college MacDonald studied the embryonic science of roadbuilding, analyzing the efficiency of horse teams on dirt roads versus paved ones. After graduating in 1904, he traveled Iowa extolling the benefits of good roads and ferreting out corrupt contractors; soon he ascended to the job of state highway engineer. He wasn't a guy with whom you would want to get a beer—biographers describe him as so painfully formal that his siblings called him "sir"—but his discipline and zeal made him the perfect corrective to Iowa's chaotic morass. By 1916 several counties had begun to whip their roads into shape, and the rest, MacDonald predicted, would soon "make the gravel fly."

All that flying gravel elevated MacDonald's profile, and in 1919 the solemn engineer was appointed chief of the Bureau of Public Roads, the agency that would become the Federal Highway Administration. He inherited a department so embattled that BPR's own staff called it the Bureau of Parallel Ruts. Despite his odd interpersonal style, MacDonald proved an effective lobbyist, extolling good roads in public speeches and smoky clubs, and in 1921 he convinced Congress to sign landmark legislation called the Federal Aid Highway Act. The law gave states funding to improve their highways and rationalized America's disjointed snarl of roads into a coherent system. Cement, brick, and asphalt manufacturers tapped the gusher of contracts, and soon forty thousand highway miles were being improved annually. Roadbuilding, once the haphazard production of hooves, feet, and wagon wheels, evolved into a sophisticated discipline; BPR's journals bulged with arcane studies of vibrolithic concretes, bituminous sealants, and the viscosity of asphalt. Mud hadn't been vanquished, but it was in fast retreat.

MacDonald's triumph transformed the country's relationship with roads as thoroughly as it molded the land itself. Before his golden age of construction, meteorology and topography limited when and where people could drive. Ungraded dirt roads became mud baths in spring, dust bowls in summer, and tongues of ice in winter. Americans perceived roads as part of the natural environment, as capricious as weather—"organic entities governed by wind and water," as the historian Christopher Wells put it. Good roads, by contrast, mastered their surroundings, transcending geography and climate to permit high-velocity transportation in all seasons. Roads, ever the conqueror, faced down nature and won.

As motorized vehicles overran the countryside, they also mustered forces along another front: the city. For centuries urban streets had been nodes of activity and commerce, as much bazaars as conduits. Yes, they were the province of carriages and electric streetcars. But they were also where kids played ball and shined shoes, where vendors flogged vegetables, where pedestrians loitered and gossiped. So nonthreatening were city streets that the Chamber of Commerce in Rome, New York, implored residents not to manicure their nails on the trolley tracks.

Cars turned streets into war zones. Early automobiles were death machines—unencumbered by safety features like turn signals and seat-belts, piloted by ill-trained drivers, and subject to only rudimentary laws. Tight curves were marked with pictures of skulls-and-crossbones or the Grim Reaper. Journalists bemoaned the proliferation of "road hogs," "speed maniacs," "Sunday drivers," "juggernauts," and the dreaded "flivverboob," the epithet for an inconsiderate motorist. The automobile struck critics as not only dangerous but depraved. The moment that "the foot touches the accelerator and the hand grasps the wheel," chided one reporter, law-abiding citizens "become afflicted with the gas rabies."

Desperate cities tried to protect their populace. In Detroit speed limits matched the gait of a walking horse, a clip so plodding that cars stalled out. New York erected rickety towers from which cops manually relayed signals along Fifth Avenue; other municipalities trialed primitive speed bumps known as "Milwaukee mushrooms." None of it helped much. In 1924 vehicles killed 23,600 people, many of them children, a per-capita death toll 60 percent higher than today's rates. The next year F. Scott Fitz-gerald published a novel at whose climax Daisy Buchanan ran down Myrtle Wilson, who "mingled her thick dark blood with the dust"—literature's most famous act of flivverboobery.

As the historian Peter Norton recounted in his book *Fighting Traffic*, the carnage incited a ferocious backlash. Cities seemed to compete over which could stage the most lugubrious anti-car demonstration. Memphis planted black flags at accident sites. Baltimore erected an obelisk inscribed with the names of dead children, while St. Louis built a cherub-festooned monu-ment titled "In Memory of Child Life Sacrificed upon the Altar of Haste and Recklessness." In Pittsburgh, a "safety parade" included a float depicting a girl crushed between cars. A demonstration in Milwaukee incorporated a trashed automobile on a trailer: behind the wheel sat Satan himself.

This was the world into which road ecology was born: one where cars were both forces of progress and unholy terrors shredding society's fabric. Automobiles were dangerous in part because their physics seemed to melt their operators' brains. Drivers accustomed to stately horses and bicycles were so discombobulated by cars' weight and speed that the American Automobile Association issued a training manual explaining why, when cars flew around a corner, centrifugal force impelled them to flip over, or "turn turtle."

And the car's mass and velocity were even more stupefying to *actual* turtles. Consider the self-defense tactics of our commonest creatures, which, like stubborn coastal dwellers facing a hurricane, hunker down rather than retreat. Skunks spray, badgers hiss, opossums play possum. Porcupines bristle. Tortoises withdraw. Armadillos, oddly, jump. Timber rattlesnakes freeze, confident in their venom. For thousands of years these stand-your-ground strategies fended off coyotes and hawks, but against automobiles they were worse than useless. Even those animals that did flee were addled by the speed of cars. Many birds depend on a "distance rule" to decide when to retreat from danger, meaning that they take flight based on how far away a threat is, not how fast it's going. This might be a reliable operating manual when your enemy is a fox skulking through the underbrush, but not when it's a Model T rattling forty miles per hour down an improved gravel road. Cars hijacked their victims' own biology, subverting evolutionary history and rendering it maladaptive.

From the get-go, the car's detractors fretted about its toll on animals. "The Haunted Auto," an illustration that ran on the cover of *Puck* magazine in 1910, depicted a swarm of creaturely ghosts—geese, pigs, cats, chickens—hounding the motorist who had run them down. Not until the Stoners conducted their germinal survey in 1924, however, did scientists attempt to quantify the problem and diagnose its causes. *Roadkill* is an apt word, for, while the vehicle is the weapon, the Stoners realized it was the road that

As the automobile proliferated in the early twentieth century, biologists and cartoonists alike fretted about its toll on animals. *Puck Magazine/Bryant Baker*

determined the casualty rate. "The surfaced roads," Dayton observed, "permit of greater speed" than dirt tracks, giving animals less time to escape. This feature doomed distance-rule adherents like red-headed woodpeckers, which, Stoner wrote, "can not quickly acquire a sufficient velocity to escape the oncoming car and so meet their death." Although the Stoners followed improved gravel roads for only one-third of their trip (the rest was on dirt), that surface accounted for half the roadkill. The better the road, the bloodier.

Roadkill was only the latest in a litany of ecological catastrophes. Nineteenth-century hunters and trappers had mercilessly persecuted bison, beavers, and wading birds for their pelts and plumage; by the early 1900s extinction, once a heretical impossibility, had become an unavoidable fact. The car posed yet another threat, and scientists took note. At that stage in the still-nameless science of road ecology, counting carcasses during a road trip and mailing your notes to a journal—a practice the roadkill historian Gary Kroll called "dead-listing"—was enough to earn publication. In California, Ernest D. Clabaugh identified fourteen car-struck birds, among them coots, red-winged blackbirds, and house sparrows. William Davis spotted eighteen dead turtles in Ohio and mourned the "colossal tragedies." In Idaho, Kenneth Gordon totted up jackrabbits and ground squirrels and "estimated that the carcasses ran very close to 100 per mile." Charles Whittle, in a paper evocatively titled "And Now the Devil-Wagon!," tallied woodcocks, finches, and warblers. "Verily, the menaces of civilisation to bird life move on apace; they multiply," Whittle wailed.

While some scientists dabbled in roadkill counts, one eccentric devoted his life to them. In 1927 an itinerant forester named James Raymond Simmons retreated to a cabin in upstate New York to decompress after years of traveling the Northeast. Soon after settling in, Simmons was visited by a friend who'd just spotted a vesper sparrow trapped in the "soft, hot tar" of a freshly paved road. The bird had broken free, leaving behind "mute evidence in the form of a tiny toenail and tail feather." Simmons, enthralled by the story, concocted a wild plan: to "salvage at least a part of the wild creatures whose lives were being snuffed out daily along the highways."

Simmons established a receiving station in a farmhouse near Albany and declared himself open to roadkill, by mail or hand-delivered. He issued no public announcement and wasn't even sure his operation was legal. "Would some of my cooperators possibly undertake to send me skunk specimens," he wondered, "and if so . . . would the postman be likely to react violently to the perfumed packages?" From the outset he was deluged with specimens. One respondent sent in a cedar waxwing "waxed eight times on the right wing and only seven on the left." A neighbor handed off a flycatcher. Observers set up satellite substations and exchanged carcasses of their own—a roadkill social network.

Over the next decade Simmons and his volunteers gathered more than three thousand specimens, many of which he taxidermied and donated to schools and museums. In 1938 he compiled his observations in a charming, scattershot book called *Feathers and Fur on the Turnpike*. Simmons was not the world's most sophisticated scientist, less inclined to statistics than flights of compassionate poetry: a bluebird against pavement reminded him of "a vanished sky that cannot return to us," while a panicky weasel, scampering between lanes, recalled a "brown leaf being blown back and forth by gusts of wind." Nonetheless, signals emerged from Simmons's noisy data. In March and April, he wrote, bird roadkill betrayed the arrival of incoming migrants. By July and August the carnage reached its zenith, a peak Simmons attributed to "inexperienced young appear[ing] on the road." Across the board, males died more often than females.* Cottontail rabbits, the most-struck mammal, were "remarkably stupid in the face of glaring headlights."

Still, Simmons mostly refrained from victim-blaming—because the true issue, as the Stoners had recognized, was speed. Below thirty-five miles per hour, Simmons claimed, cars seldom struck animals. Accelerate to forty-five, though, and they "kill rapidly." Exceed sixty miles per hour, and "you can figure on scoring a kill every ten miles or less on most of the improved roads." The Good Roads revolution had transformed cars into superpredators—heavy, impervious, impossibly fast.

⸺⸺⸺

While Simmons let the carcasses roll in, Dayton and Lillian Stoner, road ecology's power couple, continued to seek them out. The Stoners seemed constitutionally incapable of going anywhere without stopping for dead animals. In 1928, Dayton and Lillian counted the "preponderance of creeping and crawling things," mostly snakes, on their way to Florida and observed the most deaths on Illinois's "fine concrete roads." Traveling from Iowa to Albany a few years later, the couple calculated a rate of precisely 0.029 skunks per mile of improved New York road.

As the new discipline grew, it creaked beneath the weight of rivalry. In 1935, William Dreyer, an Ohio biologist whose esoteric specialty was the "water contents of ants," counted roadkill between the Midwest and Massachusetts. He found carcasses scant and calculated that cars were claiming only 7,350 wild animals each day—around 5 percent of Stoner's estimate. To the circumspect ant man, Stoner and his ilk were panicmongers who had extrapolated astronomical casualty rates from "exceptional cases of destruc-

---

* Although Simmons didn't offer an explanation, males of many species occupy larger ranges and cross more roads than females. Male puff adders, for instance, are crushed ten times more often as they search for companionship during mating season.

tion." Dreyer also implored his peers to clean up their methods: rather than noting rabbits and turtles at random, scientists should perform a "systematic statistical survey covering several seasons and various localities." Stoner, chastened, toned down his catastrophism and averaged his data with others', including Dreyer's, to generate a more modest figure. At heart, though, he suspected that everyone's research underestimated the true toll. The counts were vexed by frogs and toads, whose bodies were so "soft and yielding" that "little of the animal remains except an inconspicuous splotch of red."

Stoner's road-ecology career ended in 1941, with a series of birdkill surveys between New York and Iowa. When he died of a heart attack three years later, his obituary speculated he had worked himself to death and called his research "characteristically thorough and detailed, but unembellished by expressions of feeling." It was a curious way to describe the oeuvre of a man who had declared Iowa's roads tainted with blood. Evidently Dayton Stoner could be dispassionate about everything but cars.

Stoner's death marked the end of road ecology's first era. The moral panic around cars had faded, stamped out by a loose alliance of auto manufacturers that dubbed itself Motordom. To counter bad press, Motordom bought ads in newspapers, cozied up to editors, and defamed any anti-car activist as a "whang-doodle of a charlatan." Lobbyists blamed accidents on "jaywalkers," a derogatory neologism for a hayseed bumbling about city streets.* As the car became entrenched, animal collisions, like pedestrian deaths, came to seem inevitable. At the outset of Stoner's carcass counting, he'd believed, perhaps naively, that drivers could be persuaded to slow for wildlife: "All animals, if given a reasonable time to escape, will cause the hurried motorist little if any delay," he wrote. Simmons, too, had initially hoped that heightened awareness and tougher speed limits would reduce

---

* Animals, like pedestrians, are often laden with undeserved culpability: how often have you griped about suicidal squirrels and kamikaze deer? Roadkill victim-blaming, one sociologist observed, is a common trope in cartoons, which depict flattened animals with "a stereotypical stupidity signified by cross-eyes, buck-teeth, and sticky-out tongue"—a portrayal that both "warrants these deaths and serves in their partial denial."

the flow of roadkill to his farmhouse, but he soon abandoned his optimism. "A safety campaign on behalf of [wildlife]," he despaired, "would be laughed out of court." Road ecology seemed doomed to wither.

In the 1950s, though, a new development began to roil America's highways. A long-persecuted species had again begun to flourish in car-dominated suburbs—a large, dangerous animal whose ascent would change road ecology forever.

The deer bounded in front of Jerry Chiappetta's car at 10:05 p.m. on a May evening in 1961, a tawny streak that hurtled from the brush to slam against the hood. Chiappetta wrenched the wheel, and the car keeled sideways through the Michigan night, its interior rent by screams—his wife's, his daughters', his own. "Something soft rolled against my foot; it was the baby," Chiappetta recalled. When at last the car shuddered to a halt, he found that, though the scene reeked of antifreeze and battery acid, his family was unharmed. Even his unrestrained infant had sustained only a bump. Worst off by far was the "still-quivering" doe, who had been trapped in the grille like a giant moth.

The incident gnawed at Chiappetta, a writer for the magazine *Field and Stream*. No one, he realized, had the faintest idea how many deer were struck by cars or how many people died in those crashes. Chiappetta queried officials in every state and, from their sparse data, concluded that at least forty-one thousand deer perished each year. "You never can relax your alertness in a deer area any more than you can on a street where children are playing," he lectured. One biologist he interviewed described deer collisions as "a real national problem—one that gets worse every day."

Today, Chiappetta's crash wouldn't merit mention in a small-town police blotter, let alone a national magazine. White-tailed deer are everywhere: you can't drive through Hastings-on-Hudson, the New York village where I grew up, without spotting one elegantly devouring a homeowner's hostas. In most northeastern suburbs, deer are as controversial as they are common. Gardeners consider them defoliants as potent as Agent Orange, dendrologists accuse them of overbrowsing woodlands, and epidemiologists indict them in the spread of Lyme disease. They have an uncanny knack for polarizing communities; I'd wager that no wild animal has provoked half as many acrimonious town-hall meetings. Peter Swiderski, Hastings's former mayor, told me he ran for office on a tripartite platform: "Taxes. Development. Deer." After he was elected, Swiderski hit a deer himself, incurring $800 in damages. "My children learned a curse word that day," he sighed.

Deer preoccupy road ecologists even more than they do suburban mayors. Unlike, say, black bears, deer are so abundant they cause crashes almost constantly; a deer meets a car every eight minutes in New York State alone. Unlike frogs and turtles, deer are hefty enough to imperil human lives. In 1995 researchers estimated that deer factor in more than a million crashes each year, injure twenty-nine thousand drivers, and kill more than two hundred. (A more recent estimate roughly doubled those figures, to fifty-nine thousand injuries and four hundred forty deaths.) Cars have made deer North America's most dangerous wild animal, implicated in three times more deaths than wasps and bees, forty times more than snakes, and four hundred times more than sharks.

As Jerry Chiappetta could attest, though, the deer almost always get the worst of it. Drivers walk away unscathed from 96 percent of crashes—known, in the bloodless verbiage of highway managers, as deer-vehicle collisions, or DVCs.* Because DVCs tend to be minor wrecks compared to rollovers or head-on crashes, they haven't traditionally received much attention from transportation departments. Most DVCs go unreported, and damage calculations are scarcely more reliable than they were in Chiappetta's day. When Bridget Donaldson, a scientist at the Virginia Department of Transportation, analyzed the handwritten notes of the maintenance workers tasked with scraping up carcasses, she discovered that police records captured barely 10 percent of the state's DVCs. Deer collisions, she calculated, were

---

* Most deer-related deaths occur when drivers swerve into a tree or traffic. As the state trooper's mantra goes, "Don't veer for deer."

quietly costing Virginia more than $500 million every year. "We had no idea how bad the situation was," Donaldson told me.

Deer collisions are such fixtures of modern life they've become an inescapable cultural trope, the *cervus ex machina* that bounds from the woods to destroy sedans and advance plots. In film they've been played for thrills, as in *The Long Kiss Goodnight*; for comedy, as in *Tommy Boy*; and for horror, as in *Get Out*. Even the Simpsons have hit a deer, albeit a statue of one (Homer: "D'oh!"; Lisa: "A deer!"; Marge: "A female deer!"). Yet the DVC, for all its cinematic prevalence, is an oddly contemporary problem. Spend enough time browsing proto–road ecology studies, and you'll notice something curious: they're devoid of deer. When Dayton and Lillian Stoner drove across Iowa in 1924, they registered terrapins, bull snakes, and fox squirrels but nary a deer. One fellow who drove 77,000 miles around Nebraska in the 1940s recorded a staggering 6,723 dead animals, from sand lizards to sparrowhawks to spotted skunks. None was larger than a snapping turtle. None was a white-tailed deer.

What explained this deerlessness? Speed, or its lack, played a role: even Thomas MacDonald's improved gravel roads rarely permitted drivers to run down an animal as fleet as a whitetail. Mostly, though, early road ecologists didn't find roadkilled deer for one reason: there weren't deer to kill.

For as long as humans have occupied North America, they've depended on deer. For the Cherokee's deer clan, it was the powerful Little Deer that "[kept] constant protecting watch over his subjects, and [saw] well to it that not one is ever killed in wantonness." Wantonness, alas, followed European colonists like flies. Spanish traders launched the deerskin industry in sixteenth-century Florida; before long; merchants were exporting a half-million deerskins annually to become gloves, book bindings, and other paraphernalia. Venison fed midwestern logging camps. Skins emerged from tanneries as boot linings and aprons. By 1890 hunters and homesteaders had slashed whitetail numbers from thirty million to three hundred thousand, with most of the survivors scattered in southeastern swamps.

The whitetail's recovery proved even swifter than its downfall. In the late 1800s upper-crust sportsmen, having found the religion of conservation, convinced states to suspend hunting. Biologists relocated deer from places that had them to places with none. Regenerating forests proffered tender seedlings. Cougars and wolves had long since been poisoned or gunned down to sanitize the land for livestock. "From the islands of cover where they had survived precariously for nearly a century," one naturalist wrote, "deer pushed out in all directions."

And what sort of landscape did they push into? Mid-twentieth-century whitetails found succor in a singularly American ecosystem, as endemic to its home country as the Outback or Amazon: suburban sprawl.

Humanity's affinity for suburbs long precedes the car. A letter to the king of Persia, inscribed on a tablet in 539 BCE, praised a property as "so close to Babylon that we enjoy all the advantages of the city, and yet when we come home we are away from all the noise and dust." (All that was missing was a Whole Foods.) But it wasn't until Model Ts began rolling off the line that the 'burbs became a dominant landform. As cars permitted urbanites to escape cramped tenements for leafy outskirts, once-distinct forms of development—metropolis here, farmland there—bled together like watercolors. The mixing accelerated in World War II's aftermath, as mass-production techniques that had cranked out planes and tanks retooled to build cheap homes. Americans ate at drive-throughs, watched drive-in movies, planned last rites at drive-up funeral parlors.

Suburbia's rise supercharged the whitetail's recovery. Deer thrive in edge habitats, transitional zones where delectable meadows bump against the shelter of forests. And suburbs offered more edges than a corn maze: edges between subdivisions and weedy lots, edges between manicured backyards and second-growth oak thickets, edges between vegetable gardens and baseball fields. When Europeans arrived in North America, between eight and twenty whitetails occupied each square mile of habitat; today, many suburbs crawl with a hundred or more. For all its gifts—its acute senses of hearing and smell, its leaping prowess, its four-chambered stomach—the whitetail's greatest aptitude was its ability to live cheek-to-antler with people, to flourish in the rectilinear hash we made of once-fluid landscapes.

It doesn't take a road ecologist to figure out what happened when unprecedented driving rates met the explosive population growth of a large mammal. National DVCs leapt from around 70,000 in 1963 to nearly 120,000 just three years later—a "minimum count," observed biologists, "because injured deer may die at some distance from the highway" and thus elude roadside surveyors. In Wisconsin deerkills quadrupled in a decade; even Joseph McCarthy hit a deer, which he presumably suspected of being a Communist saboteur. In Pennsylvania, a deery matrix of farms, towns, and forests, the body count skyrocketed from 7,000 to 22,000 in seven years. Not even air travel was safe: a plane approaching the Hazleton airport in 1964 clipped a deer with its landing gear, exposing twenty-eight passengers to a "close brush with death."

The rise of white-tailed deer signaled a permanent shift in the relationship between transportation and wildlife. Road ecology had begun as the fringe concern of naturalists who feared the "devil-wagon" would overrun animals. By the 1960s, though, cars weren't invaders; they were the unchallenged rulers of a country remade for their benefit. As deer blundered into their paved paths, road ecology's wheels began to spin in reverse. No longer did cars menace wildlife; now it was wildlife that menaced cars.

In truth, deer were less dangerous than the vehicles that struck them. Early car interiors resembled medieval torture chambers, their dashboards lined with sharp edges, chrome knobs, and uncollapsible steering columns that impaled drivers like spears. Getting into a wreck, observed *Reader's Digest*, was akin to "going over Niagara Falls in a steel barrel full of railroad spikes."

Although manufacturers made some halfhearted gestures toward crashworthiness in the 1950s, the safety revolution didn't flower until 1965, the year that a young lawyer named Ralph Nader published *Unsafe at Any Speed*, an exposé that accused auto companies of bringing "death, injury and the most inestimable sorrow and deprivation to millions of people." The book became a bestseller, killed the defective Chevy Corvair, and infuriated General Motors, which hired private investigators to tail Nader and prostitutes to proposition him. Nader resisted temptation, and in 1966 Congress passed legislation requiring seatbelts and other features. Automotive safety had entered the zeitgeist—and deer stood in its way.

The deer-vehicle collision crisis flummoxed engineers. Their first attempt at a fix was to put up warning signs. If drivers paid attention and slowed down, the thinking went, they'd have an easier time avoiding animals. Early signs were eclectic and creative. Residents of Los Alamos, New Mexico, birthplace of the atomic bomb, slapped up white deer silhouettes, a scheme that newspapers covered under the headline "Atomic Deer." Transportation departments eventually hit upon a design that any modern motorist would recognize: a yellow diamond etched with a leaping black buck. Signs popped up willy-nilly—at one-time accident sites, at places where hunters noticed tracks crossing the road, at spots where locals had seen deer a time or two. The more diamonds sprouted, the less useful they became. (Do *you* slam the brakes every time you see "Deer X-ing: Next 40 Miles"?) "There are so many signs on roads now that motorists tend to become 'blind,'" Jerry Chiappetta griped. One typical experiment found that warning signs in Colorado slowed traffic by just three miles per hour and had no discernible effect on roadkill (although dumping carcasses on the shoulder did seem to catch drivers' eyes). Signs were cheap and convenient expressions of concern, a nod to the problem that didn't actually solve it—"litter on sticks," some biologists later said. The data supporting Deer X-ing signs is no more robust today than it was in 1950, when one report declared them good only for shooting practice: "These signs serve as targets for a small, undesirable segment of the hunting fraternity, but the motoring public most certainly ignores them."*

If changing driver behavior was futile, perhaps there were ways of influ-

---

* Despite their inefficacy, signs have become the default Band-Aid for road agencies around the world, as variable as cuisine: emu signs in Australia, antelope signs in Namibia, toad signs in Denmark. Traveling through Costa Rica once, I scratched my head at a sign apparently emblazoned with a brontosaurus, only to realize, as I drew closer, that it depicted a coati, a raccoon-like mammal whose curved tail resembles, in outline, the neck of a sauropod.

encing the deer themselves. In 1970 Missouri's highway department outfitted a dangerous road with dozens of reflectors, mirrors that sparkled like disco balls when struck by headlights and seemed to startle deer away from the roadside. The mirrors offered a tantalizing promise: that the burgeoning crisis could be addressed with a low-cost gimmick. American transportation agencies weren't the only ones intrigued. In Europe roe deer populations were booming, as were collisions. In 1973, Swarovski and Company, the Austrian glass company and jeweler, began manufacturing red reflectors whose strobe-like effect purportedly formed an "optical warning fence." Soon three hundred thousand Swarovski reflectors lined European highways, and John Strieter, an Illinois real-estate mogul whose niece had married into the Swarovski family, licensed the technology for sale in the United States. Strieter tirelessly promoted the concept, which he claimed could save the nation $500 million each year, and over time more than a dozen American states came to install his Strieter-Lites. An analysis commissioned by the Strieter Corporation found that the reflectors cut collisions by up to 90 percent.

Independent research, however, told a different story. Some studies found that the reflectors didn't reduce collisions at all; scientists scoffed at others for their slipshod design. One review eventually found that reflectors reduced collisions by an inconsequential 1 percent. The companies that peddled reflectors might have declared victory over DVCs, scientists archly observed, but "road planners should not take these claims at face value." The problem was habituation. Just as human drivers came to ignore warning signs, deer learned to disregard flashing lights. In Denmark, for instance, deer initially fled from reflectors, but within a week their fear wore off, and they returned to nonchalantly grazing the roadside. Reflectors were a stimulus followed by neither reward nor punishment, as if Pavlov had rung his bell without giving his dog any meat.

Ultimately, DVCs seemed almost biologically foreordained. Rush hour, for both whitetails and suburban commuters, arrived at dusk, that liminal twilight moment when neither the cones nor rods in human retinas were fully activated. For their part, deer had outstanding peripheral vision but poor depth perception, and, because their pupils fully dilated in dim conditions, they tended to be blinded by headlights—another example of cars undermining an otherwise useful adaptation. By the mid-twentieth century, the busiest human traffic had begun to coincide with the highest rates of deer movement, at the precise crepuscular moment when both parties were least able to detect and avoid each other. The DVC epidemic was a failure of both vision and timing, the rotten fruit of a landscape reshaped for cars.

As highways evolved, finding a solution to deer collisions became ever more urgent. In 1956, President Dwight Eisenhower authorized the Interstate Highway System, "a mighty network of highways" that would connect America's cities to its boondocks. Although Ike seized the credit, the interstates were born years earlier, in a report that Thomas MacDonald submitted to then-president Franklin Roosevelt. The system's mind-numbing scale reflected FDR's taste for overwhelming public works: it would span forty-eight thousand miles, cost $114 billion, and contain enough concrete to fill the Great Pyramid of Giza a hundred times. Highway builders vowed to "move enough dirt and rock to blanket Connecticut knee-deep."

The interstates would be America's safest roads as well as its biggest. Every interstate was at least four lanes wide—two in each direction, divided to prevent head-on collisions. Shoulders were broad, curves gently banked, slopes gradual. And, critically, drivers could only access them via designated ramps rather than pulling on and off at random surface roads. By controlling entries and exits, engineers hoped to keep traffic humming along at sixty-five miles per hour, eliminating the friction that provoked crashes. Manufacturers responded by outfitting cars with space-age details like tail fins and exhaust ports, features that made vehicles seem to "float like magic carpets along open roadways."

If cars were spaceships, then white-tailed deer were asteroids. Suburbs sprang up wherever interstates grasped, expanding the periurban savanna. The new highways birthed a DVC feedback loop: they fostered suburbs whose edge habitats churned out white-tailed deer, then ushered more cars moving faster than ever through those same landscapes. The deer boom was both a product of the interstates and a threat to them, a dangerous form of friction that transgressed the system's ideals. To achieve their highest purpose, interstates would have to rid themselves of deer.

Fittingly, the crisis came to a head in Pennsylvania, then America's deerkill capital. The state was split by Interstate 80, which ran from San Francisco to New York City and, by the late 1960s, had begun to funnel waves of transcontinental travelers through whitetail country. As the interstate opened in fits and starts, a bloodbath ensued. Near the village of Snow Shoe, forty-four deer died in November 1968 alone—more than a death per day, on a stretch of highway you could drive in ten minutes.

Making sense of the crisis fell to Edward Bellis, a biologist at Pennsylvania State University. Bellis and a gaggle of collaborators drove the interstate nightly in a 1963 Saab, sweeping a twelve-volt spotlight along the highway-side to catch the luminous eyeshine of deer. Whitetails were everywhere: grazing the verges, milling around roadside hayfields, lounging beneath hickories to ruminate their meals. Their activity peaked during the fall breeding season, known as the rut, when bucks moved more and seemed to think less. The main problem, Bellis concluded, was that the state had planted the highway's flanks with a buffet of clover, grass, and vetch. Deer spent their days bedded in the woods, their nights feasting on the roadside. "The rights-of-way may be visualized as long, narrow pastures bisected by high-speed highways," he lamented. Interstate 80 had created the best possible edge habitat in the worst possible place.

The solution appeared obvious: keep the deer off the road. In 1970 the Pennsylvania highway department began edging I-80 with tall wire fences. It was a sensible strategy: in Minnesota a well-made fence along I-94 would soon reduce DVCs by more than 90 percent. But Pennsylvania's fence was not well made. Whitetails army-crawled beneath it at stream crossings, wriggled through holes, and bounded over spots where fallen trees had crushed wires. The porous fence, Bellis despaired, was "probably of little value." Having deemed the fence useless, Bellis got creative. He placed deer mannequins along the highway and, like a macabre party game, pinned on tails salvaged from roadkilled animals—the theory being that because deer raise their tails as warning flags, they might take the hint and avoid the dummies. The technique, he concluded dryly, was "not recommended."

And yet, despite these failures, something mystifying was happening: deerkills were plummeting. Over two years in the late 1960s, a horrific 286 deer had perished on Bellis's segment of I-80. In a similar period spanning 1970 and 1971, though, only 22 died. Over the two years after that, only 2 deer died. The crisis had dissipated as suddenly as it began.

What explained the miraculous fix? Plenty of deer still roamed the shoulders. Whitetails, though, had all but ceased crossing the highway, repelled by traffic volumes that had "drastically increased" since I-80 opened. As the interstate caught on, denser flows of cars—fast, loud, predatory cars—were deterring deer from wandering between their roadside pastures. Roadkill declined not because the highway was safer but because it had become so heart-stoppingly dangerous that animals rarely ventured onto the asphalt. "The traffic itself forms a moving fence that inhibits deer from moving into traffic lanes," Bellis deduced.

It was a profound realization, one that would shape road ecology for decades to come. Cars created roadkill, yes. But *more* cars made it go away. And two thousand miles west along I-80's course, traffic's moving fence would soon prove deadlier than roadkill itself.

2

THE MOVING FENCE

Why highways destroy animal migrations and
how wildlife crossings revive them.

The helicopter came for the Red Desert's mule deer on a sunstruck morning in March. The herd scattered through sagebrush, deer weaving down washes and slogging through snowdrifts, the chopper in pursuit, twenty feet off the ground, then ten. With a hydraulic *pffft*, a weighted net burst forth and enveloped a doe, who capsized in a thrashing heap. A crewman bounded out to straddle the deer, blindfold her, hobble her hooves, bundle her into a sling, and clip her to a rope beneath the chopper's belly. The helicopter, a black Robinson R44, leapt skyward, two deer now dangling like cherries, wind whistling through their fluted ears, the pair thinking perhaps on some level beyond human ken, *Not this again.*

After a two-minute flight, the helicopter hovered over Crooked Canyon, a valley in southwestern Wyoming that could fairly be described as the middle of nowhere. A few junipers, made bonsai by wind, clung to the hills. The Robinson swung low and its cargo with it, silhouetted against cobalt sky. The animals touched down. The pilot detached the rope with the press of a button, and the helicopter lifted again, leaving behind a leggy load of deer.

A ground team of a dozen biologists strode out to meet the animals. I followed, shielding my eyes against stinging rotor wash. The deer lay still, calmed by their blindfolds. Steam rose from their nostrils. Leather collars encircled their necks, even now beaming data to orbiting satellites. When I touched one doe's shoulder, she tossed her head, wild and defiant. I withdrew, leaving a palm print in the thick clay of her hair.

The team worked with swift purpose, a skilled pit crew handling a racecar. They hefted the first deer onto a scale (137.8 pounds) and then a rubber mat, where a graduate student inserted a rectal thermometer to make sure the stress of capture hadn't elevated her body temperature. Others drew blood from her jugular and stretched tape measures from nose to tail. The doe brayed in objection, a honking squeal, like a squeezed balloon.

A biologist named Kevin Monteith approached the doe and, with the adroitness of a ranch hand, pinned her legs with his knee. He deftly ran an electric razor over her belly, sloughing away downy hair, then procured an ultrasound kit and pressed wand to belly. The nebulous shapes of two fetal fawns swam onscreen. Monteith tapped a keyboard. "Two-seventeen," he barked. Two, because the doe, like most of her herdmates, carried twins. Seventeen, because the diameter of the fawns' orbital sockets, a determinant of fetal size and health, measured seventeen millimeters. "Fatter animals in December tend to carry fawns with wider eye diameter in March," Monteith added for my benefit.

Monteith laid the wand on the doe's rump and squinted at the screen. We saw the white band of rib, the gray striations of muscle. "This gal has no fat on her," Monteith said. This was not unexpected. A mule deer's rump is her gas tank, the container she fills with fat during summer and draws down through the lean months. By March, the bitter tail of winter, most deer were running on empty. I patted the doe's haunch and felt angular protuberances through her hide: the spine, the sacrum, the ligament that bridges coccyx and pelvis—the hard anatomy of hunger.

Mule deer are equipped to endure privation. Although muleys, so named for their long ears, are close cousins to white-tailed deer, they're adapted to a grittier, narrower niche. Whereas whitetails are generalists whose range covers most of North America and part of South America, muleys were forged in the blast furnace of the American West and thrive in austere habitats, from Mexican deserts to Alaskan islands, that would kill a lesser cervid. For Meriwether Lewis, mule deer reliably indicated that the Corps of Discovery was nearing mountains: "We have rarely found the mule deer in any except a rough country."

Like most muley strongholds, southwestern Wyoming is rough country in the extreme—sunbaked in summer, snowbound in winter, grated by what Annie Proulx called the "great harsh sweep of wind from the turning of the earth." Over millennia mule deer have developed a sensible strategy for enduring this cyclic hardship: they move. In spring they trek into the mountains to find lush pasture; in fall they wander back to lowlands to ride out winter. They are guided by ancestral trails and their own impeccable memories. "These animals have incredible cognitive capacity," a biologist named Matt Kauffman said as we watched Monteith perform his routine. "They have the ability to maintain detailed spatial maps in their heads. They can move across any landscape—mountains, forests, rivers, plains, sand dunes—sometimes walking literally in their own footsteps from last year."

Kauffman serves as lead scientist of the Wyoming Migration Initiative, a research group that seeks to unveil the cryptic movements of deer and other mobile mammals. Wyoming is among North America's finest migration-scapes: its fickle climate rewards seasonal movement, and its scant human population allows ample room to roam. Twice a year more than a million Wyoming ungulates—a group that includes deer, moose, elk, bison, prong-horn, and bighorn sheep—take to the hoof, a mobilization of biomass that is the Lower 48's answer to Serengeti wildebeest. Along the way they sculpt the ground beneath their cloven feet. Their grazing stimulates new growth; their young feed wolves and bears. Eventually their bodies melt into the earth, fertilizing the meadows they once strode.

Among all these animals, mule deer are unsurpassed cartographic geniuses. In most mobile species, migration knowledge is hardwired. Song-birds track the stars; some moths sense magnetic fields. Deer, however, *learn* to migrate, cultivating mental maps as they tail their mothers from winter range to summer pasture. Whereas other ungulates stray widely, mule deer remain faithful to their inherited pathways. Deer migration isn't merely a movement pattern but a form of culture, transmitted from doe to fawn like family lore. And when roads thwart their treks, the loss is as thorough as the erasure of a language.

Animal lives are defined by mobility. Creatures move for countless reasons and at countless scales. Bobcats circuit their territories to chase off interlopers; young beavers abandon their home lodges to find unoccupied waters. Years ago, cruising through New Mexico at dawn, I spotted a tarantula crawling across the highway. I pulled over, nudged his hairy body onto my atlas, and ferried him to the shoulder. Moments later I came across another spider, and another, and another. I'd happened upon a "walkabout," when male tarantulas pour forth to scour the desert for bachelorettes.

This arachnid Rumspringa, however stunning, isn't a true migration. Neither is a bobcat's daily patrol nor a beaver's once-in-a-lifetime dispersal. Migration is a specific phenomenon with a narrow definition: a seasonal back-and-forth movement between points. Humpback whales that engulf Alaskan herring give birth months later in Hawaii's bathtub-warm seas. White-throated sparrows that gorge on blackflies in Canadian forests overwinter in Florida. For thousands of species and for millions of years, migration has been an ingenious way to synchronize life's rhythms with the ephemerality of our planet's resources.

You'd think that all those nomads would be hard to miss, but for decades most migrations went undetected by scientists. Animals moved under cover of darkness or through rugged terrain where they couldn't be tracked. In Wyoming biologists long suspected that ungulates migrated, but they didn't know where. Each spring herds of deer, elk, and pronghorn—a lovely mammal, known colloquially as antelope, endowed with auburn flanks and graceful horns—departed their winter range and disappeared into the mountains like loose change vanishing into a sofa. Primitive technology made them hard to follow. Although scientists began outfitting elk with high-frequency radio collars in the 1960s, they had to get close enough, either by foot or aircraft, to pinpoint the signal with an antenna. Radio monitoring revealed patches of habitat but not how animals transited between them, a connect-the-dots with no dots connected.

Biologists needed a tracking device that didn't require constant attention. In 1970, NASA deployed history's first satellite collar—a bright-red contraption that weighed twenty-three pounds and beamed its location to the Nimbus 3 satellite as it orbited overhead daily—on an animal dubbed Monique the Space Elk. Monique soon died of pneumonia, and the collar's next bearer, Monique II, was shot by a hunter. Still, the concept outlived the Moniques, and by the early 2000s, light collars with durable batteries could receive and calculate thousands of GPS points. After capturing months of coordinates, these "store-on-board" collars automatically dropped from their wearers, allowing biologists to collect them at their leisure. In Wyoming previously unknown migrations—moose that traipsed from Jackson Lake to the Tetons, elk that flowed from Yellowstone to Cody—disclosed themselves to incredulous scientists. "It was like Christmas when we got the collars back," Matt Kauffman told me. Migration wasn't a subplot of ungulate life, but the story itself.

Wyoming's most impressive migration unveiled itself in 2010, when the federal Bureau of Land Management tasked a biologist named Hall Sawyer with keeping tabs on some mule deer in the Red Desert. Sawyer was an able-bodied outdoorsman, Wyoming born and bred, who had gotten into wildlife science to commune with the critters he'd grown up hunting. He gained renown in the 1990s, when he and colleagues radio-collared pronghorn in Grand Teton National Park and followed them on foot to the Green River Valley—a 120-mile march, known as the Path of the Pronghorn, that became America's first federally protected migration corridor. Compared to that hike, tracking the Red Desert's deer, which were thought to be sedentary, seemed like a cinch. Sawyer collared forty deer in early 2010 and waited to see what they would do.

That summer Sawyer dispatched a pilot to check on his deer. The pilot flew over the Red Desert with an antenna, waiting for the familiar click of a signal. (Although store-on-board collars capture GPS points, you still have to find the collars with a radio receiver and connect them to a computer to extract their data.) The pilot texted Sawyer: *Can't find your deer.* Sawyer suggested he change frequencies. No luck. The pilot wheeled for home. As he skimmed along the Wind River Range, he tested the receiver one last time. A signal ticked through the static. He'd found the deer, a hundred miles from where Sawyer left them.

When Sawyer reclaimed his collars, the migration's full majesty became apparent. Each spring between one and two thousand deer, many of them pregnant does, depart the Red Desert. They clatter over escarpments and tiptoe along bluffs, linger in willow bottoms and bed in aspen stands, paddle across lakes and ford rivers. They intersect the Oregon trail's ruts, wild pilgrimages braided with the ghosts of human ones. Weeks after they embark, they scatter into the green folds of the Hoback Basin to give birth, then a few months later lead their fawns back to the desert to overwinter. They travel three hundred miles round-trip, the longest known ungulate migration in the contiguous United States. "We had no idea that any of them were doing this, much less the route they were taking," Sawyer told me. "It was mind-boggling how far those deer were going."

And then they reach the interstate.

The southwestern Wyoming landscape is conducive to both deer migration and human movement. While the state's northern half is crumpled by mountains, its southern plains are more rolling than sheer, a notch in the wall of the Rockies. This forgiving topography has long attracted travelers; petroglyphs depicting deer, bison, and elk attest to the ancient presence of the Cheyenne and Shoshone. In the nineteenth century the Union Pacific Railroad punched through, birthing towns wherever it laid track. Then came the Lincoln Highway, America's first cross-country road, over whose route states battled like modern cities competing to host Amazon headquarters. When the highway's founders chose southern Wyoming in 1913, the governor called for an "old-time jollification to include bonfires and general rejoicing."

After Interstate 80 supplanted the Lincoln Highway in the late 1960s, though, jollification was in short supply. Southern Wyoming might be the easiest place to cross the Rockies, but it still towers seven thousand wind-swept feet above sea level. No sooner had the interstate's concrete dried than monstrous multivehicle wrecks began piling up, eighteen-wheelers blown together like tumbleweeds. CBS News called it "the worst stretch of interstate in the nation." Truckers gave it another epithet: the Snow Chi Minh Trail.

The Snow Chi Minh Trail was as cruel to wildlife as it was to tractor-trailers. Wyoming's animal migrations generally flow north to south, in parallel with its mountains—and directly perpendicular to the east–west interstate. Where vectors crossed, disaster followed. In the decade after I-80 opened, a thousand muleys died along a single migration route.

And roadkill was merely the vanguard of a graver crisis. Back in Pennsylvania, Edward Bellis was learning that traffic deterred white-tailed deer from crossing I-80, the impenetrable wall of cars he had called the "moving fence." In the bucolic Northeast the moving fence wasn't devastating: whitetails could find food year-round, whatever side of the highway they were trapped on. In harsh western environments, however, animals had to roam to survive. If traffic stopped ungulates from migrating, they'd die.

Sure enough, the Interstate Highway System lopped off migration routes as neatly as a guillotine. Soon after I-84 opened in Idaho in 1969, biologists discovered deer herds floundering in deep snow north of the highway, "either unable or unwilling" to cross the interstate. Though the state dispensed "special pelletized ration[s]," hundreds starved, and a herd once two thousand animals strong crashed to eight hundred gaunt survivors. In Colorado, I-70 proved so forbidding that conservationists eventually dubbed it the "Berlin Wall for Wildlife." And I-80 was the worst of the bunch. In October 1971 an early storm buried Wyoming beneath deep powder, and fierce winds blew up ground blizzards for days. Herds of pronghorn amassed along I-80's north shoulder, anxiously pawing at the snow and pacing the right-of-way fences. More than three thousand ultimately perished. The new interstates spanned barely a hundred feet from shoulder to shoulder, yet they had inflicted massive habitat loss, denying animals access to millions of acres of life-sustaining land.

These early tragedies testified to the necessity of migration—and they foretold what happened in late 2010, when Hall Sawyer's deer finally returned to the Red Desert after their three-hundred-mile journey. That winter the snows fell hard and early, enveloping the bitterbrush and mountain mahogany in oppressive drifts. Muleys congregated north of I-80, outside the coal town of Rock Springs. A century earlier they would have soldiered on south, perhaps as far as Colorado, but no longer. They pooled against the interstate like water behind a dam, waiting vainly for a crack in the vehicular wall. By spring nearly 40 percent of Sawyer's deer had starved.

"They weren't killed out on the interstate—they died of malnutrition," Sawyer told me. It was the first time he'd observed such a die-off but not the last. "It doesn't happen when the snows are bad," he said. "It happens a few weeks or even a month later. Animals that are healthy move on from their winter range in April. The other ones just tip over, lay down, and die. If they could move south across the interstate during those winters, they certainly would. How far they would go, I don't know."

⸺⸺⸺

The Red Desert saga made two facts clear. First, Wyoming was rife with undiscovered migrations. Second, blockades imperiled many routes. The sagebrush steppe might look vacant, but every deer faced a gantlet that would make Odysseus blanch. Muleys had to leap fences, skirt subdivisions, negotiate oil and gas fields, and cross a welter of state highways. And always, always, Interstate 80 lurked behind them, constricting their range and depleting their reserves.

In 2014, Sawyer, Matt Kauffman, and a biologist named Bill Rudd formed the Wyoming Migration Initiative to study and protect the Red Desert herd and others like it. As the initiative deployed satellite collars and published maps on Facebook, it racked up human supporters. Adoring fans followed migrants like Mo, a beloved deer whose name stood for *mobility*, *movement*, and *momentum*. The project's maps dramatized the lives of familiar animals and made migration tangible in a way that few environmental phenomena could be: Wyomingites couldn't see molecules of carbon dioxide accumulating in the atmosphere, but they could follow their favorite deer as easily as racers in the Tour de France. By delineating pathways, the initiative also facilitated conservation. Along the Red Desert route, environmentalists purchased hundreds of acres to forestall development. "Today there's a lot of debate about what measures should be taken to protect these migration corridors," Kauffman told me, "but there's no debate about whether they exist or where they are."

I'd caught up with the Wyoming Migration Initiative in Rock Springs, near the Red Desert herd's winter range. We rolled out in a caravan of Ford F-250s and headed east on I-80, motel coffee sloshing in our veins. At Superior (population 184), we exited onto the dirt track that led up Crooked Canyon, where Kauffman's crew made camp with the practiced competence of Rolling Stones roadies. Tackle boxes and Action Packers and syringes poured forth from truckbeds and trailers, followed by clipboards, disinfectant wipes, and Equi-Phar Non-Spermicidal Vedlube. I saw a bin labeled "Disease Sampling" and another labeled "Tooth Extraction and Suture Kit."

"There's always that perpetual feeling that you've forgotten something when you're doing captures," Anna Ortega, the PhD student coordinating the research, muttered as we waited for the helicopter to arrive with the day's first deer. Kauffman rubbed pensively at the sunscreen that clung to his stubble. "I feel that way about life," he sighed.

Now, deer in hand, it appeared that all necessary gear had been mustered. The helicopter deposited two more muleys, and three biologists jogged over to restart the process. Others hustled off the first two for release. Their hobbles and blindfolds removed, the deer leapt to their feet and stotted off to rejoin the herd.* Soon their dun fur disappeared into the quilt of sage and snow.

And so the day went: deer being dropped off, deer being released, deer being weighed and shaved and ultrasounded, the scene a combination of

---

* Can you envision the bouncing gait of a gazelle, all four feet pogoing off the ground with each springy leap? That's *stotting*. Whitetails gallop, but mule deer prefer to stot, perhaps to show their predators just how fit they are as they leave them in the dust.

obstetrical clinic, phlebotomy lab, and sheep-shearing station. I assumed the least-skilled job: packing snow around overheated deer as though they were branzino at a seafood counter. Hot deer also received an ice-water enema from a bag-and-tube system that resembled a CamelBak. "You basically want to pinch her anus around it, and make sure it's all the way in," one researcher instructed an undergraduate. The nozzle writhed loose, and enema water gushed forth.

"You failed me, boss!" Kevin Monteith cried with faux outrage as the torrent soaked him.

"You're gonna want more of a slow enema drip," a biologist gently advised the chagrined student.

"That's a band name right there," someone quipped.

The banter belied the research's import. In a few weeks, Ortega explained in a quiet moment, the Red Desert herd would begin its annual pilgrimage toward summer range. Soon after departing, it would divide, amoeba-like, into smaller subherds, each of which would follow its own path. Some were homebodies, wandering only a few dozen miles. Others, as Hall Sawyer had shown, would trek 150 miles. And one legendary doe, Deer 255, ditched her herdmates and pressed on—up the Gros Ventre Range, along the shores of Jackson Lake, and across the Snake River, all the way to Idaho. Was this mere wanderlust or part of a broader survival strategy? By comparing the animals' health—revealed by their bodies, their fluids, and the size of their fetal fawns—to their migration routes, Ortega hoped to figure out why the group's units went where they did. Her hypothesis was that the deer divided to hedge against weather, like a trader maintaining a diversified portfolio. "Some migrations might do better in heavy snow years, some might do better in mild winters," Ortega said. "Over the long term they balance out. The idea is that the herd doesn't want to put all its eggs in one basket."

That intricate strategy hinted at a bigger idea. Although migration is by definition a movement between points, wild journeys are anything but simple. Wyoming's muleys, the initiative's scientists have learned, travel in puddle jumps rather than by nonstop flight. Instead of hastening between ranges, deer spend weeks hanging out at "stopover sites," newly sprouted pastures where plants are tender and nutrient-dense—"like baby greens in a fresh spring salad," as Kauffman put it. As the snow melts, grasses sprout in its wake, an emerald sweep so vivid you can detect it from space. Where plants erupt, deer follow and linger, a behavior known as "surfing the green wave." For a mule deer on the hoof, it is always early spring.

This harmony with the land makes surfing a fragile tactic as well as a profitable one. Travel too quickly, and a deer risks arriving at her stopovers before the salad is ready. Too slow, and the plants have turned brittle by the time she shows up. Roads and other impediments not only prevent animals from getting where they need to be; they stop them from arriving when they need to be there. Sawyer and Kauffman have found that migrating deer deviate from their habitual routes to avoid oil and gas drilling fields, frightened by the bedlam of compressors and trucks, then have to hustle to catch up with the receding green wave. "When you recognize their need to surf," Kauffman said, "you realize that anything that affects their ability to move can alter the long-term survival of these migrations."

The day had by now turned unseasonably warm, and the earth was churned into a slurry of snowmelt, enema water, and feces. Deer hair thatched the ground like mulch, tickled my larynx, and floated in a communal pot of chili. Kauffman stifled a sneeze.

"Time for me to take a Claritin."

"You okay?" I asked.

He laughed miserably. "I'm allergic to mule deer."

———————

The plight of Wyoming's deer illustrates a broader crisis: roads have made migration a dangerous bet. Salamanders perish on suburban streets between breeding ponds; caribou shy from Alaskan oil roads. One 2018 study, ominously titled "Moving in the Anthropocene," found that mammals from long-tailed pocket mice to African elephants are losing their *vagility*, or their ability to wander freely, as infrastructure dices their milieus. The collapse of migration transcends individual species. It is the endangerment of a biological process, the loss of a lifeway nearly as ancient as movement itself.

Precisely why roads hinder animals is a straightforward question with a complex answer. In 2016 a biologist named Sandra Jacobson and her colleagues split animals into four categories based on how they react to roads. On one end of the spectrum were the *nonresponders*, creatures that ignore roads at their peril—like leopard frogs, which hop across roads no matter the traffic—and *pausers* like porcupines and skunks, which creep onto highways and then hunker down. On the other end sat wary, intelligent *avoiders*, such as grizzly bears, who generally steer clear of even rural roads. Whereas nonresponders are fatally heedless, avoiders are imprisoned by their own innate caution.

Jacobson placed deer in a fourth category: *speeders*. Speeders are fast and alert, evolved to outrun predators. When traffic is light, speeders race between vehicles like a running back bursting through the line. As traffic intensifies, the interval between cars tightens, crossing gets riskier, and collisions increase. Crank up the traffic even further, though—to the level of, say, I-80—and roadkill plateaus, then declines. Why? The highway has become so dangerous that only the foolhardiest deer would attempt to speed across. A lonesome county road could thus have the same negligible deerkill as an interstate: the former because it's traveled by so few cars that deer easily speed between them; the latter because traffic's moving fence is so densely woven that they don't even try.

But how many cars are too many? In 2013, Corinna Riginos, an ecologist at the Nature Conservancy, took up that question. That winter she and her technicians rigged Wyoming's highwaysides with remote thermal cameras. Road ecologists had long contented themselves to bend over carcasses, the gruesome tea leaves that James Raymond Simmons called "mute evidence"; now one could watch animals make decisions in real time. Riginos's videos are surefire anxiety provokers. In one harrowing clip a doe, rendered a pale ghost by the thermal camera, edges up to a two-lane highway, fawns in her wake. She steps forward gingerly, then retreats at a car's approach. Fifteen seconds elapse. The doe reenters the frame. Her fawns follow. A pickup slams the brakes, taillights glowing; the panicked trio wheels away. Forty more seconds. The doe reappears. A semi hurtles past, then another pickup. The coast clears. She crosses alone, head up, ears pricked. Headlights flare in the distance. Her trailing fawns trot forth. The headlights brighten, the fawns hastening to their mother, death racing to intercept them, one laggard still too slow, hurry up, hurry *up*! I watch through laced fingers. And then, at the last instant, the dawdling fawn accelerates, a sprinter breaking the tape, and he's off the road and into the brush, alive to migrate another day.

In the folder Riginos sent me, that video file is labeled Near_Miss.mp4. You don't want to see BIGcollision.mp4.

Riginos's footage, however traumatizing, quantified an important metric: *gap acceptance*, an intuitive concept borrowed from pedestrian studies, road ecology's sister science. Humans, like deer, are speeders: the sparser the traffic, the safer we feel hurrying across a street. Because we're good at gauging speed, we take our chances with small gaps between cars; in India jaywalkers feel comfortable crossing at intervals as short as five seconds.* Riginos knew that deer would require larger gaps than people: to us, cars are familiar machines; to deer, they're alarming predators. Still, she was taken aback by how wide a gap deer needed. When cars rolled by every thirty seconds or less, deer refused to cross, or were killed or nearly killed, when they did. When the gap exceeded a minute, they usually made it without incident. Gaps between thirty and sixty seconds were gray areas, sometimes perilous, sometimes not. Nearly a half-century after Edward Bellis diagnosed the moving fence, Corinna Riginos identified the moment at which it formed.

A car every sixty seconds is, to be clear, not many cars. Watching the terrifying drama of Near_Miss.mp4, I was struck not by the highway's busyness but by its emptiness. This was a quiet two-laner in America's

---

* Pedestrians in more developed countries require wider gaps, perhaps because they're accustomed to stringently regulated traffic. Anyone who has braved the crossweave of motorbikes in Hanoi knows not to expect any gap whatsoever: just cross your fingers and step into the flow.

least populous state, yet it was too hectic to permit fluid migration. Most of Wyoming's highways, Riginos has calculated, exceed deer-friendly volumes, and its interstates are cursed with insufficient gaps almost twenty-four hours a day. The situation is surely worse in more crowded states, which is to say, everywhere else. "As a rule of thumb," road ecologists have written, "roads with more than 10,000 vehicles per day should be considered absolute barriers to most wildlife."

Deer, like people, lack infinite patience. Frustration makes them imprudent. "Sometimes when you see them make a risky move—maybe I'm anthropomorphizing, but it feels like they're moving out of desperation," Riginos told me. A deer that terminates four crossing attempts, she's found, has a 50 percent chance of giving up altogether. "The dead animal on the side of the road is very visible and very visceral," Riginos said. "But that's only one piece of the problem. Roads are making it hard to get to the habitat they need to survive."

In hindsight Riginos's research explains why Hall Sawyer's mule deer herd starved en masse in early 2011. By then, around thirteen thousand cars and trucks were driving I-80 each day—one vehicle every 6.5 seconds, a fraction of the gap that deer require. The interstate had calcified into a four-hundred-mile wall across Wyoming.

Yet the Snow Chi Minh Trail was not completely impermeable. Interstate 80 epitomized the problem of the moving fence, yes—but it also taught biologists how to poke holes.

When I-80's designers plowed the Snow Chi Minh Trail through Wyoming in the 1960s, they were not entirely heedless of wildlife. "Curtailed movement of some herds appears critical," cautioned one biologist. So, at the state's request, engineers installed four concrete tunnels, or box culverts, where the interstate bisected a migration pathway between Laramie and Rawlins. They would be among the first wildlife crossings in the United States.

No one had much reason to believe the passages would work. Back in

1961, Jerry Chiappetta had noted in *Field and Stream* that some states—he didn't mention which—had built tunnels beneath "particularly dangerous deer crossings" but that "deer refused to use them." At first the I-80 passages seemed likewise doomed. Wyoming's engineers hadn't designed the culverts merely for deer but also to funnel water beneath the highway, and they had egregious feng shui. They measured ten feet wide, ten feet high, and several hundred feet long—dark, echoey crypts that could backdrop an ungulate horror movie. Predictably, deer hated them. Safety, for a speeder, lay in open country, not a cramped culvert.

With deer reluctant to enter of their own volition, the Federal Highway Administration fenced off the interstate to guide herds through the tunnels. As it had in Pennsylvania, however, a supposedly deer-proof fence proved as porous as a colander. Rather than slink into the uninviting culverts, deer slithered through holes in the fence or followed it to its end, outflanked it, and died on the interstate. Evidently, someone would have to keep the deer behind the fence and coax them through the tunnels. The gig fell to a jocular Texan named Hank Henry. Henry lived in a trailer along I-80 and patrolled the roadside, shooting bottle rockets at any deer he caught trying to get around the fence. "When they got really close, it was Roman candle time," Henry told me. He also stuffed human hair, gleaned from salons, into rolls of pantyhose and hung them along the roadside, in accordance with the theory that deer fear human scent. Whether the hair worked was never clear, but the technique did make Henry popular among Laramie's barbers. "They loved it when I showed up. 'Oh, you've come to clean my floor again.'"

The real key, Henry figured, was to make the tunnels less sinister. Mule deer wouldn't probe the fence for gaps if they had an appealing place to cross. In 1978, Henry and his supervisor, Lorin Ward, began to pile the culverts with hay, apple pulp, and vegetables scavenged from Kroeger. "We would go out with two or three big boxes of cabbage and lettuce and beet tops," Henry recalled. "Anything to entice 'em." Deer took the bait. Animals that had loitered near the tunnels for weeks now breezed through in an evening. Some got so comfortable that they bedded down in the under-passes.* "Hell, they had a roof over their head," Henry said.

But the veggies were a stopgap, not a solution. That summer, at Ward's and Henry's urging, the highway administration extended the fence another mile to the east. There, it reached a machinery crossing—a broad, dirt-floored underpass intended for tractors, bulldozers, and other equipment. No longer did deer find themselves on the pavement when they reached the end of the fence. Now the fenceline guided them to a crossing that, though not intended for their use, felt more natural and attractive than the eerie concrete culverts. Soon hundreds of deer were ambling through every sea-son, and roadkill had fallen some 90 percent. "Each year it just got better and better," Henry said.

Through trial and error, Lorin Ward and Hank Henry had developed road ecology's most important formula, one that future generations of sci-entists would accept as gospel. Neither fencing nor crossings worked in iso-lation: fences without underpasses prevented roadkill but blocked animals' movements, while underpasses without fences didn't get used. In combi-nation, though, they permitted migrations to continue—but only in safe, predictable places beneath the highway's surface. Fences kept ungulates off the interstate. Underpasses got them across it.

The success on I-80, alas, proved fleeting. Once Henry was no longer around to bribe deer with veggies, they mostly shunned the culverts. Wyo-ming's transportation department took all the wrong lessons from the experience. In 1989 the state slapped up roadside fencing along Highway 30 in Nugget Canyon, a valley where more than a hundred deer perished every year as they migrated down the spine of the Wyoming Range. This

---

* Road ecologists still bait crossing structures today. When an elephant underpass opened in Kenya, biologists left a mound of pachyderm dung near its entrance; an elephant named Tony crossed that very night. In Costa Rica, jaguars have been lured through underpasses with Obsession, a Calvin Klein perfume that purportedly sets a "sexy, provocative mood."

time, though, engineers didn't bother adding underpasses. Instead, they routed deer across the highway through a narrow fence gap, which they marked with a few useless flashing lights and warning signs, like a muley crosswalk. "That just collected all the dead deer in one place and made it easier for the maintenance guys to pick them up," a state engineer named John Eddins told me.

When Eddins began to lobby for underpasses at Nugget Canyon in the late 1990s, he met with incredulity. His colleagues remembered the I-80 culverts as failures. "People were saying we needed the money elsewhere. The bridge department was saying our bridges were failing, our roads were going to hell," Eddins recalled. But he caught the ear of a sympathetic supervisor, and in 2001 the state cobbled together $400,000 for his pet project. "Truckers, men in the street, guys having coffee: everyone said it wouldn't work," he chortled.

Eddins—square in build, stoic in manner—struck me as every bit the engineer when I met him at his office in Rock Springs. But he was also a hunter with a hard-earned understanding of deer behavior. What his underpass needed, he decided, was a high *openness ratio*. A tunnel's openness is a crude formula that anyone can compute on a napkin: width times height divided by length. Think of it as the Scariness Index. Long, narrow, creepy underpasses had low openness ratios. Wide, well-lit, appealing ones had high ratios. For a mule deer, inching through the I-80 tunnels was something like crawling down a mineshaft.

The concrete culvert that Eddins installed at Nugget Canyon was superficially similar to the original I-80 passages, but it had a far higher openness ratio. Eddins's underpass was twice as wide—twenty feet instead of ten—and much shorter; a deer standing at one end could see clear to daylight on the other. From the moment the culvert went in, Eddins said, "just outrageous numbers of deer were using it." The state, finally convinced, built six more underpasses nearby, with thirteen miles of fencing between them. Roadkill plunged, from almost ten victims monthly to fewer than two. The deer proved steady learners. When Sawyer set up cameras in the passages, he saw that muleys strolled through more confidently each year as

In Wyoming, biologists discovered that the combination of roadside fences and underpasses allowed mule deer to cross I-80 safely. *Gregory Nickerson*

they acclimated to the culverts and incorporated them into their routes—ungulate culture and human technology together forging a new pathway. Within three years nearly fifty thousand muleys had trekked under Highway 30. Mule deer didn't need lettuce trimmings to enter an underpass. They were simply claustrophobes.

The Nugget Canyon underpasses were also backed by compelling fiscal math. Historically, wildlife crossings had drawn from the same limited pots of funding as basic maintenance, like lane repaving and bridge repairs. Pitted against crumbling infrastructure, animals almost always lost. In 2009, however, a group of eminent road ecologists, led by Marcel Huijser, tallied up the many costs of wildlife crashes: the hospital bills, the car repairs, the loss to hunters, and so on. The average DVC, they calculated, set society back $6,600. Bigger animals inflicted worse damage: elk collisions cost $17,500, while moose came in at a whopping $31,000.* All told, ani-

---

* Adjusting Huijser's figures for inflation in 2021, researchers at the Center for Large Landscape Conservation calculated that the average collision now costs around $9,100 for deer, $24,000 for elk, and $42,200 for moose.

mal crashes were costing America more than $8 billion per year. By that calculus, engineers could save the public money by building underpasses and fencing along any mile-long highway segment that killed at least three deer, one elk, or one moose per year—in other words, pretty much any road that bisected important habitat or a migration path. The Nugget Canyon underpasses, which prevented ninety-five crashes annually, paid for themselves within five years. Crossings weren't lavish expenditures; they were downright thrifty. "I honestly think the cost-benefit analysis is the most important piece of work I've done in my career," Huijser told me in 2013. "We could keep ourselves busy for a long time mitigating road sections where it's beneficial not only to safety or conservation but to our wallets."

As cost-conscious engineers warmed to crossings, fences and underpasses proliferated in Montana, Colorado, Arizona, Idaho, and other western states. Road ecologists tinkered with the formula. Engineers installed more capacious crossings, building dirt-floored, open-span underpasses that mimicked natural conditions better than box culverts. They nailed down the optimal spacing between crossings (no farther apart than a mile so that animals could find them without too much trouble), the ideal fence length (at least three miles so that ungulates couldn't easily outflank them), and the design of "jump-outs," one-way ramps that allowed deer to escape the highway if they broke through the fence and found themselves trapped on the road. Some biologists installed "skylights," open grates that illuminated the passage; others eliminated overhanging ledges from which mountain lions might have launched ambushes. The tweaks varied, but the big picture didn't: where crossings and fences went in, crashes ceased and migrations continued. High-quality remote cameras, activated by movement or mammalian body warmth, captured parades of deer and elk striding through underpasses, dispelling any lingering doubts. "We could finally show the engineers that structures work, and suddenly they believed us," a road ecologist named Patricia Cramer told me.

As wildlife crossings caught on, a geographical rift cleaved America's transportation departments. Western mule deer and elk migrated so faithfully that they formed unmistakable roadkill hotspots. Driving through

Idaho one March, I narrowly avoided a band of muleys as they darted from a draw, across the highway, and up a ridgeline. When I pulled over, shaken, I noticed a half-dozen of their fallen companions, necks twisted grotesquely and ribcages open to the sky. I didn't need satellite data to identify a promising underpass site. The ravens said it all.

Meanwhile, eastern states continued to struggle with their old nemesis: white-tailed deer. Unlike muleys, whitetails didn't migrate along consistent routes; instead, they seemed omnipresent. When everywhere was a roadkill hotspot, nowhere was. "I have photos of deer lying down along the grassy side of the interstate, sleeping, breeding, browsing—I mean, *everything*," Bridget Donaldson, of Virginia's transportation department, told me. "Deer seem ubiquitous. People just throw their hands up and say, what can we possibly do? We can't have wildlife crossings absolutely everywhere."

But Donaldson refused to surrender. Virginia's whitetails didn't migrate, true, but they did use certain predictable habitats, like stream corridors and woodland edges. In 2016 she strung up a mile of fencing around two existing underpasses beneath Interstate 64: a bridge that allowed a creek to trickle under the highway, and a culvert through which dairy farmers once herded cattle. The crossings weren't built for deer, but Donaldson suspected that whitetails, with guidance from fences, would adopt them. As she'd expected, roadkill declined by more than 90 percent, and deer—as well as black bears, skunks, and foxes—adopted the underpasses, often with fawns, cubs, and kits in tow. Within two years the fences had paid for themselves. It was a crucial lesson: whitetails, too, could be cajoled into crossing beneath interstates, even via underpasses that weren't expressly built for them. The world was rife with wildlife crossings, just waiting for fences to put them to use.

Environmentalists tend to deal in controversy. There are agencies to sue, pipelines to protest, polluters to prosecute. Talk about wildlife crossings with any proponent, though, and you're guaranteed to hear the phrase "win-win." Often, it's extended to "win-win-win"; with especially enthu-siastic interlocutors, you might get to "win-win-win-*win*." Consider the healthy herds, the averted hospital bills, the car-insurance savings, the fore-stallment of animal suffering. Wildlife crossings, with rare exceptions, have become one of the country's few truly bipartisan environmental interven-tions, equally beloved by hunters and the Humane Society.

When I visited Wyoming, I found a deep-red state that couldn't get enough of costly government projects. The transportation department was riding high on a grant to install underpasses near Big Piney. Activists in Teton County had pushed through a tax that would raise $10 million for passages. John Barrasso, a Republican senator not known for his sympa-thy to environmental causes, had sponsored a highway bill that earmarked hundreds of millions of dollars for crossings, and the Trump administra-tion had enjoined Wyoming and other states to protect migratory corridors. The legislature had even approved a special mule-deer license plate whose sales would fund crossings and fences. "A license plate fits perfectly to the cause," Josh Coursey, president of a conservation-minded hunting group called the Muley Fanatic Foundation, told me. "You're driving around out there generating brand awareness."

The Muley Fanatics, who had campaigned for the license plates, endorsed wildlife crossings because they worked. But crossings also seemed attrac-tive for reasons of political expediency. When the Fanatics first lobbied the legislature to commission the plates, Coursey said, the politicians "weren't even lukewarm. So how do we get their attention? We needed to rope in some powerhouses." The campaign took off when the Fanatics secured endorsements from two of Wyoming's most powerful forces: the Stock Growers Association and the Petroleum Association. "We saw an immedi-ate change in momentum," Coursey said.

I asked Coursey why ranchers and oilmen had joined the cause. "They recognize their vehicles contribute to the problem," he said. But I won-dered whether a shrewder calculus was also at play. Wyoming's mule deer population was crashing, and while highways had surely contributed, they weren't the only culprit. There were the burgeoning subdivisions, the clamorous oil and gas pads, the ranchers who fettered the land with fences. (More than once, Sawyer's cameras had captured deer and prong-horn flayed by barbed wire, blood painting their legs as they stumbled through an underpass.) Compared to roads, however, those crises were covert: we see the carcass on the shoulder, but never the doe who loses her fawn because a remote drilling operation disrupted her green-wave surf-ing. By endorsing crossings, Wyoming's most prominent industries could cast themselves as conscientious stewards and deflect blame from their own sins. Wildlife crossings were, as Matt Kauffman put it, a "safe, neutral space where everybody can work"—a nonregulatory approach in a deeply conservative state.

If crossings were political pawns, they were also subject to economic forces. Transportation departments tend to build crossings along moder-ately trafficked rural highways with high collision rates, where cars smash deer so routinely that underpasses pay for themselves. Along I-80, how-ever, the math wasn't so clear-cut. The interstate's moving fence might be disastrous for ungulates, but, perversely, it protected drivers from colli-sions. If crossings had to pay for themselves, I-80 might not qualify for one. "My impression is that the DOT hasn't plunked down millions of dollars on projects just to connect historical seasonal ranges," Kauffman said as we rolled back to Rock Springs after his crew's long, allergy-inducing day of muley research. The fact that deer rarely breached the interstate also made it hard to know where they wanted to cross. Satellite collars could tell Kauffman where muleys approached the highway, but there was no conspicuous heap of carcasses. "We can't rely on the animals themselves to tell us where they'd like a crossing," Kauffman said.

This raised other troubling questions: Even if Wyoming did install cross-ings along I-80, how would ungulates learn to use them? How do you revive a migratory culture that's been dormant for decades? "My thought is that it's a trial-and-error thing," Kauffman mused in the darkness of the truck's cab. "In a string of normal years, they might not discover the crossings. But you get a really harsh winter, and their traditional winter range has four feet of snow, so they push down another twenty miles. They would eventually figure it out."

This isn't conjecture. Kauffman and his colleagues have found that ungulates relocated into unfamiliar landscapes fail to migrate, likely because they lack detailed mental maps of their surroundings. As generations pass, though, they acclimate to their environs and pioneer new pathways. Fawns, calves, and lambs perfect their mothers' routes, knowledge flowing meme-like through the herd as social learning and lived experience form a migration where none had been. Individual routes may be fragile, yet migration itself is resilient—if our infrastructure permits it.

The next day I left Rock Springs and turned north on Highway 191 into Wyoming's rugged heart. Blue skies turned iron; poplars waved against winter's canvas. A couple hours' driving put me in Pinedale, where nineteenth-century fur traders once mustered to swap pelts. Just past town, rivers cinched the land into a ridgeline called Trappers Point, a mile-wide funnel through which thousands of deer and pronghorn pass like sand slipping through an hourglass. This bottleneck lies along the Path of the Pronghorn, the well-trod migration route that Hall Sawyer and others detailed in the 1990s, and it has served both ungulates and humans for millennia. Archaeologists here once unearthed the six-thousand-year-old bones of pronghorn slain with stone-tipped spears, among them eight fetal fawns—proof that Indigenous hunters had intercepted the animals during spring migration, when their bellies bulged with young.

In the early 2000s, Pinedale attracted another kind of migrant: roughnecks capitalizing on a fracking boom. The region's population doubled, and traffic exploded. Herds of pronghorn and deer approached Highway 191 and retreated, approached and retreated. Trappers Point was an obvious candidate for wildlife crossings, but there was a hang-up. While mule deer readily adopted culverts, pronghorn disdained most underpasses. Pronghorn are North America's fastest mammals, endowed with enlarged lungs and shock-absorbent toes that permit sprint velocities of more than fifty miles per hour. If muleys preferred open country, pronghorn, the ultimate speeders, demanded it. In addition to underpasses, then, Trappers Point also called for bridges, whose long sightlines would in theory allow pronghorn to cross the highway in comfort. Overpasses, though, would run $2.5 million each, six times more than culverts. Skeptics grumbled, but in 2011 the state completed a network of two bridges, six underpasses, and twelve miles of fencing.

Yet again a leap of faith paid off. Hall Sawyer's cameras captured nearly sixty thousand mule deer and more than twenty-five thousand pronghorn within three years, and the overpasses prevented enough crashes to recoup their costs. Most gratifying was the spike in "nondirectional crossings": rather than simply darting across the highway and carrying on, animals *recrossed* to feed on either side. The passages not only permitted migration; they allowed herds to linger, forming another layover on their flight path.

I reached the first overpass in late afternoon. Snowflakes flicked against my windshield, silent as moths. The bridge, faced in muted brown, leapt over the highway, massive and abrupt in the featureless landscape. The wind had scoured the snow from its deck to expose tufty yellow grass. The light faded as I rolled under the bridge, each modular arch curving above me like a whale's rib. Then I was out the other side.

A quarter-mile downhill, I pulled over to examine the roadside fences. Slush trickled into my boots. If there was a green wave to surf, I couldn't see it. But when I looked toward the overpass, a dozen mule deer stood poised on a rise, gray shapes against gray snow. I stepped back, afraid to spook them, but it was too late—I'd been heard, or scented, or seen, or felt—and they took off, stotting in unison like a murmuration of starlings. Near the overpass they paused, etched against sky. I watched them move, in awe of their wildness and wariness, their exquisite choreography with the slow dance of spring.

It was a reminder that bridges, despite their expense, fill a vital niche in the road-ecology ecosystem. Yes, their open decks better serve pronghorn and other skittish species, but they play a communicative role as well. Underpasses are the field's cheap, invisible workhorses; you could drive over a box culvert without ever knowing it. Overpasses, by contrast, are billboards that advertise the preservation of movement, reminding us that we share the land—even with very large carnivores, in our very largest cities.

3

HOTEL CALIFORNIA

*Can America's largest wildlife bridge
save an isolated cluster of cats?*

The mountain lion came down from the hills after midnight in February 2009, in the small brown hours when humans sleep and the wild world moves. He padded through bunchgrass on paws the size of saucers, angular shoulders shifting beneath tawny hide. His tail floated behind him like an independent creature. He flowed south and downslope, following the earth's folds toward the freeway. Earlier that evening he'd crouched in lion-colored brush and watched rush-hour traffic ooze along U.S. 101, a snarling wall advancing west from Los Angeles. But now the freeway's roar had dulled, and the lion neared its north shoulder. His amber eyes flashed, lit by the headlights that swept the scrub.

And then—what? Likely he slunk beneath the freeway via a residential street called Liberty Canyon Road and glided past the houses huddled on its south side. Or perhaps he bounded over the highway's surface in a few leaps, each thrust of his haunches lofting him over forty feet of pavement, unnoticed by anyone save an overcaffeinated trucker who surely knuckled his eyes, thumbed his CB radio, and muttered, "I just saw the *damn*edest thing."

Millions of animals negotiate roads every day—or attempt to. Yet few crossings had ever been so consequential. Months earlier, biologists had captured the two-year-old lion in the Simi Hills, the low mountains north of U.S. 101, some forty miles west of L.A. They'd affixed him with a tracking collar and dubbed him P-12: P for *Puma concolor*, his species' scientific name, and 12 to denote that he was the twelfth cougar they had collared. (The mountain lion has more than forty names, per the *Guinness Book of World Records*; depending on where you live, you might know them as cougars, pumas, panthers, painters, catamounts, or mountain screamers.) No one expected P-12 to cross the 101; no lion was known to have managed it. But young male cougars, like young males of most species, are impetuous and risk-tolerant, and P-12 survived his crossing. In so doing, he entered a new domain: the Santa Monica Mountains.

Although P-12 didn't know it, he'd broken into a prison. The Santa Monicas are a tiny cell of wildness delimited by obstacles: the Pacific Ocean to the south, sprawl to the west, the 101 to the north, Interstate 405 to the east—and beyond that, Los Angeles. Just 240 square miles of chaparral and oak woodlands lie within these borders, not much more than the home range of a single male cougar. And this fiefdom already had a ruler, an iron-pawed despot known as P-1. Trapped in his palace, P-1 had gone mad, or so it seemed to the humans who tracked him. In 2005 he'd killed his longtime mate, P-2, in a brutal struggle soundtracked by the combatants' unearthly screams. The next year he slew two of his own children, then sired four more kittens with another daughter. Granted, violence and incest are natural features of cougar life, yet the intensity of the strife suggested a diseased feline society—too many cats trapped in too small a cage. Something was rotten in the Santa Monica Mountains.

Now P-12 had joined the fray, like Macduff coming to Inverness, and tensions seemed certain to boil over. But the tyrant had grown long in the tooth, and scientists soon found P-1's collar near blood-splashed rocks. P-1's decadent reign was over. P-12, conqueror from the north, ruled the Santa Monicas.

At first P-12 was a savior. He quickly mated with one of P-1's daughters, infusing the inbred population with fresh DNA from the Santa Susana

Mountains, his birthplace. As the years passed, though, P-12's bloodline curdled. No new cats followed him across the freeway, and his coupling options dwindled. Finally P-12 did what P-1 had done before him: he mated with his own daughter, twice. Then he bred with his granddaughter and great-granddaughter. Soon his family tree was as tangled as a grapevine. Freeways had made the Santa Monicas his personal Hotel California: he'd checked in, but he could never leave.

Few roads have influenced their surroundings more than Los Angeles's freeway system, a five-hundred-mile labyrinth that structures the city's layout as surely as any mountain range or river. Los Angeles began building freeways in earnest in the 1950s, superimposing its concrete colossi atop ancient Spanish military trails. Although their ostensible purpose was to relieve surface-street traffic, California's freeways were agents of social as well as civil engineering, designed to advance "a long-established pattern of decentralized, low-density development"—essentially, sprawl. Some exalted them for their convenience; others reviled the "dehumanized quality of life" they imposed on the citizenry. But no one denied their dominance. Freeways, observed one critic, shaped Los Angeles so thoroughly that they created a discrete ecosystem: "a single comprehensible place, a coherent state of mind, a complete way of life."

Among the system's grandest components was the Ventura Freeway, a sixty-five-mile section of U.S. 101, the coastal highway that ran from Los Angeles to the Bay Area and, from there, all the way to Washington State. Just west of L.A., the Ventura passed through the San Fernando Valley, an agricultural enclave that by 1960 was morphing into a "sprawling complex of suburban housing developments, space age industry, and sparkling commercial centers." Reporters predicted that the freeway would halve downtown travel times and catalyze the valley's growth. In that, it succeeded too well. By 1985 the Ventura Freeway had become the most-traveled road in America; by 2016 it hosted the nation's worst commute. During morning rush hour, cars crept along at an average speed of seventeen miles per hour on a freeway designed for sixty-five.*

Less obviously, the 101 had also clogged *animal* mobility. The freeway ran near the northern border of the Santa Monica Mountains National Recreation Area, the rugged tract that is effectively America's largest urban national park. In the process the 101 drove a wedge between Southern California's interior mountains and its coastal scrublands, effectively splitting the region in two. It was a textbook case of habitat fragmentation, the breakup of once-contiguous ecosystems—or, more aptly, *Landschaftszerschneidung*, a German compound word that means "landscape dissection." The term, popularized by a road ecologist named Jochen Jaeger, is roughly synonymous with *fragmentation* but with an added specificity. While fragmentation can be inflicted by all manner of forces—deforestation, farming, development—dissection connotes the surgical precision of a road slicing through land, clean as a scalpel.

The dissection of the Santa Monicas disrupted practically every species that called them home. Bobcats and coyotes descended into social chaos as the freeway distorted their territories. Wrentits, pudgy birds that prefer hopping between bushes to flitting over highways, saw their breeding curtailed. But one creature, by virtue of its rarity and expansive requirements, was more harmed than any other: the mountain lion.

If you've spent much time on your feet in the American West, odds are good that you've been surreptitiously observed by a 140-pound cat with a bite-force capable of shattering an ungulate's skull. Mountain lions are the most widely distributed large mammal in the Americas, capable of thriving everywhere from South American rainforest to Arctic tundra. Exterminated east of the Mississippi in the early twentieth century, their populations are stable or growing in the West. Today cougars steal through

---

* For the record, no modern Angeleno has ever referred to it as the Ventura Freeway. Instead it's known, in Californians' endemic highway vernacular, as *the* 101. Call it otherwise at peril of mockery.

the deer-filled outskirts of towns like Boulder and Bozeman, exploiting the edgy world we've engineered for their prey.

Once, the thickets and shrublands that cloak greater Los Angeles afforded top-notch cougar habitat. In 1892 the *Los Angeles Times* described lion-hunting as "exhilarating sport, lightly spiced with danger." But as the city sprawled and freeways webbed the valleys, the cats vanished, or seemed to. Hikers claimed to glimpse them near Los Angeles, but no one produced evidence. Still, in 2002 a biologist with the National Park Service set up motion-activated cameras and scoured the region for lion scat. The effort seemed only slightly more promising than a Bigfoot hunt, but the cameras soon snapped photos of a huge male cougar—the animal that would become P-1, imperious king of the mountains.

That apex carnivores were dwelling on Los Angeles's doorstep raised questions: Did they have adequate space? Food? Sexual partners? The Park Service tasked two biologists, Seth Riley and Jeff Sikich, with ascertaining the answers. The duo had divergent hobbies and personal styles—Sikich was a voluble surfer, Riley a reserved ballroom dancer—but complementary professional interests. Riley was an urban wildlife specialist who had tracked raccoons in Washington, D.C., and foxes in Marin County; few people better understood how carnivores navigated cities. Sikich's talents lay in handling predators—wolves in Minnesota, bears in Virginia, cougars in Montana. While Riley coordinated the project's logistics and research, Sikich would lead the nitty-gritty of capturing lions in foot-snares and box traps to affix GPS collars. (Roadkill deer made a handy bait.) Success was hardly assured. "These animals are constantly moving and have huge home ranges," Sikich told me, "and we're trying to get them to step in an area the size of a dinner plate."

But step in the plate they did. In July 2002, Sikich and Riley captured and collared P-1, followed soon by his ill-fated mate, P-2, and their offspring. By 2004 their study had come to include eight animals; by 2012, twenty-six; by 2022, 112. They took blood samples and analyzed DNA, constructing lion genealogies of Tolstoyan complexity. As their project's profile grew, cougar sightings became almost routine. Like L.A.'s human celebrities, its lions were seldom spotted, but liable to make news whenever they set foot in public. A cyclist photographed one devouring a deer on Mulholland Drive. Another appeared on *Sixty Minutes*. Backyard security systems and doorbell cameras composed a citywide surveillance network that one journalist called "lion TMZ." When Will Smith's cameras captured a lion stalking his estate, he displayed the images on *Ellen*.

All that notoriety suggested that L.A. was brimming with mountain lions. In reality, Riley and Sikich were learning, the cats were few and embattled. Just ten to fifteen adult cougars prowled the Santa Monicas at any given time: a dominant male or two, four to six females, and some juveniles. These animals lived fast and died young. Several ate rat poison. One burned his paws in a wildfire and starved. Another broke into a courtyard and got shot by a jumpy cop.

Mostly, though, lions perished in two ways. The first was on the road. Carnivores require a lot of space, and cougars are especially demanding. "The female bobcats that we've tracked, their home range is maybe a couple of square kilometers," Riley told me. "The average home range for one of our female mountain lions? *One hundred and thirty-four* square kilometers." Inevitably, such peripatetic habits led cougars across highways. P-23 met her end on Malibu Canyon Road, P-32 on I-5, and P-39 on the 118 freeway—followed, heartbreakingly, by her kittens. Highways were taking a chainsaw to the cougar family tree.

Nearly as common as death by car was death by fellow lion. In wilder areas young cougars stake out territories far from dominant, filicidal adults. Southern California's freeways, however, stymied dispersal. Juvenile cougars approached the 101 and bounced away like ping-pong balls, desperate to escape their birthplace but unwilling to brave traffic. With flight impossible, they were forced to fight, often with their own fathers. These Oedipal conflicts rarely ended well for the upstarts. "Most of the males born in the Santa Monica Mountains haven't lived beyond the age of two," Sikich said. Sometimes the twin terrors of cougar life—cars and other cougars—augmented each other. Such was the case for P-61, a male lion who somehow crossed the 405 freeway and settled near Bel-Air, a feline fresh prince.

Two months later P-61 was attacked by a rival, a tussle revealed by security footage. Hounded by his foe and with nowhere to flee, P-61 braved the freeway one more time. It would be his last attempt.

Landscape dissection, Sikich and Riley realized, had turned the Santa Monicas into an island—an isolated chunk of land encircled by a freeway sea. This was both an apt metaphor and a foreboding one. For two centuries biologists from Charles Darwin to E. O. Wilson have sought nature's truths on islands, whose small size and simplified ecosystems make them easier to study than mainlands. Their cumulative research has revealed some discouraging principles. Islands are weird, volatile places where resources are scarce, populations are small, and reinforcements are an ocean away. Around three-quarters of all human-caused extinctions have occurred on islands, among them infamous cautionary tales like the dodo and the Tasmanian tiger. "Islands," the writer David Quammen has observed, "are where species go to die."

In Darwin's day a scientist had to venture to the Galápagos to observe island effects. Today, landscape dissection has created islands everywhere. Even national parks have become too isolated to support most predators. In 1987 a bombshell study in *Nature* revealed that parks were hemorrhaging fauna: lynx had vanished from Mount Rainier, otters from Crater Lake, red foxes from Bryce Canyon. They hadn't been hunted into oblivion or squeezed out by development; they simply needed more room than protected areas afforded.

If one axiom about islands is that they're too small for large carnivores, another is that they're hard to reach. Freeways had left the Santa Monica cats marooned, cut off from the rest of cougar society by straits of traffic. And without immigrants to freshen it, their gene pool had turned as stagnant as a roadside puddle.

When the automobile permeated American culture in the early 1900s, it revolutionized human social contact. We have cars to thank for dating: before autos, one historian observed, a gentleman caller would sit primly in his sweetheart's parlor, sip tea with her mother, and "[listen] politely while the young woman displayed her skills as a pianist." Empowered by a Chrysler or Nash, he could take his girlfriend out on the town, free from supervision. Not for nothing did moralists denounce cars as "brothels on wheels."

Even as cars facilitated human sex, however, they thwarted wildlife from doing the deed. Traffic's moving fence deterred animals from crossing between populations, and cars crushed would-be lovers who dared the trip. By stymieing life's most fundamental act, highways scrawled their signature into its molecular code. In Switzerland roads distorted the genes of species from roe deer to bank voles; in the Mojave Desert they pared the genetic diversity of bighorn sheep. In the Northern Rockies grizzly populations are so disunited by highways that researchers can tell, from the merest snippet of DNA, on which side of which road any bear was born. *Abax parallelepipedus*, a flightless European beetle, disperses so feebly that biologists once found a genetically distinct population encircled by a highway exit loop.

And even crossing the highway doesn't guarantee sexual success. Seth Riley, who tracked bobcats and coyotes as well as mountain lions in the Santa Monicas, found that these smaller carnivores sometimes crossed the 101 through drainage pipes, yet they seldom reproduced on the other side. The explanation lay in which animals were crossing. Strong, virile adults were loath to cross and risk relinquishing their hard-won roadside territories. The animals who braved the freeway were hapless teenagers who, on finding the best real estate claimed by their elders, fled without mating. The freeway forced a turf war that favored entrenched animals over low-caste nomads and skewed genes on both sides. Coyotes north of the 101, Riley found, were so distantly related to southsiders that they might as well have lived hundreds of miles apart.

For medium-sized carnivores, such distortions were worrisome but survivable: a male bobcat or coyote trapped south of the 101 still had good odds of finding a mate who wasn't his mother or daughter. The rarity of mountain lions, however, made them especially vulnerable to genetic harm. Every lion killed on a freeway drained genes from the pool, a winnowing known as genetic drift. Even worse, traffic prevented cats from mingling and forced dominant males like P-1 and P-12 to mate with their daughters. It was only a matter of time until the Santa Monica lions, like Egypt's pharaohs, suffered the repercussions of inbreeding. In 2016, Riley, Sikich, and their colleagues calculated that the Santa Monica lions had some of the lowest genetic diversity of any cougars in the West. Their odds of dying out were as high as 99.7 percent. The cats, they wrote, had been sucked into an "extinction vortex."

In that regard they reminded biologists of a disturbing precedent. In the mid-1990s the Florida panther, a cougar subspecies that roams the Southeast's cypress swamps, had suffered its own brush with extinction. After decades of development and roadkill cut the panther's population to around thirty animals, the survivors had no choice but to breed with their relatives, and genetic anomalies cropped up. Some mutations, like kinked tails and cowlicked fur, were benign. Others were life-threatening. More than 60 percent of males developed undescended testicles, and 20 percent of all panthers suffered atrial septal defects—holes in the walls of their heart. After biologists introduced eight female lions from Texas, the defects abated, and the panther's numbers ticked upward. But the saga demonstrated an alarming truth: without drastic intervention, a small, inbred cougar population was doomed.

Now, in California, history was repeating itself. On March 4, 2020, Sikich captured and tranquilized a subadult male, the study's eighty-first cougar. When he inspected P-81, he found an undescended testicle and an L-shaped tail—the same symptoms of inbreeding that had plagued the Florida panther. Researchers who examined sperm from five Southern California mountain lions found that nearly all specimens were abnormal. The vortex had the cats in its grip.

The Florida panther had shown that the vortex, once entered, could be escaped. But translocating more lions into the Santa Monicas, the strategy that saved the panther, wasn't an option; if anything, it would only intensify the fighting. Besides, importing lions wasn't necessary. North of the 101, potential reinforcements roamed the Simi Hills and the Santa Susana Mountains; beyond that lay the Los Padres National Forest, a cougar-filled paradise

Although outwardly healthy, P-81 had the kinked tail and undescended testicle that were characteristic of inbreeding in the Florida panther. *National Park Service*

that Sikich called the promised land. Just one immigrant from these north-lands every two years would save the Santa Monicas' cats. Sikich and Riley simply had to coax a stream of lions across one of the busiest roads on earth.

- - - - - - - -

Identifying the inventor of the wildlife overpass is like identifying the inventor of the internet. It's a technology with many parents, some celebrated, others obscure. The first overpasses appeared in France, which built around 150 *passages à faune sauvage* in the 1960s to appease deer hunters. Although France's narrow "game bridges" weren't rousing successes—they conveyed as many farmers as deer—they inspired engineers elsewhere in Europe to build grander structures. Wildlife overpasses, known also as ecoducts, green bridges, and landscape connectors, arced over roads in Germany, Switzerland, and Austria. The undisputed leader was the Netherlands, home to the world's largest overpass, the Natuurbrug Zanderij Crailoo, a half-mile-long crossing that swooped above a highway, a railroad, and a sports complex.

As Europe innovated, North America lagged. The first American overpass popped up in Utah in 1975 to usher mule deer across I-15. A decade
later New Jersey erected a hundred-foot-wide crossing, dubbed the "Bunny
Bridge," where an interstate split a park. But few states followed suit. By
the early 2000s, Europe sported dozens of overpasses and North America
a paltry six. "You would go to engineers and say, hey, can you try this, and
it was always, no, no, that won't work," Trisha White, an American conservationist who toured Europe's overpasses, told me. "And then we saw
a crossing in the Netherlands that had a living, working wetland on top of
it—not just some dirt and brush, a *functioning wetland*. They had to drag us
off because we wanted to stay all day. Our minds were blown."

Yet American engineers were not entirely bereft of close-to-home inspi-
ration. Just across the Canadian border, the world's most renowned wildlife
bridges had begun to prove their worth in Banff National Park.

Banff, which encompasses 1.6 million acres of Alberta's Rocky Mountains, is, from an animal's standpoint, a deceptive landscape, at once vast
and restrictive. Like many parks, it protects mostly glaciers, talus fields,
and windy crags—the kind of sublime scenery that makes mountaineers'
hearts sing but doesn't offer much sustenance for an elk or deer. Animals
gravitate instead to the fertile comforts of the Bow Valley, the river corridor
that wends through the park's bottomlands. Good wildlife habitat—flat,
hospitable, close to water—is almost by definition a good place to build a
road. In the 1950s the valley floor was slit by the Trans-Canada Highway,
a continent-crossing behemoth that came to funnel twenty-five thousand
cars through Banff daily. By the 1970s so many animals were dying on the
Trans-Canada that conservationists called it the Meatmaker.

And the Meatmaker was about to start making more meat. In 1978 the
Canadian government announced plans to "twin" the Trans-Canada, a
process that would expand the highway from two lanes to four and make
it even more impermeable. Engineers installed fences and underpasses to
avert the incipient crisis, and elk roadkill plummeted. Yet Banff's early
underpasses did little to help another park resident: grizzly bears. Grizzlies were textbook "avoiders" who rarely crossed the Trans-Canada to

forage or breed; highways made them so leery that they seldom lumbered through standard deer underpasses. Large, road-shy carnivores needed crossings of their own.

What kind of crossings would do the trick? That was less obvious. "In the mid-nineties, there was really no such thing as the internet," Terry McGuire, the engineer tasked with figuring it out, reminded me. "You were stuck with anecdotal information—you know, somebody went over to Europe and saw a structure, or had a brief conversation with someone at a conference." After some thought McGuire's team decided to route grizzlies over the highway rather than under it. He had recently constructed a vehicle bridge using a technique known as the "buried precast concrete arch"— imagine two massive arcs, each 90 degrees, snapping together to create one semicircular span—and assumed that bears could walk across a precast arch as readily as cars could drive over it. By 1997 he'd built two wildlife bridges, each more than 150 feet wide, over the Trans-Canada. Pines, shrubs, and grasses soon rooted in the soil that overlaid the concrete— graceful sutures of habitat sewing up the highway's wound.

McGuire was uncomfortably aware that his bridges were at heart experiments. In addition to cutting-edge overpasses, Banff also called for a comprehensive research program. A biologist named Tony Clevenger took the job. Clevenger was a Californian by way of Spain, where he had followed brown bears around the Cantabrian Mountains. He had no road-ecology experience and few resources at his disposal. At the time, the primary method for studying animal movement was the trackpad, a sandy pit that registered the prints of the animals who walked over it. It was a rudimentary technique—a herd of elk could churn a trackpad into an illegible muddle overnight—but a reasonably effective one. Clevenger set up trackpads at the Banff passages, then returned every few days to record the hoof and paw marks stamped in the sand, estimate how many animals had strolled through, and rake the pit clean. "It was the cheapest fieldwork in the world," Clevenger recalled. "All we needed was a pickup truck and an iron garden rake."

Clevenger's research began slowly. Grizzly tracks rarely appeared in his sand pits; in town, folks whispered that McGuire's overpasses were expensive failures. "The first time I met Tony, I said, 'I heard these things don't work for grizzly bears,'" a road ecologist named Adam Ford told me. "Tony's a pretty calm guy, but if he could've thrown his coffee mug across the room, he would've." In reality, grizzlies were simply acclimating. Bear movements began to climb around 2000, as they summoned the courage to use passages and taught their cubs to follow. Within the decade grizzlies were traversing the highway nearly two hundred times each year. It was another vital discovery: smart, cagey carnivores didn't immediately warm to wildlife crossings. Road ecology required patience.

By 2008 the Banff crossings had become, by any standard, among the most successful ever built. The Meatmaker had ground to a halt, and more than eighty thousand wild commuters had used the park's under- and over- passes. Networks of animal trails descended from the forest to convene at the passages, the product of countless hooves and paws: a collective exter- nal memory that guided successive generations of animals to the crossings their forebears had pioneered. "It was like the land was learning together how to get across the highway," Ford recalled.

Even so, Clevenger wasn't satisfied. He had begun to grapple with a fun- damental road-ecology dilemma: what does it mean for a wildlife crossing to work?

This might seem a strange question, if only because the answer appears obvious: a crossing works when animals use it. Just because critters tra- verse the highway, though, doesn't ensure that they'll breed on the other side or enhance the gene pool when they do. In California, Seth Riley had shown that bobcats and coyotes who crossed the 101 were mostly low- status juveniles who seldom reproduced. The hundreds of grizzly tracks in Clevenger's sandpits could likewise have been left by a few frisky young males scampering back and forth in futile search of a mate, or a single ter- ritorial boar who had enfolded Banff's crossings within his kingdom and didn't permit rivals to use them. As some dour researchers concluded in 2009, "There is no evidence that wildlife overpasses do or do not efficiently address genetic issues."

Clevenger's questions had outgrown his glorified sandboxes. Fortunately, he now had access to more sophisticated technology. By the early 2010s it had become possible to identify bears from the DNA in strands of hair. Clevenger and a biologist named Mike Sawaya stretched threads of barbed wire across the entrances to Banff's passages, which snagged the fur of any grizz who squeezed past. "With a bit of hair, it's like leaving a pad and pencil," Clevenger said. "And every individual that comes through writes down their species, their sex, who their mother and father is, and where they came from." At last he could tell how many individual bears were crossing and how grizzlies on either side of the highway were related—the sort of detailed genealogy that Riley and Sikich were compiling for Los Angeles's lions.

Thanks to Banff's crossings, Clevenger learned, its bears weren't cleaved by the genetic divides that splintered other populations. Eight males and seven females crossed the highway during his and Sawaya's three-year study, and several produced cubs. (That may not sound like many, but most grizzlies live in the distant backcountry and never encounter the road at all.) One courageous sow crossed eighteen times, mated with a male who crossed thirty-four times, and birthed three daughters who crossed frequently themselves. Like pronghorn, grizzlies preferred bridges to underpasses, a choice rooted in their past: whereas black bears are forest creatures who tolerate tight spaces, grizzlies were historically plains dwellers who overwhelm their opponents with power and speed in the open. All-important females, especially sows with cubs, particularly favored overpasses, whose sightlines allowed them to watch out for infanticidal males. Overpasses were thus the most "family-friendly" structures, capable of transporting bears regardless of age or sex.

As Clevenger's research rippled through the scientific community, Banff's influence grew. A visit to the park became a rite of passage for ecologists, who expressed such admiration that "Banff envy" became a well-known condition. Emissaries arrived from Japan, Korea, and Israel, and Banff-inspired bridges popped up in countries as far-flung as Argentina and Singapore. The self-effacing Clevenger became road ecology's *éminence grise*,

Banff's wildlife overpasses are today the most celebrated crossings in the world, and likely the best studied. *Ben Goldfarb*

enlisted to provide guidance on crossings around the world. "Tony is basically the Brad Pitt of wildlife biologists," one of his colleagues told me.

In 2014, Seth Riley and Jeff Sikich approached Clevenger about yet another project, perhaps the most ambitious in road ecology's history. In Banff, Clevenger had shown that wildlife crossings could reconnect grizzly bears sundered by a highway; now Riley and Sikich wanted to know if a bridge might help California's inbred mountain lions. Clevenger thought it could. But Los Angeles posed problems that road ecologists had never confronted: while the Trans-Canada Highway was four lanes wide, U.S. 101 stretched a jaw-dropping ten. Clevenger, intrigued, flew to California in January 2015 to give his input. Along with other experts, he spent a day touring prospective wildlife-crossing locations and offered recommendations. "And then there was kind of nothing, for three or four years," Clevenger said.

Back in California, though, conservationists were already laying the groundwork for an overpass that would dwarf the Banff structures, a crossing grander than any in North America. It would be a gargantuan project, both in scale and cost, as difficult to fund as to build. And it would have as its representative the most famous wild cat in the world.

Miguel Ordeñana sat before his computer in early March 2012 and clicked idly through camera-trap photos. Ordeñana, a biologist at the Natural History Museum of Los Angeles County, had stationed sixteen motion-activated cameras around Griffith Park, an amalgam of wildlife habitat and urban amenities in northern L.A. Ten million people annually visit Griffith Park, which contains a golf course, an observatory, a zoo, and the Hollywood sign. Yet much of the park is curiously wild, an ecotone where the outcroppings of the Santa Monicas shoulder into L.A.'s sprawl. Mule deer browse its draws; bobcats hunt woodrats in neon twilight. Ordeñana's cameras had revealed coyotes entering the park over the same bridges that drivers used during the day, nocturnal commuters whose lives were negatives of our own.

Still, Ordeñana wasn't prepared for what he saw as he flipped through photos of coyotes and deer: the rippling haunch of a mountain lion.

Ordeñana stared at the screen. Sure, Seth Riley and Jeff Sikich had documented lions in the Santa Monicas, but those animals haunted L.A.'s exurbs, not the city proper. Ordeñana had grown up barbecuing in Griffith Park, playing catch with his mother, eating *pollo loco* beneath the Hollywood sign. It seemed unfathomable that an apex predator stalked the park now. "It was like seeing Bigfoot or *la chupacabra* for the first time," he told me.

But the Griffith Park cat was real. Sikich soon captured the lion, collared him, and designated him P-22. Blood tests revealed that he was the son of P-1, the half-mad dictator of the Santa Monicas. This meant that P-22 had crossed both the 405 *and* the 101 freeways en route to Griffith, where he'd settled in an eight-square-mile patch of habitat—the smallest territory of any known cougar.

There P-22 flourished. By day he napped in oak thickets; by night he ambushed deer. He had a preternatural ability to avoid people—even when he broke into the zoo and mauled a koala, his victim wasn't discovered until the next day—yet his star rose quickly. A photographer captured him striding beneath the Hollywood sign, as confident and muscular as Schwarzenegger. Paparazzi attended his rare misadventures: when he wedged himself in a crawlspace, news trucks flocked to the scene. (Like any savvy celebrity, P-22 stayed put until the media left.) He ingested rat poison and contracted mange, which seemed to melt his haughty features; Sikich recaptured him to administer medicine. Over time he attained legendary status: P-22 the action hero, the roguish bachelor, the unrequited lover, admired for his pluck and pitied for his isolation. He urinated on leaf piles and chirped in the night to entice females who never came. "For those of us in L.A.," a conservationist named Beth Pratt told reporters, "having a romance prospect quashed by traffic is something we can all relate to."

Pratt was P-22's self-appointed publicist and agent, and one morning I met her in the Hollywood Hills to emulate her client's journey. We rendezvoused on the narrow bridge over which Mulholland Drive crosses the 405. Pratt arrived forty minutes late after getting stuck in traffic, a condition as ambient in Los Angeles as weather. "It's the only cliché about L.A. that's actually true," Pratt said.

I'd come to join Pratt on the last leg of her annual P-22 hike, a fifty-mile pilgrimage that traced the cat's path to Griffith Park. Because P-22 hadn't received his tracking collar until reaching the park, his precise route through the Santa Monica Mountains was a matter of conjecture. Within city limits, though, the green space shriveled to a strip, and his trail became easier to follow. Pratt fished out her phone and launched a livestream on P-22's Facebook page. "This bridge is likely where he had to cross the 405," she told her audience, gesturing at Mulholland Drive. "Most of the hike, we're within a mile or two of where he walked. Right here, we're in his literal footsteps."

Pratt and I trudged up Mulholland, past wrought-iron fences and gates with keypads. The road's hairpins lacked shoulders and sidewalks; at one curve a sign read Pedestrians Prohibited. Drivers honked behind tinted SUV windows. Even in this autoscape, though, wildness persisted. We passed deer huddled between mansions, heard frogs cheeping in washes, watched an acorn woodpecker sip from a fountain. Los Angeles is riddled with secret wooded pockets known only to rattlesnakes and rabbits; today it's one of two megalopolises with big cats within its borders. (The other is Mumbai, where leopards hunt feral pigs in back alleys.) "Anywhere else, P-22 would've been shot or removed immediately," Pratt said. "Here he's a folk hero."

Although we passed homes rumored to belong to Clooney, Sheen, and Gaga, Pratt worshipped a different megastar. She hiked in a tank top, revealing a tattoo of P-22's whiskered countenance on her left bicep, rendered in the hagiographic style usually reserved for dead relatives. A stuffed P-22 toy dangled from her backpack. (She used to hike with a life-sized cardboard cutout, but it kept getting tangled in poison oak.) She even wore one of Jeff Sikich's tracking collars around her neck to illustrate how unobtrusive it was. She had turned the GPS off, though she'd activated it for past hikes. "You could track me to the pizza place at the end of the day," she said.

As we dodged traffic, Pratt told me how she had become P-22's most ardent booster. She'd grown up in the Boston suburbs, a heritage that tinged her accent ("cars" occasionally lost its *r*), then moved to California after college. Pratt bounced around western national parks, working in Yosemite and Yellowstone before returning to California in 2011 to direct the National Wildlife Federation's state office. A year later she learned about P-22's existence the same way everyone else did, in a newspaper article featuring Miguel Ordeñana's photos. After the shock wore off, she began to wonder how her organization might assist this stranded cat and his relatives.

Pratt was a newcomer to the cause. Riley and Sikich had studied the lion population for years, and a group called the Santa Monica Mountains Conservancy had been buying up land since the 1980s, stringing together patchy corridors of protected habitat. The issue was the freeway that sundered these linkages. It had become evident to Riley and Sikich that their lions couldn't persist without a wildlife crossing, and Pratt agreed to aid the campaign to build one. "I naively said, sure, I'd be happy to help," Pratt recalled. "How hard could it be?"

The answer, she realized, was hard indeed. The 101 dwarfed even interstate highways, and its dimensions frustrated conventional wildlife-crossing designs. An underpass was out, since any tunnel beneath a ten-lane freeway would be too long and dark to reliably entice animals. Like Banff before it, Southern California needed an overpass. But where to put it? Roadkill data wouldn't help: the 101 was so heavily trafficked that cougars almost never braved it. Riley and Sikich had to figure out where mountain lions would cross if they could. Fortunately or not, there weren't many places an overpass could feasibly slot within L.A.'s sprawl. The obvious location was Liberty Canyon, where strips of public land converged at the 101 and formed a natural funnel between the Simi Hills and the Santa Monica Mountains. P-12 had traversed the freeway at this very pinch-point in 2009, and radio-collared bobcats and coyotes sometimes drew within a stone's throw of traffic as they sought vainly to cross. "You don't need a PhD in wildlife movement," Pratt said. "You look at a map, and there ain't too many options."

An overpass it would be, then, at Liberty Canyon, the site of the cougars' own liberation. Pratt just had to figure out how to pay for it. Elsewhere, wildlife crossings had drawn from various pots of money—lotteries, gas taxes, license plates—but mostly their funding came from state and federal transportation budgets. While other crossings had recovered taxpayer dollars by preventing crashes, that fiscal logic didn't hold on the 101. "As far as I know, this was going to be the first wildlife overpass intended to prevent the extinction of a population," Pratt said. That goal was noble, but trying to fulfill it with public money wouldn't fly with politicians. The bulk of the funding would have to come from private sources, and Pratt would have to raise it.

Pratt's campaign had one crucial asset: P-22, California's most sympathetic animal. She organized a festival on his behalf, created social-media accounts in his name, and gave life-sized lion cutouts to museums and schools so that Angelenos could pose with him, like tourists at a wax museum. Small donations trickled in—$20 from London, $50 from Australia—as well as the occasional large one. Leonardo DiCaprio chipped in; so did Tom Petty's widow and Mel Gibson's divorce lawyer. Rainn Wilson recorded a public service announcement. Viggo Mortensen posed with a stuffed animal. As Pratt's campaign snowballed, she became a minor celebrity herself. Drivers cat-called the Cougar Lady during her hikes, and admirers accosted her in coffee shops. Her predilection for humanizing her feline subject attracted skeptics, too. "Some people thought I was committing heresy—'Oh my god, mountain lions don't date!'" Pratt said. "I'm like, I think the public knows that. When I say P-22 is longing for a girlfriend, they don't think he's literally at Griffith Park with a bouquet of roses."

As the Liberty Canyon overpass wended through the bureaucratic maze, its price tag ballooned. There were feasibility studies to conduct, workshops to convene, designs to design, consultants to consult, and North America's largest wildlife bridge to build. It would cost ten million dollars; no, thirty; no, fifty; no, eighty-seven. Yet Pratt's fundraising kept pace. An inflection point came in 2019, when a CBS News spot about the campaign prompted a call from California's governor, Gavin Newsom, and eventually a twenty-five-million-dollar foundation grant. Like a university library, the Liberty Canyon overpass—make that the Wallis Annenberg Wildlife Crossing—would be named for a well-heeled patron.

By the time I visited Los Angeles in late 2021, the overpass was a *fait accompli*, and Pratt's P-22 hike had become a victory lap. Her entourage grew throughout the day; by late afternoon we'd been joined by a YouTube vlogger, a reporter for a British magazine, and an Italian filmmaker. Celebrity-spotting bus tours slowed to photograph our posse. As the light faded, we trudged into a pullout that overlooked Griffith Park. The blocky *H* of the Hollywood sign jutted from behind the hillside's rim. Unseen below it, P-22 awoke in the chaparral, languidly arched his back, and prepared to hunt.

One curious fact about P-22, not obvious to outsiders with a shaky grasp of California geography (like me), is that he had become the mascot for a wildlife crossing that wouldn't help him. Liberty Canyon is in Agoura Hills, a suburb thirty miles west of Griffith Park. Even after the overpass was built, P-22 would remain trapped on his islet. As Pratt was fond of saying, the structure would instead benefit his cousins—not his cousins in the vague taxonomic sense but his literal blood relatives. "If the crossing had been there, he probably wouldn't be in the plight he's in," Pratt told me. "He's taking one for the team."

One afternoon I drove out to Liberty Canyon to see the site for which P-22 had sacrificed himself. I escaped the 101's bumper-to-bumper congestion at Liberty Canyon Road and parked beneath gnarled oaks on the south side. Soon the overpass would break ground, and this spot would crawl with groaning machinery, hard-hatted workers, and politicians come for shovel-in-hand photo ops. I clambered into the foothills and tried to apprehend the land as a lion might. Above me the Santa Monicas jutted into the southern sky; to the north the Simi Hills rolled along the horizon, the dusty bronze of lion fur. Between them ran the 101, its hiss audible a half-mile away. The clonal subdivisions of Agoura Hills encroached on all sides; to the east lay Calabasas and its prime Kardashian habitat. The overpass would slot into a gap within this feckless sprawl like a key in a lock, opening a door between abutting mountain ranges.

It was obvious, too, that it would have to be gigantic. The structure, more than two hundred feet long and nearly as wide, would span both the 101 and a parallel residential road. Engineers would also rebuild the steep "approach slopes" with fill and topsoil, smoothing the transition between the crossing and its surroundings. Less than a bridge, one planner opined, the crossing was "architecturally solid terrain through which vehicles must pass."

The assignment of molding that terrain had gone to a firm called Living Habitats and its principal, a landscape architect aptly named Robert Rock. In addition to working with state engineers on the overpass design, Rock's team was responsible for sculpting the nine acres of land that lay atop and around the overpass—essentially, for directing wildlife's user experience. This was not something to which engineers historically gave much thought. "The default has been to dump some soil on top and throw some grass seed on it, and just hope it works," Rock told me. But the 101's enormity demanded maximalism. The freeway's noise was deafening, its light blinding. Animals would avoid a crossing that was too loud and bright. Rock's task was to dull sensory pollution through artful walls, berms, and vegetated screens—to forge topography that masked America's busiest freeway.

Making this more difficult was that, while cougars were Liberty Canyon's emblem, the overpass would serve the entire ecosystem. In Banff animals could browse a buffet of several dozen prospective crossings and choose the one that met their needs. At Liberty Canyon, however, one structure had to cater to every species. The overpass required woodlands for lions, grasslands for mule deer, rockpiles for lizards, logs for mice, pools near its base for frogs. (Rock described these features as "microclimatic moments"; others call them, delightfully, "faunal furniture.") Every organism would experience the structure differently. Bobcats might dart across it in minutes; harvester ants would call it home. No design detail was too minor. Rock's team planned to inoculate the overpass with symbiotic fungi and plant it with trees grown at an on-site nursery. They'd held lengthy debates about the orientation of individual boulders. "Some people may think this is a little overzealous," Rock said. "But we're trying to create a unique ecological stitch that unites two disparate places."

That Liberty Canyon was "overzealous" was an objection I had heard before. More than one road ecologist whispered that its outsized cost and dimensions could establish a problematic precedent; after all, not every crossing can drum up $87 million on the back of a celebrity cat. If Liberty Canyon's antithesis existed, it was in Utah, where engineers built a $5 million wildlife bridge above a stretch of I-80 with such outrageous roadkill that it was known as Slaughter Alley. The overpass was a sparsely landscaped catwalk that measured 320 feet long and just 50 feet wide, about one-third of a wildlife bridge's ideal width; one ecologist dismissed it as "a piece of spaghetti." Two years after the bridge opened, though, Utah's transportation department posted a video of moose, elk, bears, and bobcats ambling along it, a tantalizing hint that even minimalistic overpasses might do the job. And you could build seventeen Slaughter Alley overpasses for a single Liberty Canyon.

When I mentioned Slaughter Alley to Beth Pratt during our hike, she admitted that she, too, had been astonished by Utah's footage. "People say, why can't it be more like that one?" she told me. But the two crossings existed in different contexts. "Their traffic is 10,000 cars per day. We have 300,000 to 400,000 cars per day. We have light and sound problems that nobody else has." In Utah animals crept over I-80 in the dead of night, largely undisturbed. On the 101, the traffic never slept.

Liberty Canyon, to Pratt's mind, was both precedent-setting and *sui generis*. The overpass would be the culmination of decades of road-ecology art and science: if you could entice an apex carnivore over a California freeway, you could do anything. No other crossing would ever cost so much or require so much private funding. "When this is over," Pratt added, "the idea is that I'm out of a job doing fundraising for wildlife crossings, and these become embedded in budgets." She sounded, I thought, a little wistful.

The next morning I met up with Pratt, Miguel Ordeñana, and a few hardcore cougar fans for the final stage of the P-22 hike, through Griffith Park itself. The morning was damp and mist-wreathed; wrentits purred in the brush. We trekked up the road behind the Hollywood sign, letters blurry in the fog. Pratt herded us for a photo. "This is important," she called. "I raised $87 million based on perfecting my selfie."

As it did every year, the hike concluded at the P-22 festival, a collection of information kiosks, merch stands, and food trucks near the park's eastern border. A camera crew filmed us marching into the festival grounds to the applause of volunteers. Amid the fanfare it was hard to believe that P-22 himself was out there. Griffith Park's isolation had thwarted his love life, but it had also protected him. He was eleven years old, geriatric for a lion, yet he had the babyface of a cat who'd never known a rival's claws. Never had he menaced a hiker; rarely had he even been sighted. He was L.A.'s Boo Radley, the harmless recluse who dominated local imagination.

As P-22 had learned to live with humans, Los Angeles's humans were learning to live with lions. A population of cougars (or wolves or bears) might have all the contiguous terrain they need, but they won't survive if people shoot them on sight. Researchers call this principle "anthropogenic resistance"—the notion that human attitudes as much as habitat determine whether animals can navigate the land. Perhaps P-22's greatest accomplishment was lowering Los Angeles's anthropogenic resistance. Over the years, Pratt had invoked him as spokescat for all sorts of conservation causes: a ban on rat poisons, an ordinance that would compel homeowners to install wildlife-friendly fencing. Coexistence had become a buzzword.

The city's tolerance imbued me with hope. Large cats are among the planet's most road-afflicted organisms—in part because their immense territories are easily dissected, in part because their small, slow-growing populations can't withstand many collisions. A few untimely roadkills can condemn a population of jaguars, ocelots, or Iberian lynx. Against that dispiriting backdrop, it was incredible that the Santa Monica lions had endured. "You know, whenever you read articles about our study, it's always like, oh, another lion dies from rat poison, or another one gets hit on the freeway," Jeff Sikich told me later, as I pestered him with questions about roadkill and inbreeding. For every cougar who died on a highway, though, another was quietly doing her thing. When I asked Sikich to name some cats whose biographies epitomized the population, I expected he'd mention P-81, whose tail was deformed by inbreeding, or the mad king P-1. Instead, he brought up P-19, a canny matriarch who had reared five litters. "She's reproducing, raising her young, killing and eating mule deer," Sikich said. "She's behaving and acting like a lion in a more natural area."

This behavior was encouraging for reasons that transcended Los Angeles. By 2030 cities will surface nearly 10 percent of the planet; if wild animals are to survive, they and humans must learn to cohabitate. Many creatures are holding up their end: Chicago's coyotes have become so car-savvy that they look both ways before crossing the street. Wildlife crossings are a way of returning the favor, of deliberately inviting nonhumans into our cities. In Singapore an overpass known as the Eco-Link ferries pangolins and civets across an expressway; in Brandenburg, a German state with more towns and farms than forests, wolves have used seven green bridges to recolonize the region. Crossings foster what the geographer Jennifer Wolch has described as *zoöpolis*: a "renaturalized, reenchanted city" that integrates humans and wildlife.*

One benefit of zoöpolis, Wolch added, is interspecies empathy: when you share a city with wildlife, you can't help but "grasp animal standpoints or ways of being." That seemed especially true of P-22, a protean symbol who had come to represent all things to all people. One gay-rights activist donated to Pratt's campaign because he considered P-22 a fitting emblem for Griffith Park, which hosted Los Angeles's earliest pride rallies. Warren Dickson, a hip-hop artist from Watts, has written songs linking P-22's plight to his own neighborhood's, hemmed in by freeways that planners plowed through communities of color.

At the P-22 festival I attended a talk by Alan Salazar, a storyteller of Chumash and Tataviam descent. Salazar was tall and stern, with a stentorian voice and a ponytail that flowed from his fedora; he had worked as a juvenile probation officer, and it was easy to picture him setting a wayward kid on the straight and narrow. As the festivalgoers listened, rapt, Salazar related his people's origin story. The Chumash, Salazar explained, came into being on Limuw, the island known today as Santa Cruz. As the Chumash grew, they needed more space, but the roiling ocean surrounded them. So the Earth Mother stretched a rainbow from Limuw to a mainland mountain, and the Chumash people crossed safely onto the North American continent.

Salazar does some consulting in his spare time, and in 2019 the Liberty Canyon design team had hired him to provide input on vegetating their overpass. As he scanned the blueprints, an epiphany dawned: *this is a modern Rainbow*

---

* We also have self-interested reasons to welcome predators into our midst. Cougars are the ultimate agents of deer population control, and, per one study, reintroducing them to the Northeast could prevent 700,000 deer-vehicle collisions and 155 driver deaths over thirty years. In Wisconsin wolves already avert enough DVCs to save nearly $11 million annually. Even better, while wildlife crossings only target collision hot spots, deer-eating carnivores make highways safer everywhere they tread.

Bridge. "The Rainbow Bridge is a metaphor for how you get across something that's dangerous," he told me. "It's about how we safely get from here to there." The overpass would convey animals, yes, but it was also laden with story, a structure that spanned meanings as well as freeways, a bridge to zoöpolis.

Liberty Canyon's own Rainbow Bridge would not be completed during the lifetime of its feline muse. On November 9, 2022, P-22 killed a leashed chihuahua in the Hollywood Hills; over the weeks that followed, the formerly introverted cat loitered near houses and a Trader Joe's. In early December he seized another chihuahua and tussled briefly with its owner. The confrontations, so out of character for the hermit of Griffith Park, struck biologists as "signs of distress" that warranted a health exam. In mid-December, Jeff Sikich tracked P-22 to a suburban backyard, darted him with a tranquilizer, and took him into custody. His time in the wild was over, but perhaps he could live out his final years in the comfort of a sanctuary.

Like so many famous Angelenos, though, P-22 was destined to burn out rather than fade away. When veterinarians examined him at the San Diego Zoo, they discovered the extent of his ailments. His kidneys were failing, he had suffered injuries to his head and right eye, and he had lost thirty-five pounds. No wonder he'd exchanged deer for miniature dogs. Most of his maladies likely stemmed from a single recent trauma, awful and inevitable: P-22, this symbol of freeways' harms, had been hit by a car.

Euthanasia, the vets agreed, was the only option. The day before P-22 was to be put down, Beth Pratt went to see the cat she loved for the first and final time. He lay in a kennel the size of a living room, separated from Pratt by a row of steel bars, and hissed when she entered, wild to the last. She sat on the floor, so close she could feel his hot breath on her face. He settled into a burlap-sided shelter, shoulders hunched, paws before him like a sphinx, silently nursing his injuries. For fifteen minutes Pratt gazed into P-22's amber eyes and wept. "He was probably the longest serious relationship of my life," she told me later.

In the months after P-22's death, his acolytes would debate how best to memorialize him: perhaps a statue, or a postage stamp, or a star on the

Walk of Fame. Surely his most durable legacy, though, would be his impact on the planet's infrastructure. A year earlier, when Pratt and I had walked in P-22's pawprints, she'd claimed that she would quit raising money for crossings once Liberty Canyon was paid for. But road ecology had roped her back in. Pratt had recently agreed to lead the Wildlife Crossing Fund, an effort to raise an astonishing half-billion dollars; at P-22's memorial service, Wallis Annenberg, the philanthropist who had ponied up for the 101 overpass, announced that she would chip in the first $10 million. The campaign would pay for new crossings in California and someday around the world.

"How can I morally say, 'One and done, great, I'm retiring now?'" Pratt said. "It's not in me." There was no better way for Pratt to honor P-22, in the end, than to spend her own life advancing the cause for which he'd given his.

4

IN COLD BLOOD

How toad tunnels, frog shuttles, and turtle passages
avert biological annihilation.

The night began with a cat-eyed snake, as long as a forearm, slender as a finger, curled in a lopsided crescent, a scrawled *C*, absolutely dead. His markings were dapper, crisp cream stripes against a slate body, as though he'd been bound for a black-tie gala at the moment of his demise. His scales were flawless parallelograms, thousands of them, a bewitching miracle of geometry. With the toe of her rubber boot, Daniela Araya-Gamboa nudged him onto a metal hook and lowered him into a jar of clear preservative. He slumped in a loose coil, the last strand of spaghetti clinging to the pot.

We pressed on, a party of four: three scientists, one journalist, cruising a highway on the Papagayo Peninsula, the crooked thumb of Costa Rican dryland forest that extends into the Pacific Ocean. We inched past fields and thickets at fifteen miles per hour, our gaze fixed on the pale cone cast by our headlights. A third light atop our Nissan Navara flashed amber, lending us the mien of a maintenance crew, which in a way we were. The night was unpolluted by streetlights; when a dive bar loomed out of the void, spilling its neon interior onto the highway, it felt like passing a satellite in deep space.

Every few minutes we came upon a snake—*Trimorphodon quadruplex*, the lyre snake; *Leptophis diplotropis*, the parrot snake; *Leptodeira nigrofasciata*, another cat-eyed snake. The team fell into a routine honed over countless nights of collection. The driver, Esther Pomareda-García, pulled into the scrub, and a biologist named Esmerelda Arevalo-Huezo set out traffic cones to cordon off our worksite. *Carro, carro, cuidado,* she called, as pickups wove around us. The night reeked of cow patties and salt water. Arevalo-Huezo and Araya-Gamboa fed snakes into the jar while Pomareda-García scribbled their scientific names. On the data sheet, they were an incantation, a recipe for ethanol stew.

It would have been a doleful scene were it not for Araya-Gamboa—alert, gregarious, jubilant with purpose. "Your heart starts to get hardened," she said as she ladled another snake into the jar. "At the beginning it's very hard, but you need to do it."

Few Central Americans had handled more roadkill. Araya-Gamboa worked for Panthera, a nonprofit devoted to protecting ocelots, pumas, and, rarest of all, jaguars. Like cougars, jaguars roam widely, and roads disrupt their travels. "We started thinking, how do we save the jaguar?" Araya-Gamboa recalled, her face sallow in the dashboard's glow. She and her colleagues began to survey the country's roads—tallying animal corpses, setting out cameras, touting wildlife crossings to government ministries whose interest waxed and waned. Sometimes their surveys turned up dead felines, but more often they encountered the middle tiers of the ecological pyramid: skunks, rabbits, raccoons, birds. One night they found thousands of crushed crabs that had been migrating seaward to spawn, a chitinous scramble of claws and carapaces. "It was like a horror movie," Pomareda-García said.

"We said, wait, we have to save the *prey* of the jaguar," Araya-Gamboa said. "And the *prey* of the prey." Although the project's official name was "Wild Cats Friendly Roads," its scope came to encompass the totality of the Costa Rican ecosystem, as Borges's perfect map inflated to the size of the world itself.

As Friendly Roads grew, two groups of animals occupied more and more of its focus: reptiles and amphibians, the cold-blooded masses that compose much of the planet's biodiversity. Snakes and frogs, the crew realized, were perishing in numbers that beggared belief. The most traumatizing evenings fell during the wet season, when amphibians emerged to breed. "When it starts raining, you say, oh my god, it's going to be a long night," Araya-Gamboa said.

Pomareda-García nodded grimly. Recently, she said, "we had the field trip of the toad." She leaned forward to peer at another sinuous shape. "And this—this is the night of the snake."

We stopped again, and again, and again. *Imantodes cenchoa*, the blunthead tree snake. *Porthydium ophyromegas*, the slender hognose viper. *Scolecophis atrocinctus*, the black-banded snake. We scooped up snakes with candy-cane stripes, snakes as black as coal, snakes whose mottled camouflage resembled leering skulls—the totality of Central America's reptilian splendor arrayed before us on the pavement. Often we slowed to inspect snaky-looking sticks. "Sometimes we stop, and it's dead tomatoes or carrots," Araya-Gamboa added: the flung remains of somebody's lunch masquerading as roadkill.

Occasionally we came upon a living snake soaking up the asphalt's lingering warmth—*¡Está viva!* went the call in the truck. We shooed them into the grass with our boots. "We are always trying to help them cross the way they were going," Araya-Gamboa said. A month earlier the team had escorted a fer-de-lance, the ill-tempered pit viper that is Central America's most dangerous snake. Araya-Gamboa once nearly stepped on a live rattlesnake curled on the shoulder. The obliging rattler hadn't budged.

After a time—it could have been two hours, or four, or six—we reached the road's end, and Pomareda-García pulled a U-turn. We returned the way we'd come, faster now, hastening to bed. In the backseat sloshed the jar of snakes, treasure that other biologists would sift for disease or pollutants, like beachcombers working the tidal wrack.

The shifting focus of Wild Cats Friendly Roads, from large mammals to reptiles, paralleled the evolution of road ecology itself. The field's earliest interventions had been intended to prevent life-threatening collisions with deer, as in Wyoming, or to protect rare carnivores, as in Banff. Yet many of the worst-afflicted species aren't mammals at all. Instead, reptiles and amphibians—our scaly, slippery, shell-clad fellow travelers—are among cars' most frequent victims. These are animals to whom we don't readily relate, when we notice them at all. Happening on a road-killed snake in New England, the poet Mary Oliver initially shuddered at the creature's inanimacy, how he lay "looped and useless as an old bicycle tire" and "cool and gleaming as a braided whip." His life force extinguished, a snake was only an object, inert as rubber or rope.

The great contradiction of roadkill is this: its most conspicuous victims tend to be the species least at risk of extinction. Simple probability dictates that you're more likely to collide with a common animal—a squirrel, a raccoon, a white-tailed deer—than a scarce one. Most of the roadside dead are culled from the ranks of the urban, the resilient, the ubiquitous: the animals furthest from extinction's door.

As a result, roadkill is an overlooked culprit in our planet's current mass die-off, the sixth major extinction in its history. In the United States at least twenty-one critters are existentially threatened by cars, from the Houston toad to the Hawaiian goose. If the last California tiger salamander ever shuffles off this mortal coil, odds are it will be on rain-slick blacktop on a damp spring night.

Focusing exclusively on the wholesale disappearance of rare species, moreover, risks ignoring a subtler cataclysm: the diminishment of abundant ones. We are losing not only species but individual animals. Since 1970 the world's animal populations have dwindled by an average of 60 percent; pull back to 1900, and a third of our vertebrates have declined in numbers and range. Species endure as shadows of themselves, in minuscule fragments of their erstwhile domains. Commonness itself has become, well, rare.

The poster groups for this diminution are reptiles and amphibians, known collectively as "herps." (Herpetology is the scientific field concerned with their study.) The small bodies and secretive habits of herps conceal their dominance: many forests in the eastern United States support more salamanders than mammals and birds combined. But this exuberance is failing. Around one in five reptile species and two in five amphibians are endangered, and many more are on their way. Snapping turtles, spotted salamanders, and leopard frogs aren't near extinction; I've caught them in ponds across New England. Yet they've become rarer and more isolated, retreating from our landscapes and our lives. Biologists refer to localized extinctions—a pond emptied of frogs here, a pool robbed of salamanders there—as extirpations, many small losses that over time can amount to a very big one. The wetlands advocate David M. Carroll has lamented "the silence of the frogs," a hush as disquieting as the one that terrified Rachel Carson.

The forces muzzling the frogs are many—habitat loss, fungal disease, pollution—but it's not coincidence that herps are predisposed to become roadkill. Reptiles and amphibians move slowly and, being cold-blooded, gravitate toward warm surfaces, whether limestone or asphalt. They're surprisingly wide-ranging: turtles lumber across lakefront streets to deposit their eggs; snakes slither over highways to huddle in hibernacula. Worst of all, most herps aren't speeders like deer or avoiders like grizzlies. Instead, they're nonresponders: animals who are unfazed by traffic, even when prudence would serve them well.

Amphibians, whose name means "double life," are especially susceptible. Frogs, toads, and salamanders belong to two worlds: the water in which they're born and the upland forests where many species, upon swapping gills for lungs, spend their adulthood. When you have a toe in two realms, you must migrate between them. Amphibians move most frenziedly on spring nights, when rain refills the ephemeral pools that dimple forest floors. Wood frogs that spent the winter as cold and hard as popsicles, preserved by their own natural antifreeze, thaw and stir. Salamanders clamber from their catacombs. Peepers trill with a vehemence out of all proportion to their thumbnail-sized bodies. Thousands of minuscule lives go on the march, called by wetlands that will soon be cloudy with gelatinous egg masses. In some places the emergence occurs over weeks; in others, in a bacchanal known as the Big Night. And a salamander on a Big Night will cross any road in her path—come hell, high water, or Honda.

When an aggregation of libidinous herps boils over a road, the outcome is what biologists call, none too scientifically, a "massive squishing." Squishing statistics are both appalling and abstract, in the way that astronomical death tolls often are: nearly 28,000 leopard frogs killed over four years on a Lake Erie causeway; 10,000 red-sided garter snakes slain in one season in Manitoba; 2,500 toads flattened on a French country road. In Indiana scientists who counted 10,000 crushed animals found that *95 percent* were reptiles and amphibians. You've probably never heard one pop beneath your tires, but in many places it's herps—not deer, not squirrels—that make up most vertebrate roadkill.

Intuitively, any species subject to massive squishings would seem likely to suffer. But for a long time, few considered roadkill a threat. Cars might be crushing millions of frogs each year, but millions more crawled in the wings. Even Henry David Thoreau considered wagon-flattenings cause for perverse celebration. "I love to see that Nature is so rife with life that myriads can be afforded to be sacrificed . . . that tender organizations can be so serenely squashed out of existence like pulp," Thoreau raved in *Walden*, "tadpoles which herons gobble up, and tortoises and toads run over in the road."

Many biologists concurred. Roadkill was widely considered "compensatory mortality," a form of death balanced by the scales of life. If more frogs were flattened, the thinking went, maybe fewer would get eaten by predators, or there would be more food for their tadpoles. "When you deal with a group like amphibians, which have a high reproductive rate, people just think, 'Well, they'll be able to compensate for road mortality,'" a Canadian biologist named Lenore Fahrig told me. "The idea that roads actually have effects on populations, I don't think that was on anybody's radar at all."

Fahrig put the lie to that cavalier attitude. She grew up in Ottawa and in 1991 returned to take a job at a local university. Her parents still lived nearby, and she visited them often. The drive to their house took Fahrig along two roads—one with a lot of traffic, the second with only a little. One spring evening she noticed that her route was littered with dead frogs. This wasn't surprising in itself, since she was passing along riverfronts and through spongy fields. What was strange was where frogkill was concentrated. Because frogs readily cross busy streets, she should have seen the most dead herps on the road with the most cars. Yet the carnage was worst on the quietest road—the opposite of what she expected. Fewer cars had somehow produced more death.

Fahrig puzzled over the conundrum and came up with a hypothesis: the busy road didn't have many frogkills because there weren't many frogs left to kill. Traffic, she suspected, had already wiped out the local population. Curiosity piqued, Fahrig devised a study to test her hunch. In the spring of 1993 she and colleagues drove around Ottawa, scanning roadsides for dead frogs and toads and stopping to listen for the trills, croaks, and squeaks of live ones. Sure enough, the busiest roads had the poorest remnant amphibian communities. Given enough time and traffic, roadkill could indeed diminish a population, even extirpate it.

Other studies soon confirmed Fahrig's discovery. In Ontario just nine roadkills per year could rub out a population of black rat snakes. In Massachusetts a roadkill rate above 10 percent was enough to eliminate any given group of spotted salamanders; by that standard, up to three-quarters of the region's populations might be doomed. Roadkill was not exclusively a source of compensatory mortality: it could also be *additive* mortality, death that never came out in nature's wash.

One of roadkill's cruelest aspects is not how many animals it culls: it's which ones. Wild ecosystems weed out the sick and the old—the diseased fawn devoured by wolves, the senescent moose who collapses in a snowdrift. Roadkill, by contrast, is an equal-opportunity predator, as apt to eliminate the strong as the feeble. In Canada, for instance, vehicle-killed elk are healthier than animals slain by wolves and cougars. The same dynamic plagues amphibians. In New York researchers discovered that roadside ponds held unusually small salamander egg masses, likely because young females were getting crushed before they could grow into big, ripe breeders. Cars not only kill animals, in other words: they crush the very individuals who would otherwise help populations recover.

In a vacuum, reptiles and amphibians might be able to endure all of this. But they're also bedeviled by "synergistic threats," dangers that compound in pernicious ways. The same suburbanization that drains marshes also funnels more traffic through wetlands, piling roadkill atop habitat loss. As populations bow beneath comorbidities, they become more vulnerable. Healthy animal communities ride natural fluctuations like gulls bobbing in the surf, buffered from extinction by sheer abundance. Plunge too deep into the wave's trough, though, and a few mishaps—a dozen SUVs on a soggy night, say—can be fatal. And landscapes, once drained of their herps, rarely refill. The roads that run between wetlands and uplands also sever the link between these realms, destroying any frog or newt brazen enough to strike out toward an empty pond. Roads disunite land and water, short-circuiting the experience of amphibiousness.

The diminishment of herps is a hard problem to grasp. Utter extinction, the fate that befell the passenger pigeon and the Carolina parakeet, is a concept universally understood, as clean as a broken bone. A species existed; now it's gone. But the gradual ebbing of abundance strains language. Some researchers have called such losses *defaunation*; others know them as *biological annihilation*. The biologist E. O. Wilson favored *Eremocene*: the age of loneliness, a near and desolate future in which humankind bestrides an empty world, or perhaps drives over it.

In North America it took until the 1990s for biologists to comprehend the vehicular threat. But in Europe the epiphany arrived decades earlier, courtesy of a long-maligned creature: the common toad.

Common toads are archetypal in their toadliness. Theirs is the squat, belligerent visage that pops into your head when you think *toad*. Few creatures have received more abuse. Superstitious Britons considered toads "bred from the putrefaction and corruption of the earth" and believed them capable of such devilish feats as transforming wine to vinegar. Shakespeare judged them "foul." One seventeenth-century philosopher who kept a "large live toad . . . in a vessel of water" was accused of perversion and executed for his amphibious heresies.

Bufo bufo's life cycle lends itself to supernatural associations. Toads emerge from their winter hideaways each spring like a bumbling army of the undead, then, on reaching water, enter what George Orwell described as a "phase of intense sexiness." Males clasp females from behind, an awkward-looking position known as amplexus, and fertilize their eggs. "Frequently one comes upon shapeless masses of ten or twenty toads rolling over and over in the water," wrote Orwell, "one clinging to another without distinction of sex."

Before toads can satiate their lust, they must cross roads. Roads tend to pass through bottomlands, which also happen to be where water collects and toads congregate. One naturalist, writing in 1919, described watching a toad migrate "right out in the middle of the road, a most dangerous place for him"; so slow was his progress that it recalled the Via Dolorosa, Christ's path to the crucifixion. The danger that vehicles pose to amphibians may have inspired Kenneth Graeme, author of *The Wind in the Willows*, whose protagonist, Mr. Toad, is a horrid driver: "Toad the terror, the traffic-queller, the Lord of the lone trail, before whom all must give way or be smitten into nothingness and everlasting night." Nothing good can happen when toads and cars converge.

Outside the realm of children's fiction, toads weren't doing the smiting, but were being smitten. Massive squishings plagued Europe. Some twenty tons of toads were crushed in Britain every spring. Scientists calculated that one car per minute—traffic that most mammals would negotiate with ease—could flatten 90 percent of the toads crossing a Dutch street. In Germany, claimed the *Wall Street Journal*, "several people were killed or injured when a motorist skidded on rain-soaked toad bodies."

In the 1960s and 1970s toad rescue projects arose in the Netherlands, Switzerland, Austria, and Germany. These were simple operations: picture a handful of locals, outfitted with slickers, pails, and flashlights, dashing along wet roads on spring nights, scooping up toads, and carrying them to safety. Before long, toad patrols reached the United Kingdom. "There were actually quite a lot of people doing toad rescues, here and there and everywhere," Tom Langton, Britain's unofficial toad czar, told me. "But they weren't really connected or networked or organized."

Langton was well qualified to coordinate the United Kingdom's toad-movers. He'd grown up beside Hampstead Heath, a London park whose ponds abounded with toads. In the literary tradition of *The Secret Garden*, he'd had his own childhood grotto, a mossy recess at the base of his house where an enormous toad reliably lurked. At age eight he began ferrying toadlets across the road and into his garden. "You just had to be careful you didn't get run down yourself," he recalled. In the 1970s Langton, older but no less toad-crazy, joined a nonprofit called the Flora and Fauna Preservation Society, first as a volunteer and later as staff herpetologist, where he formalized the toad-carrying he'd done in his youth. Langton pulled together around four hundred patrol groups under the society's aegis, persuaded the government to post Toad X-ing signs at hot spots, and distributed "Help a Toad Cross a Road" stickers and shirts. Soon crossing guards were conveying a quarter-million toads each year.

While carting toads around in buckets made for cute photo ops, it was woefully inefficient. Langton began to look into toad tunnels: miniature underpasses, between ten and eighteen inches in diameter, that had first been installed in Switzerland and later proliferated in Germany. In 1987, at Langton's urging, Britain opened its first toad tunnel, custom-made by a German concrete company, in a village called Hambleden. An undersecretary named Lord Skelmersdale snipped a tiny ribbon. "The toad is an inoffensive and often misunderstood creature," the lord intoned, toad in hand. As if testing his goodwill, the toad, one of Langton's pets, peed on Skelmersdale's noble palm. "They usually do when you hold them," Langton told a reporter, a bit defensively.

Despite the urine incident, or perhaps because of it, the Hambleden tunnel instigated a media frenzy. Enthusiasm also swelled across the pond. That same year, the Flora and Fauna Preservation Society's American wing built a pair of eight-inch-wide underpasses for spotted salamanders in Amherst, Massachusetts. On the first night of the 1988 migration, fifty Amherstians gathered in the rain to watch a single salamander trundle through a passage—"an historic event," declared one attendee. In the nights that followed, biologists glued color-coded paper dots, clipped from IBM punch cards, onto migrants to determine what proportion of the population was using the tunnels. The answer: around two-thirds. Salamanders crawled through the tunnels most avidly when volunteers held up a flashlight at the far end, attracted perhaps to a facsimile of the moon. By 1991 the once-disdainful media was reporting rapturously on Amherst's vernal rite. "Spring is in the air," exulted one columnist, "and salamanders are in the tunnels."

As more herp tunnels sprouted in Europe, design principles cohered. The basic recipe—a combination of fences and underpasses—was the same that engineers had developed for deer. But the finicky biology of reptiles and amphibians posed distinct challenges. Compared to large mammals, herps were homebodies. A moose could pace a fenceline for a mile looking for an underpass, but a frog might quit, or desiccate, if crossings were farther apart than two hundred feet. Opaque fencing or walls were preferable to transparent mesh, since frogs and toads—not the world's most ingenious problem-solvers—didn't always perceive clear barriers and would stubbornly try to power through them. Tunnel design mattered, too. Because air didn't readily circulate through small passages, tunnels often developed the cool, dry microclimate of a wine cellar. Amphibians, however, preferred conditions warm and damp, more like a sauna. Biologists rectified the problem by building larger tunnels with open grates that permitted the entry of light and water. But even effective passages fell into disrepair. Fences buckled and frayed; tunnels flooded and clogged with sediment. All infrastructure requires maintenance, but a 150-foot-wide overpass can't get plugged by a wad of leaves. Toad tunnels, Langton said, "are living systems, as vulnerable as any type of habitat."

Few American engineers got the memo. Aside from the Amherst salamander passages, the United States remained mostly tunnel-less, and the herp crossings that did get built were so shoddy they cast aspersions on the cause. In Davis, California, western toads refused to crawl through a six-inch tunnel beneath I-80; when the city added lights to entice them, some allegedly baked to death. Stephen Colbert, then a *Daily Show* correspondent, visited Davis in 1999 and observed that the city had no idea how many toads had been squished before the tunnel's installation or whether any were using it now. "Probably on a subtle psychic level [the toads] know that the people of Davis care about them and that they're being looked after," the mayor told Colbert, surrendering a few more credibility points. The tunnel, wrote one local historian, made Davis a "national laughing stock."

When herp passages weren't being mocked by Comedy Central, they were running afoul of penny-pinching politicians more enamored of Newt Gingrich than of actual newts. In Michigan a $318,000 turtle fence drew the animus of a Republican lawmaker who, in one unhinged speech, invoked it as a metaphor for profligate health-care spending. A federal grant for an amphibian culvert in Monkton, Vermont, provided fodder for salamander sex jokes on Fox News. In Texas a columnist described a proposed passage for Houston toads as a "toad welfare scam."

Although some projects survived the attacks, the paucity of passages bespoke a greater truth: a society that cared only for monetizable costs and benefits would never protect herps. Few people had ever totaled their truck running over a garter snake; toad tunnels and turtle culverts were thus easy to shrug off as fiscal frivolity. This dismissive attitude irked road ecologists. When I met him in 2013, Marcel Huijser told me that American transportation agencies had largely ignored herps and other animals that didn't endanger drivers. Huijser's own cost-benefit analyses had done so much to popularize wildlife crossings, yet he regretted including only those factors, like hospital bills and car repairs, that could be easily quantified. "Our well-being partly depends on having natural spaces and wild animals around us," he said. "That wasn't part of our economic analysis. But we must ask what it is worth to have these animals in the landscape." He and other researchers had since tried to answer that question; one survey of Minnesotans found that every crushed turtle cost the state around $3,000 in "passive-use value"—econospeak for how much the public liked having turtles around. Still, the silence of the frogs was a mostly incalculable loss in a country obsessed with the calculable.

For herp passages to go mainstream, they would have to activate the sympathy of a callous public. Fortunately, in Florida, they would gain a savvy advocate.

- - - - - - -

What Tom Langton is to toads, Matthew Aresco is to turtles. Like Langton, Aresco is an earnest, tenacious biologist with a lifelong affinity for herps. He came of age in Connecticut, near a wetland that abounded with turtles. To the young naturalist, the ornate, high-domed shell of a box turtle or the golden constellations that adorned a spotted turtle's keratinous scutes were nature's loveliest self-expressions. Aresco found something poignant, too, in the futile optimism of retreating into a shell against overwhelming threats. "They always seemed so vulnerable, like they needed protection," Aresco told me.

After college Aresco moved to Tallahassee to study turtles' role in food webs. One afternoon in February 2000, his girlfriend returned from a shopping trip with disturbing news: she'd found a turtle massacre on Highway 27. Aresco hastened to the scene and deduced the situation. Highway 27 was a relic of an earlier, anarchic era of development, built decades before laws, like the National Environmental Policy Act, that checked ill-conceived infrastructure. The highway had bisected a wetland and split it into two lobes, known as Lake Jackson and Little Lake Jackson. Months of drought had reduced big Lake Jackson to mud, and its turtles were now fleeing toward the smaller lake, which still held water. Between the lakes, however, lay Highway 27, the axe stroke that had cleaved them in the first place. Few of the refugees even made it past the shoulder.

That first afternoon Aresco scraped up ninety dead turtles and piled them onto a tarp. The reptiles looked volcanic in death, pink meat glowing like magma through fissures in their shells. He snapped a photo and went home. "I thought that all you had to do was document a problem, and somebody would do something," Aresco recalled with a sour chuckle. "I was naive."

Aresco began visiting Highway 27 daily. Lake Jackson was still drying up, and the turtles—softshells and sliders, cooters and snappers, gopher tortoises and box turtles—continued their hopeless exodus. Aresco spent eight hours a day intercepting turtles, plopping them into Rubbermaid bins, and lugging them between lakes. Some days they emerged faster than he could wrangle them; sometimes car-struck turtles shot through the air like hockey pucks. The situation had the grim efficiency of a monstrous factory designed expressly for industrial turtle destruction. At night Aresco lay awake, shattered shells kaleidoscoping through his mind.

Aresco had come upon a quintessential road-ecology scene, a conflict that predated the internal combustion engine. Turtles are the ur-roadkill victims: "When they get into a rut," Thoreau observed, "they find it rather difficult to get out, and, hearing a wagon coming, they draw in their heads, lie still, and are crushed." In *The Grapes of Wrath*, Steinbeck evokes a turtle's harrowing journey as a metaphor for the trials of westbound Okies. As the animal shuffles across a highway, a trucker veers to hit him: "His front wheel struck the edge of the shell, flipped the turtle like a tiddly-wink, spun it like a coin, and rolled it off the highway." The turtle miraculously opens his "old humorous eyes," unscathed.*

Turtles are vulnerable to cars because—and this will astound you—they're slow. They're built for basking on logs, not skittering through traffic. But turtles aren't just slow moving. They're slow everything. A field mouse starts cranking out pups within two months of her own birth; a grizzly cub reaches sexual maturity at age five. A Blanding's turtle, meanwhile, may spend twenty years paddling around her swamp before she deposits her first clutch. Some species can live for more than a century. Scientists have found turtles wandering suburbs in search of wetlands that were paved over decades earlier, chasing their own dim memories into cul-de-sacs and parking lots. "They've lived here longer than the neighborhood has existed," one biologist told me.

For 220 million years this leisurely life history was wildly successful. But the laidback timescale of turtle life has proved incompatible with the Anthropocene's breakneck pace. Even a 3 percent increase in additive mortality, a few crushed elders each year, can send turtle populations into a

---

* Research vindicates Steinbeck's bleak view of human nature: one study found that nearly 3 percent of Canadians swerved to flatten rubber turtles and snakes.

A box turtle makes the long and perilous trek across a Georgia road. *David Steen*

tailspin. In one Canadian marsh a population of snapping turtles that numbered nearly a thousand fell to 177 in less than two decades, pulverized by traffic. Highway 27, Aresco calculated, was crushing enough turtles to "potentially cause irreversible declines." Biological annihilation was happening in real time.

Aresco knew that he couldn't move turtles by hand forever. Fortunately, a corrugated metal culvert ran beneath the highway and connected the two lakes: a ready-made wildlife passage, just waiting for some fencing to guide turtles through. Aresco begged Florida's transportation department to intervene. The response was tepid. One day a maintenance guy pulled up in a truck, dumped some mesh fencing on the roadside, and told Aresco to go nuts. Aresco spent the next four days scratching at the earth with a shovel until the fence was up.

The barrier worked to an extent. Turtle tracks soon crosshatched the culvert's muddy floor. But the fence was a nightmare to maintain. The sun fried its fibers; rodents undermined it; termites ate the posts. Aresco found himself visiting Highway 27 as often as ever, fixing fences and moving turtles ad infinitum, each carapace a burden as eternal as Sisyphus's rock. When rain finally refilled Lake Jackson, he carried the turtles back. That year he rescued 4,177 turtles. In the three years that followed, he saved four thousand more.

As time wore on, Aresco noticed another troubling trend: turtle-kill discriminated between sexes. Chelonian society is marked by an unequal division of labor. During nesting season males loaf in the pond, while females heave themselves overland in search of egg-laying sites, including sandy roadside soils. On one Texas causeway roadkilled female turtles outnumbered males by an astonishing 345 to 5. A population that loses too many females is doomed, even when its fate is masked by what scientists call the "perception of persistence." The turtles basking in your local pond look content enough, but they may be the patriarchs of failing lineages, ancient kings without heirs.

If Lake Jackson were to avoid demographic disaster, it needed dedicated turtle underpasses, with barriers to match. Aresco began touting a design called an EcoPassage, a chest-high concrete wall punctuated by four box culverts. In the late 1990s Florida had built a similar system in a roadside wetland called Paynes Prairie, where cars hit so many herps that tires sometimes skidded on the pulp. The first EcoPassage had reduced deaths more than fifty-fold, and critters from rat snakes to alligators had readily adopted the culverts. (Perhaps too readily: one of the passage's few drawbacks was that Floridians occasionally climbed its wall to feed the gators.)

Aresco figured the state wouldn't build another EcoPassage without a push. He had proven that Highway 27 had the world's worst turtle-kill, yet the transportation department hadn't lifted a finger. Aresco was a scientist, not an activist; still, if he didn't take up the cause, who would? He founded a nonprofit, rallied supporters, and pleaded his case to politicians. In tribute to the unlikely magnetism of turtles, thousands of chelonophiles rushed to his aid. The EcoPassage lobby—Big Turtle, you might call it—became a force. Turtle lovers elected a pro-passage candidate to the county commission and swamped the state with so many letters that officials begged Aresco to call off the dogs. Aresco told them to build the passage.

After nine years of panels, reports, studies, and other rigmarole, the EcoPassage's moment arrived. In 2009 the Obama administration disbursed a massive stimulus package to ease the recession it had inherited. Florida, perhaps to get Aresco off its back, gave the EcoPassage a skinny slice of the pie. In that era of bailouts, credit markets got $27 billion, the auto industry $80 billion, and banks $250 billion. Turtles, one of nature's most troubled assets, got a piddling $3.4 million stimulus of their own.

Even that was too much. Tom Coburn, Oklahoma's tight-fisted senator, issued a report excoriating the stimulus, the EcoPassage included. Why waste millions on an underpass, Coburn opined, when Aresco's temporary fence "saves a lot of our four legged friends"? The stunt generated a wave of negative coverage; Sean Hannity, smelling blood, invited Aresco on his show. (Aresco wisely declined.) More hurtful than outrage was derision. On CNN, Anderson Cooper found the imbroglio more amusing than tragic. Aresco had spilled blood and sweat, saved eleven thousand lives, and broken his hand with a sledgehammer while pounding posts for a crappy fence that Tom Coburn insisted was adequate. His reward was mockery.

To Aresco's amazement, though, the transportation department didn't cave. In late 2009 the state granted a contract to a construction firm, and by August 2010 the EcoPassage was complete, wall, culverts, and all. There was no ribbon-cutting, no sign. "I think they just wanted to hide from it," Aresco said. Today you could drive across Lake Jackson without an inkling of the EcoPassage's existence or of the machinations that created it.

Humans didn't notice, but the turtles did. The torrent of roadkill on Highway 27 shut off, neat as a turned spigot. During Aresco's years of turtle relocations, he'd clipped metal tags, stamped with his name and phone number, to hundreds of shells, a low-tech tracking system. After the EcoPassage's completion, sightings trickled in—*saw number 137 nesting in the neighborhood*—like postcards from forgotten friends. When a trapper found a tag in an alligator's stomach, even that was a comfort: proof that turtles were again playing every part in Lake Jackson's ecological drama.

"They were still out there doing their thing," Aresco said. "Laying eggs, getting eaten, being turtles."

These days, there's no longer doubt that a well-built herp crossing does its job. American roads sport tunnels for tiger salamanders, culverts for wood turtles, and in California an elevated stretch that accommodates the Yosemite toad. Yet the most common strategy for getting herps across the road hasn't changed since Tom Langton's childhood: pick 'em up and move 'em. In the Northeast alone, some thirty bucket brigades schlep thousands of herps, and chaperones from British Columbia to Belgium move hundreds of thousands more. It's a phenomenon both heartwarming and mysterious. What coaxes people into the freezing rain to assist creatures that are the antithesis of warm and fuzzy? The botanist and salamander escort Robin Wall Kimmerer has posited that carrying herps relieves our "species loneliness—estrangement from the rest of Creation." Perhaps it's our yearning to alleviate the Eremocene, E. O. Wilson's lonesome age, that explains herps' gravitational tug, though even practitioners don't fully understand it. "I never sought to be a frog rescuer—I never even had a particular affinity for frogs," a frog-mover named Shawn Looney told me. "I just couldn't . . . not do it."

Unlike Aresco and Langton, Looney fell into her avocation by accident. One evening in January 2013, she'd eased her Prius onto Harborton Drive, a hairpin road that winds through a quiet neighborhood in Portland, Oregon, to find the pavement alive with frogs. They trampolined through her headlights and down the street, beelining toward traffic on Highway 30. Looney swerved back and forth, trying to straddle the frogs rather than crush them. When she returned later that night, the wave had passed like an eerie dream. The only trace was a fishy reek, the lingering odor of frogkill.

Looney had discovered the vestiges of a once-mighty migration: the last northern red-legged frogs in Portland. She'd encountered them during their winter walkabout, when hundreds depart a big block of land called Forest Park, bounce down a ferny cliff, and make for a wetland tucked next to an oil depot. Along the way they cross Highway 30, a railroad, and two residential streets: a real-world game of Frogger, with the highest conceivable stakes. When Looney stumbled upon—or drove through—this journey, she felt moral clarity. She and a friend resolved never to permit massive squishings again.

And so the Harborton Frog Shuttle was born.

One damp March day I drove to Portland to observe the shuttle in action. I arrived at the wetland at dusk, just as the rain stopped and purple cracks fractured the sky's gray ceiling. A few flatcars sat idle on tracks; cars hissed along Highway 30 behind a maple curtain. Looney, a retired speech pathologist who radiated good cheer, turned up with her volunteers a few minutes later. She distributed headlamps and buckets, then explained the situation to me, the neophyte. Months earlier, patrollers had intercepted hundreds of frogs on Harborton Drive and ferried them to the wetland to breed. Now the marsh was strewn with egg masses, and the frogs, sated, were ready to head home. The Frog Shuttle would chauffeur them back to the forest, like parents picking up teenagers after a night of revelry.

Looney's team busied itself setting up a flimsy, knee-high strip of fencing outside the wetland. With luck the frogs would hop from their marsh, bump into the mesh, and sit bewildered until a volunteer gave them a lift. When conditions aligned, the emergence could be stunning. On a Big Night a year earlier, so many frogs had bubbled forth that frantic patrollers snatched them up in fistfuls. "One night in seven years I go on vacation," Looney groaned. "Six hundred-plus frogs. Broke every record. And I missed it."

Darkness fell, and the frog people began to walk, buckets in hand, headlamps aimed at feet. They had a cultish, ritualistic air. Some whistled. Others clasped their hands behind their backs, like Sherlock Holmes pacing his study. "I've logged a lot of steps on my FitBit," one volunteer whispered. I shuffled to the end of the fence, turned, shuffled back. End, back, end, back, through my own condensed breath. A circuit took ten minutes. I settled into a focused trance; I felt mindful and embodied in a way that I rarely do. The world shrunk to the diameter of my headlamp's beam.

And then, a male red-legged frog, resting obediently against the fence. He was slim-waisted and lithe, his umber skin peppered with spots. I'd expected him to require corralling, but he seemed to give himself willingly. He sat tranquil in my palm, cool as a riverstone. I laid him in the bucket, took a few steps, and found a gorgeous female, nearly twice as large as the male. She glowed a warm, lambent ocher, as though candlelit from within. Her eyes were black as wells and rimmed in gold.

Soon my bucket was floored with frogs. I could hear them thumping hollowly off its walls.

I rendezvoused with Looney, who offered effusive praise and transferred my bucket to two other shuttlers. They climbed into their SUV, swung onto Highway 30, and veered off at Harborton Drive. There they parked, lugged the bucket to the forest's edge, and coaxed out its contents. The frogs crouched in the dark, motionless as leaves.

"See you next year," someone whispered.

Escorting frogs was spiritually nourishing work. I felt grateful for having communed with such lovely creatures, inordinately proud of the minor aid I'd bestowed, and pleased to learn from Looney that the population was trending upward. But even Looney knew that the shuttle was a flawed bulwark against extirpation. Every year some unknowable number of red-legged frogs venture out after the volunteers have gone to bed, or elude them in the darkness. This is a familiar problem for herp chauffeurs. In Italy patrollers moved more than a million toads in the 1990s and 2000s, yet populations still declined. Some brigades in England have been discontinued for want of toads. This is not to imply that shuttles are futile, merely imperfect. "We so desperately want not to have to do this," Looney said. "That's the challenge: how do we work ourselves out of a job?"

But the Frog Shuttle's job has proven hard to replace. Although scientists had discussed the prospect of building an underpass, a tunnel beneath Highway 30 would have to be twice as long as any other amphibian crossing in the world, and no one was certain whether frogs would use it or where the funding would come from. Even if Portland did install a passage, there was no guarantee it would save the frogs. In one much-publicized case, a toad population in the Netherlands collapsed after biologists replaced volunteers with tunnels, perhaps because the passages were too far apart for toads to find them. The only fail-safe measure would be to close the road altogether during migration. The city of Marquette, Michigan, annually shuts down overnight traffic near a park in deference to blue-spotted salamanders; in Illinois biologists cordon off "Snake Road" to protect copperheads, rattlers, and cottonmouths. Those are local roads, though, not a federal highway. Portland would no sooner close U.S. 30 than it would reintroduce wolves to Forest Park.

Lately, Looney told me, the city had toyed with the idea of digging artificial ponds in Forest Park, which would give the frogs a safer place to breed. That proposition struck me as both sensible and depressing. As wetlands define frog habitat, frogs define a wetland; lacking amphibious life, what was a marsh but a glorified puddle? The thought of this one falling into frogless destitution seemed wrong, a capitulation to the highway's primacy. Red-legged frogs weren't mule deer; their journeys were measured in feet, not miles. Yet even a modest migration couldn't persist in a roaded world.

The chief irony of biological annihilation is this: we are both its cause and its solution. We so thoroughly dominate the earth that our conservation interventions, too, are necessarily heavy-handed. We introduce exotic species, then stamp them out; we poison California condors and breed them in captivity; we drain the planet's wetlands and excavate ersatz ones. "If there is to be an answer to the problem of control," the journalist Elizabeth Kolbert has observed, "it's going to be more control." Having curtailed wildlife migrations with our roads, we micromanage the survivors: we shuttle them in buckets from here to there, fence off roadsides and direct animals to the few narrow crossings we permit them. We built roads to subjugate nature, and by the time we learned the consequences, we had no choice but to subjugate it further. We are forever at war with the world we've built.

This was not a productive line of thinking, and so I did what felt right: I walked. Along the fence the patrollers ambled, heads bent as though in prayer, headlamps bobbing like shiplights in a harbor. And then a light descended, and I knew someone had found a frog.

PART II

MORE THAN
A ROAD

5

ROADS UNMADE

*Why the U.S. Forest Service operates the world's largest
road network—and how it's being undone.*

National Forest Road 5621 is a dirt furrow scratched into the mountainous face of northern Idaho, a road merely in the word's loosest, most euphemistic sense. It is unadorned by traffic lights, by signage, by guardrails, by pavement, by lane lines, by lanes. Fewer people drive it in a year than glide along an interstate in a day. It is neglected and desultory, barely more defined in places than ski tracks in snow, a pair of suspension-busting ruts that wend aimlessly into what William Least Heat-Moon called the American Outback. A road, wrote Heat-Moon, is "a beckoning, a strangeness, a place where a man can lose himself." With every mile, National Forest Road 5621 loses its passengers further.

I drove NF 5621 one afternoon with a retired engineer named Anne Connor and two of her colleagues, a hydrologist named Rebecca Lloyd and an ecologist named David Forestieri. The road was precipitous and ill-maintained, and the green chasm of the forest yawned beyond its crumbly shoulder. Alder and maple slapped our Ford Explorer; boulders grated the undercarriage. We tacked to and fro up NF 5621 like a sailboat against a stiff wind, switchbacking out of dense cedar into airy pine and fir, climbing ever higher into the Nez Perce–Clearwater National Forest. We limboed under tilted spruce and skirted windthrow. Forestieri edged the Explorer along a cliff, and the tires spun. My knuckles blanched in my lap. "David could probably drive this with his eyes closed," Lloyd said placidly.

No doubt she was right. Connor, Lloyd, and Forestieri had spent twenty-five years jouncing together down Idaho's forest roads; likely they knew NF 5621 better than any three people alive. If the road was a strangeness to me, it was a friend to them, or a frenemy. As we rumbled along, they reminisced about their years in the Clearwater: the ancient case of dynamite they'd found in a streambed, the dead moose that stunk up a campground. The drive felt like a reunion, with both land and people. "Our idea of a family day is to go look at an old road," Forestieri said.

At 4,300 feet we crested a pass and the road leveled. Across the Lochsa River, green mountains rose into pale sky. We got out and gaped. The landscape was surreal, simultaneously wild and cultivated: nary a building in sight, but roads, roads, everywhere roads. Roads striping the slopes like rice terraces, roads as clustered as topographic lines, roads spiraling into the blue. On one ridge the roads were so thick that they did not crease the mountain; it seemed that they had *become* the mountain. In some corners of the Clearwater, Connor said, sixty miles of road squirmed over every square mile of land—a road density that exceeds New York City's. Forestieri unfolded a map. On the page the forest looked parasite-ridden, dirt tracks writhing like tapeworms. Compared to the self-explanatory logic of a highway, these were roads that demanded questions: who built them, and why?

This perplexing snarl was the responsibility of the U.S. Forest Service, the federal agency that manages 193 million square miles of American land. National forests cover a boggling diversity of ecotypes: sunlit sweeps of southeastern longleaf pine; moss-garbed Douglas firs that scrape Washington's low clouds; pinyon-and-juniper shrublands that barely resemble forests at all. Few categories of land are more contested or paradoxical. The Forest Service's holdings are at once industrial stomping grounds for loggers and cattlemen who generally pay minimal rent to extract profit from public lands—*your* lands—and egalitarian wonderlands on which any person can backpack, cut firewood, hunt elk, catch trout, or camp for two weeks and drink beer around a firepit. They are reservoirs of wood and water, playgrounds for motorheads and critical habitat for wolverines. Many contain clear-cuts and wilderness virtually side by side.

Although the Forest Service refers to its holdings as "Lands of Many Uses," it might be more accurate to call them Lands of Many Roads. On a zoomed-out map of America, forests appear as green blocks mostly unmarred by interstates, creating the illusion of roadlessness. No impression could be more wrong. Although few of its roads show up in your Rand McNally atlas, it's almost certainly the U.S. Forest Service—not the Federal Highway Administration, not the Chinese Ministry of Transport—that is the largest road manager on earth. You could make it to the moon on Forest Service roads, and most of the way back.

If interstates are America's arteries, it's tempting to call the Forest Service's 370,000 road miles its capillaries—its closest-knit, most minuscule vessels. Whereas a circulatory system is essential to a body's survival, however, the vast majority of forest roads have outlived whatever dubious purpose they once served; these days their primary function seems to be backdropping Hummer commercials. Like the ruins of ancient churches, they are half-forgotten artifacts that reveal cultural values, reminders of who we are and who we were.

The Nez Perce–Clearwater National Forest is the poster land for road obsolescence. The Clearwater's roads were born with good intentions, midwifed by Forest Service managers who considered themselves the land's enlightened caretakers. Roads didn't desecrate nature; they helped men steward it. As Bud Moore, a ranger who began working in the region in 1934, put it,

> Improving access was prerequisite to managing the area's natural resources. The idea prevailed that once we had access we could get everything done. The elk could be cared for. Natural stream fisheries could be improved. Log jams could be moved from the outlet of lakes to create more room for fish to spawn. The timber could be harvested

properly, with vigorous, young stands replacing the old. Insect epidem-
ics could be forestalled. . . . Everybody would be happier and better off.

Moore's road-building campaign accelerated in the 1950s, when an infesta-
tion of spruce beetles struck the forest. He and other officials authorized
a frenzied round of roading and logging to excise blighted trees. Roads
pierced meadows, befouled streams, and allowed contractors to blaspheme
the woods with "the junk of American industrialization": cans, broken
cables, feces. "None of us had the wisdom to foresee the consequences of
the program we had devised," Moore wrote in his memoir. Rainstorms liquefied dirt tracks, churning roads into coffee-colored slurries that hurtled
into pellucid creeks and smothered fish eggs. "Our eyes were not on wild
things to be appreciated," the regretful ranger lamented. "Instead, we saw
obstacles in the way of the forthcoming road—obstacles to be moved or
smashed or circumnavigated."

With its major roads in place, the pillaging of the Clearwater began in earnest. The forest was overrun by "jammer" logging, whereby timber companies
used cranes and cables to haul logs onto waiting truckbeds. Loggers exploited
Moore's roads and bulldozed hundreds of rudimentary new ones. After jammer operations slowed in the 1980s, unused roads sank into the brush like
Mayan ruins. Once, Connor told me, she'd bushwhacked a mile from a ridgeline to a creek. Along the way she crossed the vestiges of thirty-three roads.
"We joke that they must have driven to every tree," Forestieri said.

While forest roads are physically immutable, or nearly so, their meaning is less fixed. "The road has its own reasons," Cormac McCarthy wrote,
"and no two travelers will have the same understanding of those reasons."
To Bud Moore and his ilk, forest roads were tools of conservation; to timber
companies they were instruments of profit. To many modern environmentalists, roads are syringes that inject humans and our poisons into nature's
veins. Countless logging roads have been appropriated by outdoorspeople—
hunters, fishermen, all-terrain-vehicle enthusiasts—who, having grown
accustomed to vrooming through the woods in Polarises and Arctic Cats,
don't readily surrender the privilege. If roads denote freedom, any attempt

to curtail their use is tyranny. In 1998, after the Forest Service blocked roads in Idaho with giant pits called "tank traps" to protect grizzlies and elk, an unlit gasoline bomb turned up on a ranger's doorstep. At a public meeting I attended once in Idaho's Fremont County, a map of nearby forest roads hung between portraits of Washington and Lincoln, an unsubtle conflation of access and liberty.

Decades after their haphazard construction, America's forest roads have become proxy battlegrounds in a cultural war: What are public lands for, and who gets to decide? Is the highest purpose of a forest to be wilderness or a source of timber? Do pickups and ATVs belong everywhere, or should we shut them out for bears and bull trout? Should federal agencies determine how forests are managed, or should local governments—which usually favor industry over preservation and access over exclusion—call the shots? The most-pitched public-land battles these days tend not to be over cattle grazing or timber harvest but over "travel management," jargon that means, basically, who gets to use which roads and for what purposes. That might sound like a lot of portent to pile onto some two-tracks in the middle of nowhere. But roads are used to carrying symbolic weight: as one historian has noted, they are as much "expressions of ideology" as physical features. A road is never just a road.

~~~~~~~~

Overseeing the world's largest road network was not the Forest Service's original intent. Instead, its infrastructure was an unhappy accident spawned by thousands of foresters, engineers, and timber companies operating bulldozers and excavators in America's most obscure corners—the wildland equivalent of sprawl.

Theodore Roosevelt signed the Forest Service into existence in 1905, an action that aligned his dual passions for conservation and trust-busting. No longer would timber barons raze America's woods, Roosevelt vowed. Instead, a new agency would govern the people's forests, carefully doling out lumber for what Gifford Pinchot, the service's first chief, called "the greatest good for the greatest number in the long run." It was a mission both noble and nebulous. Roosevelt and his lieutenants delineated the forest system by spreading maps across the White House floor and outlining acreage almost at random. "Oh, this is bully!" the president cried. "Have you put in the North Fork of the Flathead? Up there once I saw the biggest herd of black-tailed deer." Meticulous, he was not.

Surveying this *terra incognita* fell to a cadre of young foresters whom newspapers mocked as "Teddy's Green Rangers." Among them was Elers Koch, the Montana-raised son of Danish immigrants and a fresh graduate of the Yale School of Forestry. Two years after the Forest Service's founding, Koch, then twenty-seven years old, was appointed supervisor of two million acres in western Montana—"all of it," he wrote, "practically unknown to me, and much of it unmapped and unexplored." It was a thrilling and labor-intensive job. There were ranger stations to build, telephone lines to install, and roads to carve. Koch was up for it. He was an indefatigable woodsman and a paragon of rectitude; decades after his death, foresters still referred to a tall, straight pine as a "Koch tree." He scaled mountains that had never seen a white bootprint, scampered up trees with bears in pursuit, busted wildcat miners and brothels.* Gradually he brought his realm to heel. "When one has spent many arduous hours painfully toiling up a steep timbered creek bottom, obstructed by brush, rocks, swamp, and down timber, and a few months later rides a horse easily up a cleared and graded trail—that is a real satisfaction; that is opening up the wilderness," he enthused.

Teddy's rangers took an especially proactive approach when it came to fighting wildfires. The early Forest Service was the bane of timber moguls and their Senate toadies, who wanted to defund the agency and privatize its trees. In response the service justified its existence by saving lumber from the perceived wastage of fire. Whenever smoke arose, Koch rounded up laborers from mining camps and saloons, armed them with saws and pickaxes, and dispatched them to cut firebreaks. "At least six or eight men

---

* "Two undesirable prostitutes established on government land," one of Koch's colleagues wired his supervisor. "What should I do?" Came the reply: "Better get two desirable ones."

Like many early Forest Service employees, Elers Koch came to regret "opening up the wilderness" with roads and trails. *K. D. Swan*

have been killed by burning snags crashing down on them," Koch wrote, "and it is not a pretty sight to see a man with his head smashed and the brains spilling out."

Mobilizing all that personnel required infrastructure. Armies throughout history have built new roads and appropriated old ones, from the Roman legionaries who trod the Appian Way to the American soldiers who poured into Fallujah via Iraq's Highway 10. The Forest Service's quasi-military approach to firefighting was no different. The agency cut roads with abandon, abetted by millions of dollars in appropriations. Surveyors infiltrated the forests, lugging maps and theodolites "along the face of precipitous slopes and rocky cliffs where a misstep or a loosened rock would be disastrous." Much of the construction was performed by the Civilian Conservation Corps, a New Deal program that kept workers busy scratching out roads, trails, and firebreaks. Koch considered the corps's laborers inept urbanites, but you can't employ three million grunts without getting something done. All told, the corps erected 48,000 bridges; strung 360,000 miles of telephone line; and cut 125,000 miles of roads. By 1936, wrote the environmentalist Rosalie Edge, the forests were chockablock: "roads parallel, roads crisscross, roads elevated, roads depressed, roads circular . . . a madness of roads, too many of which will be left untended to fall into disrepair and disrepute."

Where roads perforated the woods, motorists followed. Like Bud Moore, who came to regret roading the Clearwater, Koch decided that opening the wilderness had been a mistake. Time-worn footpaths were debauched by "the print of the automobile tire in the dust," and horse-packing camps became "no more than a place to store spare barrels of gasoline." Making matters worse, Koch realized that snuffing out backcountry wildfires was "a practical impossibility." Rather than lancing the woods with roads and firebreaks, he declared, the Forest Service should let them burn. The idea was decades ahead of its time, and it backfired badly. Earl Loveridge, a rival forester, countered Koch's laissez-faire philosophy with his own draconian one: the agency should stifle every conflagration by the morning after it arose, a strategy later known as the "10 a.m. policy." The Forest Service sided with Loveridge and redoubled its roadbuilding.

The service's road madness only grew more acute over time. In the late 1940s logging companies, having clear-cut private timberlands, cast their gaze toward the national forests to supply fodder for the postwar housing boom. The Forest Service, once the adversary of loggers, became their handmaiden. "Roads in the western forests," wrote the agency's chief in 1952, were the "key to attaining full timber harvest and net growth." America locked into a roadbuilding cycle: interstate highways drove sprawl, which drove lumber demand, which drove the construction of forest roads. Between 1946 and 1969 the service's road network doubled in size. Over the next thirty years, it nearly doubled again.

The furious roading amounted to a massive subsidy. Taxpayers splurged on logging roads that wormed into public lands and allowed private firms to liquidate the people's timber. Where roads went in, trees—Douglas firs so straight they made fine sailboat masts, larches that glowed gold in autumn, Sitka spruce whose heartwood was stronger pound for pound than steel— came out. Roads fueled deforestation's business model. Often, the Forest

Service paid timber companies to build roads on its behalf, "essentially trad[ing] national forest trees for a forest transportation system," as the geographer David Havlick put it. "Some loggers admit that they make their profits only on building roads, losing money on the trees," one ex–Forest Service employee charged.

Elers Koch didn't live to see the road boom's end. In 1954—afflicted by chronic pain, despondent over the untimely deaths of his wife and eldest son—he took his own life. The Forest Service's industrialization surely hadn't helped. "I would that I could turn the clock back and make a plea for preserving the area as it was twenty-five or even five years ago," Koch sighed in one essay. "Alas, it is too late. Roads are such final and irretrievable facts."

Koch wasn't the only horrified observer. A new generation of conservationists was rising, united by their distaste for roads. In Montana a socialist forester named Bob Marshall inveighed against "propaganda spread by the Automobile Association of America." In New England, Benton MacKaye, founder of the Appalachian Trail, denounced "development of the Coney Island type." The leader of the antiroad movement was Aldo Leopold, a perspicacious ecologist who observed sarcastically that "There is No God but Gasoline and Motor is His Prophet"—an adage that could form the ethos of *Mad Max*. "To build a road," Leopold wrote, "is so much simpler than to think of what the country really needs."

Leopold came by his road antipathy honestly. His journey paralleled Elers Koch's: after graduating from Yale in 1909, he moved to the Southwest to work for the Forest Service and found the region in a state of collapse. Goats and cattle had "badly cut up" the range, and "an incalculable amount of very fine soil" had blown away. Like Koch, though, Leopold reserved his fiercest ire for the motorized hordes pouring into the woods on roads that his colleagues built. To counteract the spoilage, he proposed a novel concept: the creation of wilderness "kept devoid of roads, artificial trails, cottages, or other works of man." Wilderness areas, Leopold believed, would safeguard the rights of a threatened minority, that dwindling class of outdoorsmen who preferred to travel by foot and horse. He didn't resent car campers, he insisted. He just wanted to stave off "the impending motorization of every last nook and corner of the continental United States." In 1924 the Forest Service heeded his call by designating a 750,000-acre wilderness, the country's first, in New Mexico's Gila National Forest.

While the Forest Service was amenable to piecemeal wilderness declarations, it opposed the idea of a broader Wilderness Act, a law that threatened to put millions of acres off-limits to roadbuilding and logging. Despite the service's objections, Congress, won over by environmental advocates, passed the act in 1964, sixteen years after Leopold's death. In its most famous lines, the law defined wilderness in gauzy terms that philosophers have puzzled over since: a wilderness was an area "untrammeled by man, where man himself is a visitor who does not remain." Thankfully, the law's authors also threw in some solid guidelines about what could qualify: any roadless block of public land that measured at least five thousand contiguous acres. For all the lofty rhetoric that attended it, wilderness was defined by a material absence—the absence of roads.

In retrospect, one oddity of the Wilderness Act is that it said little about the nonhumans who called wildlands home. This omission reflected the concerns of the movement's early proponents. Roads irked Leopold not because they degraded nature but because they degraded our experience of nature—made us shallow, hasty, oblivious. "Recreational development," he wrote, "is a job not of building roads into lovely country, but of building receptivity into the still unlovely human mind."

In the 1980s, though, an ecological case against forest roads joined Leopold's spiritual one. Much of America's private acreage had been devoured by sprawl, and many parks were too small to permanently sustain large animals. By contrast, large, contiguous swaths of national forest functioned as a hybrid habitat: well trafficked by loggers, cows, and ATVs, yet vast and wild enough to support skittish, wide-ranging critters like wolverines and lynx. Or at least they would have been, were it not for roads. In North Carolina's Pisgah National Forest, black bears hid deep in the woods, far from marauding hunters and the roads they stalked. When the Forest Service opened a logging road on Washington's Mount St. Helens to the public in 1987, elk fled, scared off by noisy bipeds. In Wyoming's forests, streams crisscrossed by roads held fewer trout; in Minnesota people shot or trapped nearly half the wolves in a densely roaded section of Superior National Forest. Worst off were grizzly bears, who perished in the conflicts that ensued when roads ushered humans into their domains. Some were shot by poachers or frightened elk hunters, while others were killed after they habituated to human food. Even a measly half-kilometer of road set inside 250 acres of forest could send bears into a tailspin. Roads, scientists declared, had fragmented many forests more egregiously than had clear-cuts—a revelation that wouldn't have surprised Leopold, who once opined that he would "rather see cut-over lands than Fords any day."

The Forest Service wasn't blind to the crisis. In the late 1980s the agency, having belatedly resolved to study the impacts of its timber bonanza, fenced off twenty-five thousand acres in the Starkey Experimental Forest and Range, a quilt of pine stands and grassy hills in eastern Oregon. The Starkey Project became a living laboratory, a closed system in which animals could move and reproduce, like a petri dish the size of a city. Biologists affixed tracking collars to several dozen elk and erected radio towers around the forest, which paged the collars every few hours and relayed their positions to bulky desktop computers. By 1995 the ungulate panopticon had generated more than a hundred thousand locations. "No one else in the world was that data-rich," a biologist named Mary Rowland told me.

Rowland was particularly curious about how the forest's elk would interact with its roads. She wasn't sure how bothered they would be. Although twenty-seven miles of open road ran through the Starkey, none were paved. The public rarely visited, save for the occasional mushroom forager, mountain biker, or hunter. Outside of hunting season, a half-dozen vehicles in a day constituted a rush. Yet when Rowland and her colleagues analyzed the data beamed back by the collars, the results were unequivocal: elk detested roads, preferring to steer more than a mile clear. Given the sparse traffic, it seemed unlikely that the animals feared cars themselves. Instead, the elk had learned to connect vehicles with the hunters they carried. "They asso-

ciate traffic with being pursued in some way, and learn that it's something
to be avoided," Rowland said. Roads brought cars, cars brought people, and
people brought metal sticks that went bang and killed you. "No matter how
you slice it," Rowland added, "roads are always a strong predictor of where
elk are on the landscape."

What was true for elk also held for bears, wolves, and bull trout. More
than any other factor, roads—or their lack—determined whether animals
would survive in peace or be rousted by humans. This was a profound discovery and a disturbing one. Early road ecologists had focused primarily
on major highways, yet "low volume" roads made up 80 percent of America's transportation grid. If even forest roads caused serious problems, their
cumulative influence must be catastrophic indeed. Roads, opined the biologist Michael Soulé, were "daggers thrust into the heart of nature." The
forests were bleeding out.

The Forest Service was ill-equipped to cope. Its personnel reflected its
mania for infrastructure: at one point in the 1970s, the agency employed
24 fisheries biologists, 104 hydrologists, 108 wildlife biologists—and 1,081
civil engineers. Teddy Roosevelt had established the service partly to safeguard the headwaters of America's rivers, yet foresters and engineers bris-
tled when their colleagues implored them to protect streams and wetlands.
"There was no limit on clear-cut size," Mike Dombeck, who joined the
agency as a fisheries scientist in 1978, told me. "There were no riparian zone
protections. There was no concern about ephemeral springs."

And there were certainly no curbs on roads. Dombeck had learned to
disdain roads through wild places at an early age. He'd grown up in Wisconsin and often played hooky to catch brook trout in the nearby forest.
One year he found a new road leading to his favorite creek. "Lo and behold,
it got fished out," he said. When Dombeck joined the Forest Service years
later after a stint as a fishing guide, he and like-minded colleagues cast
themselves as "combat biologists"—rebels against their employer's roadbuilding frenzy.

Despite his frequent internecine clashes, Dombeck climbed the Forest Service's ranks, and in 1997 Bill Clinton appointed him its chief. Although the agency held conservation in higher esteem than it once had, it remained a house divided, torn between protecting habitat and facilitating logging. Dombeck's predecessor warned him that he had inherited "a rudderless ship with a muddled mission." At the muddle's heart were roads, which had become an environmental and political liability. The agency's maintenance backlog exceeded $8 billion, and Congress was threatening to strip its budget. If the service couldn't afford its existing roads, it seemed ludicrous to build more. In 1998, Dombeck announced one of the most consequential actions since the Wilderness Act: an eighteen-month pause on road construction.

Dombeck knew the moratorium would be polarizing. "Fasten your seatbelt," he told a colleague. One senator in Idaho, where timber and mining reigned, declared it a "hand grenade rolled under my door." Environmental groups, hoping to make Dombeck's suspension permanent, rallied their members with banner ads and listserv blasts, one of the first digital campaigns. Activists sent the White House so many emails they crashed its servers.

After a year of bombardment, Clinton caved. The administration instituted the Roadless Rule, a policy that permanently secured the national forest system's large, unroaded blocks of land. Disgruntled lobbyists fought it, and the Bush administration tried to undo it, but the rule endured and today protects nearly sixty million acres of roadless forest. The Roadless Rule represented both a management change and an ideological one. When Congress foisted the Wilderness Act on the Forest Service decades earlier, the agency had to be dragged along. The Roadless Rule, by contrast, had come straight from its own chief. National forests, the rule implicitly declared, had a loftier mission than churning out board feet. "It's pretty tough to harvest timber without a road," Dombeck told me.

And Dombeck did not merely proscribe new roads. He also unmade old ones. In 1997 he requested $22 million to destroy around 3,500 miles of logging roads. Granted, that was an infinitesimal fraction of the service's total network, but it still felt momentous. For the first time, the Forest Service's road system would shrink.

"If you didn't know there was a road here," Rebecca Lloyd said, "you probably wouldn't believe me if I told you."

I was back in the Nez Perce–Clearwater National Forest in Idaho, standing on—or sinking into—a hillside with Lloyd, Connor, and Forestieri. The slope steep, the earth spongy. A riot of wildflowers: strawberry, trillium, the season's last wilted glacier lilies. A creek sang below us; in fall, we would have seen Chinook salmon skinning up its riffles. Without Lloyd's guidance, I couldn't have known that twenty-five years earlier this slope had been ribboned with logging roads, parallel lines as dense as pinstripes on a suit.

When the Forest Service began to obliterate roads in the 1990s, the Clearwater's system was among the first to go. Closing forest roads wasn't a new idea. The service had erected berms and gates across logging roads for years, often to protect grizzlies and elk from hunters and joyriders. But even a closed road exerted a magnetic attraction. Sometimes rangers forgot to lock gates; sometimes ATVers circumvented the barriers, adding miles of what the service generously called "user-created roads" to its ledgers. In some forests more than a third of closed roads showed signs of illicit travel.

Besides, closures only addressed one of the road's ills. "Even an abandoned road changes everything—the carbon cycling, the soil nutrients, the hydrology," Lloyd said as we squelched along the hillside. The Clearwater's old logging roads hosted little traffic; they did not produce noise pollution or roadkill. Yet the Clearwater was a timebomb, roads its tangled wiring. In 1995 the bomb detonated. A succession of winter storms drenched Idaho, destabilizing the shoddily packed dirt that bulldozers once shoved aside. Like poorly constructed dams, logging roads intercepted runoff, then burst, the torrents swelling with each cascading failure. More than nine hundred landslides roared down the slopes, over half of which stemmed from derelict roads. Deep runnels gashed the earth, as though the mountains had split their seams. Examining the aftermath of one slide, Connor heard a series of explosive pops—tree roots snapping as the mountain's bulk shifted around her. She turned to her companions.

"Uh, guys, we need to get out of here," Connor said. Roads had destabilized the ground beneath her feet.

Connor was then relatively new to the Forest Service. She had joined the agency a few years earlier, an engineer whose bosses didn't share her passion for restoration. "The attitude seemed to be, 'What do we with a woman engineer?'" Connor recalled. They shunted her toward facilities: building bathrooms, offices, and the like. Connor refused to be shunted. In the early 1990s she'd scrounged up grants to tear out some derelict roads. But the destruction proceeded at a crawl, hamstrung by a puny budget. Then the landslides hit. "That was the a-ha moment," Connor said. "An overgrown road wasn't benign." A gate might deter poachers, but it wouldn't prevent a logging road from melting in the rain like a scoop of ice cream.

Galvanized by disaster, Connor set to destroying what her forebears had created. Roads had stair-stepped the forest, superimposing their angular geometry atop the mountains' graceful slopes. Connor sought to smooth the terraces, to restore natural contours to disfigured hillsides. Backhoes and excavators trundled through the Clearwater, clawing up compacted roadbeds and dragging woody debris onto tracks where logging trucks once rolled. Downed trees and rocks furnished microhabitat; technicians spread wildflower seed and transplanted saplings. The Nez Perce Tribe, the land's ancestral stewards, joined the effort as an equal partner, contributing money and personnel. The result was not initially lovely—one visitor called it "as ugly as a war zone"—but the sites recovered fast. By 2005 more than five hundred road miles had been expunged from the Clearwater's network.

The obliteration shaped rural economies as well as land. Once, Connor's daughter came home from school with a question: *Mom, how come all my friends' parents know you?* Connor leafed through her daughter's yearbook and realized she was connected to more than half the class. Parents had worked for the Forest Service; grandparents had operated excavators; uncles had been contractors and mechanics. Road decommissioning filled the void that logging opened, a small jobs program for the earth.

Despite their labors, Connor and her crew weren't certain what they had accomplished. No one had obliterated roads at such a scale, let alone studied the outcome. "We didn't know if a recontoured road still functioned like a road," Rebecca Lloyd said. Lloyd dug up hundreds of soil samples from the roadbeds that Connor had torn up, picking through layers of earth hardened by tires and then loosened by machinery. Obliterated roads, she learned, were rich in nutrients and organic matter, and plants plunged deep roots in the churned dirt. Even the microbes shifted, from a bacteria-dominated community to a fungal web more typical of the forest. Catastrophic erosion virtually ceased, as rainfall and snowmelt soaked into the earth rather than running over it. Black bears feasted on new thimbleberry and huckleberry, turning once-perilous roads into pantries. Decades earlier, roads had defined the Clearwater as a human environment, the ultimate manifestation of our dominion over nature; now their destruction redefined it as essentially wild. Roads, and their absence, changed the land and the meaning of the land—changed who and what it was for.

As worthy as road decommissioning proved itself, however, it has proceeded only in fits and starts since the Clearwater's heyday. Although ten thousand road miles have vanished from the agency's rolls in the last twenty years, that represents just 2 percent of its overbuilt labyrinth. The Forest Service's road network is a protean object whose dimensions are easily altered through official legerdemain. It's overloaded with "ghost roads," a clandestine lattice of undocumented routes—ATV trails, mining tracks, temporary logging roads that no one bothered ripping up—whose scale is unknown but immense. Wildfires have a knack for burning away brush to reveal forgotten roads, which return to the fold like long-lost family members.* Thousands of roads linger in "storage," just waiting for a timber sale, a prescribed burn,

---

* While the Forest Service's road network was built largely to fight fires, these days it mostly causes them instead. Blazes can be ignited by hot mufflers in grass, sparks flying off wheel rims, or smoldering campfires along dirt roads. In Southern California the proximity of a road—not the forest type, the slope, or even the temperature—is the greatest determinant of when and where a fire will erupt.

or some post-fire salvage logging to justify their reactivation. "The timber guys believe they'll need the roads at some nebulous time in the future," one conservationist told me. "The fire managers think they're going to be needed for suppression. And, of course, the recreation specialists feel like they need them to offer recreation opportunities."

The stronger the arguments against roads, it seems, the harder it is to get rid of them. Mary Rowland, the biologist whose research suggested that elk avoided roads in the Starkey forest, told me that her findings initially met broad local support. In recent years, though, her studies have been derided in county meetings and dismissed as one-offs—even though, as Rowland observed, "I could count fifty road-ungulate studies that all point to the same conclusions." Today the forest where Rowland conducted her research is among the only ones without a travel management plan, and recreators have largely driven elk off public lands and onto private farms and ranches. "There's some irony there," Rowland said with admirable restraint. "Some of the best road research was conducted in your own backyard, and you refer to it as junk science."

━━━━━━━

At one point during my visit to the Clearwater, Connor, Lloyd, and Forestieri took me to Packer Meadows, a lovely, sunlit grassland near a mountain pass. The Nez Perce have gathered at Packer Meadows for millennia to harvest camas, whose edible, onion-shaped bulbs are among the tribe's traditional staples. In the 1930s the Forest Service drilled a road through the area, which diverted a stream and dried up the meadows, to the tribe's displeasure. Nonetheless, the agency has declined to demolish the road. "The Forest Service believes the road itself is a cultural resource," Forestieri said.

"What is its cultural value?" I asked.

Lloyd shrugged. "That it's old," she said.

Forestieri, who worked for the Nez Perce, added, tongue half in cheek, "And built by white people."

Roads as history, as legacy, as proof of our presence: this, too, explains our attachment. America's road-craziest state is Utah, Idaho's neighbor, where the question isn't just how roads should be used but what they *are*. There, county officials are forever "discovering" roads on federal land and asserting their right to govern them, the legacy of a nineteenth-century statute, known as R.S. 2477, that granted miners, ranchers, and home-steaders the right to build roads pretty much anywhere in the course of colonizing the West. Many of these routes—"hoax highways," environmentalists call them—strain any common-sense understanding of what constitutes a road. A cowpath through the sage? Call it a road. A sand-stone scar carved by mining trucks? Road. Ruts left by the wagons of Mormon pioneers? Definitely a road. In Utah, indeed, roads are quasi-religious objects: physical evidence, the historian Jedediah Rogers has suggested, of how early Mormons transformed the redrock desert from "wasteland to 'productive' Eden." Roadbuilding was perceived as a divinely ordained sacrifice, the means through which settlers would "subdue the land and make it useful to humans," an admixture of holy creation and stubborn labor. Identifying and naming archaic roads validates ancestral connections, an infrastructural genealogy that ties present-day people to forefathers and land.

Of course, there's also a more prosaic explanation for Utah's road fervor. The state—like Idaho, Wyoming, Nevada, and others whose lands disproportionately fall under federal management—has, to put it mildly, an antigovernment streak. Utah was the original staging ground for the Sagebrush Rebellion, a loose coalition of miners, loggers, ranchers, and right-wing politicians who had grown sick of environmental laws like the Wilderness Act and demanded more control over the public lands in their backyards. The rebels' first act was to bulldoze a road into a proposed wilderness in Grand County on July 4, 1980, accompanied by much flag-waving and speechifying. (In fact, the rebels misread the map and ended up a half-mile short of the wilderness boundary.) Politicians in various Utah counties have since bladed roads into parks, pried open gates that bar motorized access, and led rowdy ATV rides down canyons that the feds had closed to protect archaeological sites. Roads remain cudgels against conservation: if wilderness is inherently roadless, the logic goes, a land webbed by roads cannot be wilderness. As one federal official put it to me, "If they can control the roads, they control the public lands."

And the struggle for control has only grown more pitched. In 2019 scientists advised the world's governments to adopt an ambitious new goal to stave off ecological collapse: the protection of 30 percent of the earth's surface by the year 2030. The target, known pithily as 30x30, became conservation doctrine and, after Joe Biden assumed the presidency, federal policy. Converting roadless forests into permanent wilderness, researchers wrote, was "a relatively easy and cost-efficient step" toward achieving 30x30. When it comes to roads, however, nothing is easy. No sooner had Biden touted 30x30 than anticonservation forces mobilized against it, chiefly American Stewards of Liberty, a far-right group whose leaders insisted, all evidence to the contrary, that 30x30 was a plot to trample property rights and cause "the destruction of our nation." The group concocted a boilerplate resolution condemning 30x30 and distributed it to county governments in Utah, Colorado, Nebraska, and other states, which dutifully ratified it.

And what were the American Stewards fighting against? The usual bogeymen: wilderness designations; restrictions on logging and grazing; and, not least, any "road closures, road decommissioning, [and] moratoria on road construction" that might result from 30x30. Like abortion or critical race theory, roads had become shibboleths, entrenched cultural markers that distinguished warring factions. The idea of roadlessness is in some quarters so antipathetic that it's unfathomable. When an adviser briefed Donald Trump on a proposal to open unroaded parts of Alaska's Tongass National Forest to loggers, the president was flummoxed by the very premise. Blurted the former real-estate developer, "How the hell can you have an economy without roads?"

And yet I'm reluctant to unilaterally pillory roads' defenders. I drive forest roads to fish remote lakes, reach trailheads, and spend nights in ancient, rickety fire towers. I don't consider this hypocritical, for access and obliteration are not mutually exclusive. The vast majority of travel within national forests occurs on just 20 percent of their roads; some units' networks are so tortuous, redundant, and laced with dead-end "cherry-stem" roads that the Forest Service could demolish most of its mileage without diminishing access in the slightest. Using the right roads has never precluded destroying the wrong ones.

Since Aldo Leopold's day, a funny thing has happened to the concept of wilderness: it has taken as many fusillades from the left as the right. Wilderness, critics have charged, is a colonial construct that ignores how artfully Indigenous people managed pre-European landscapes, a romantic idyll propagated by "well-to-do city folks," and a displacement of nature that blinds us to the splendor of our backyards. It's also a fiction: how can any place remain untrammeled when climate change is trammeling everything?

As an abstraction, wilderness is passé. But as a physical place—a landscape category defined by roadlessness—it's more vital than ever. Roadless areas are kingdoms through which grizzlies can wander without fear of bumping into hostile humans, where salmon can spawn without their eggs being choked by mud. Soon after the Roadless Rule was passed, scientists calculated that it had protected more than two hundred endangered species and some of the "most important biotic areas in the nation." What's true in the United States holds globally. In Russia, Amur tigers live longer in roadless areas than outside them; in the Congo persecuted elephants adhere to a "siege strategy," hunkering in roadless strongholds. Roadless areas are planetary buffers against extinction: a species that resides within a wilderness is more than twice as likely to survive as one in the hellscape beyond. Decommissioning just *1 percent* of roads through public lands each year for a quarter-century, scientists have calculated, would increase wolf habitat by around 25 percent. Road networks, like economies and tumors, are predisposed to grow; it's crucial that ours begin to shrink.

This, granted, seems impossible, both socially and physically. We're surrounded by evidence of roads' immortality: the Roman Empire's military highways crease Europe's fabric; the Oregon Trail scars Wyoming's sagebrush. David Farrier has observed that our roads and bridges will solidify into "future fossils" at which archaeologists will someday wonder, "written into the earth like speech marks around a lost quotation." Elers Koch, too, deemed forest roads permanent features: "final and irretrievable facts."

And yet, in the Clearwater, decommissioning had proved that roads can yet be retrieved. Forest roads are a treatable disease, debilitating but not necessarily fatal. Occasionally, Anne Connor told me, she soars over Idaho on commercial flights and gazes on her former domain, its green carpet threaded by silver rivers, the old reticulation of logging tracks faint as pencil lines. One day, she thinks, she will fly over the forest and not see any roads at all.

6

THE BLAB OF THE PAVE

Why road noise pollution disrupts animals' lives
everywhere, even in our national parks.

The essential insight of road ecology is this: roads warp the earth in every way and at every scale, from the polluted soils that line their shoulders to the skies they besmog. They taint rivers, invite poachers, tweak genes. They manipulate life's fundamental processes: pollination, scavenging, sex, death.

Among all the road's ecological disasters, though, the most vexing may be noise pollution—the hiss of tires, the grumble of engines, the gasp of air brakes, the blare of horns. Noise bleeds into its surroundings, a toxic plume that drifts from its source like sewage. Unlike roadkill, it billows beyond pavement; unlike the severance of deer migrations, it has no obvious remedy. More than 80 percent of the United States lies within a kilometer of a road, a distance at which cars project twenty decibels and trucks and motorcycles around forty, the equivalent of a humming fridge. Traffic noise is so inescapable that it has become our aural frame of reference: "By the time I was ten," the British writer Helen Macdonald has observed, "I could stand by Europe's second largest waterfall, listen to it roar, and think, simply, *it sounds like the motorway when it's raining.*"

Like most road impacts, noise is an ancient problem. The incessant rumble of Rome's wagons, griped the poet Juvenal two thousand years ago, was "sufficient to wake the dead." To walk in New York, Walt Whitman wrote in *Song of Myself*, was to experience the "blab of the pave," including the "tires of carts" and "the clank of the shod horses." If Whitman seemed more delighted than dismayed by the hubbub, it's because he was never subjected to the din of internal combustion. As car traffic swelled in the twentieth century, road noise became a public-health crisis, a not-so-silent killer. Noise exposure deprives us of sleep, impairs our cognition, and triggers the release of stress hormones that lead to high blood pressure, diabetes, heart attacks, and strokes. A 2019 report by one French advocacy group calculated that noise pollution shortens the average Parisian life span by ten months. In the loudest neighborhoods, it truncates lives by more than *three years*.

To a road ecologist, noise is pernicious precisely because it isn't confined to cities. It also afflicts national parks and other ostensibly protected areas, many of which have been gutted by roads to accommodate tourists. A vehicle inching along Going-to-the-Sun Road, the byway that wends through Glacier National Park, casts a sonic shadow nearly three miles wide. "Once you notice noise, you can't ignore it," a conservation biologist named Rachel Buxton told me. "You're out in the wilderness to have this particular type of experience where you escape and relax, and noise ruins it. It totally ruins it."

And road noise isn't merely an irritant: it's also a form of habitat loss, a repellent that evicts wild creatures from environments in which they would otherwise thrive. In Glacier the distant grind of gears startles mountain goats; in one Iranian wildlife refuge a highway drives a forty-decibel wedge through a herd of Persian gazelles. The bedlam directly challenges the purpose of parks, places that exist in large part to safeguard the animals within them. Consider Devils Tower National Monument, the surreal Wyoming butte made famous by *Close Encounters of the Third Kind*, which annually hosts a group of travelers even more disruptive than aliens: bikers from the nearby Sturgis Motorcycle Rally. The roaring fleet of Harleys and Yamahas, Buxton has found, pushes prairie dogs back into their burrows, drives off deer, and deters bats from feeding near the monument for weeks. Had Spielberg's extraterrestrials arrived mid-Sturgis, Richard Dreyfuss would have missed their five-note musical greeting.

In hindsight road noise was a predictable problem in national parks. Nearly from their inception, America's parks were "windshield wildernesses," in the historian David Louter's memorable phrase—road-trip destinations designed to be experienced from behind glass. In 2019 nearly 330 million people visited national parks, almost all in cars. That parks are being "loved to death" has become a truism, repeated as often as Wallace Stegner's claim that they're our best idea. More visitation means more roadkill, more erosion along shoulders requisitioned as parking spaces, and, most of all, more noise. Nearly half the total area under the National Park Service's jurisdiction today suffers at least three decibels of sonic pollution.

From the car's earliest days, its critics worried that its clamor would overwhelm nature. Soon after the Model T's release, John Burroughs, an influential writer and proud Luddite, denounced the automobile as a "demon on wheels" that would "seek out even the most secluded nook or corner of the forest and befoul it with noise and smoke." The screed caught the attention of Henry Ford, who well understood the symbolic power of converting the car's detractors. In 1912, Ford gifted Burroughs a placatory Model T, and soon the septuagenarian was flivverboobing merrily around the Catskills, plowing through barns and into trees. (He was easily distracted by birds.) "Out of that automobile," Ford recalled, "grew a friendship."

In 1914, Ford and Burroughs traveled to Florida to visit another famous friend, Thomas Edison. After greeting fans in Fort Myers, the trio and their families struck out for the Everglades. The jaunt devolved into a comedy of errors: their cars foundered in mud, snakes invaded camp, and a thunderstorm forced the drenched men to cower in the women's tent. Despite these mild hardships, the self-proclaimed "Vagabonds," joined later by the tire magnate Harvey Firestone, made car camping a semiannual tradition, visiting California, the Adirondacks, and the Appalachians in convoys of Packards and Model Ts. They bought apples from farmers' daughters, talked chemistry around campfires, and disavowed shaving—no trouble for Burroughs, whose trademark was a sternum-length beard. "We cheerfully

The Vagabonds—from left, Thomas Edison, John Burroughs, Henry Ford, and Harvey Firestone—helped make car camping an American pastime. *The Henry Ford*

endure wet, cold, smoke, mosquitoes, black flies, and sleepless nights, just to touch naked reality once more," Burroughs rejoiced. Not that they were roughing it: a passel of bowtie-clad servants waited on them, and their 1919 glamping trip included a kitchen car complete with stove and icebox. Still, the press was enthralled by the notion of Ford and Edison, these champions of progress, returning to nature, or a facsimile of it. After Burroughs died in 1921, he was replaced by an even more august celebrity: President Warren Harding.

The Vagabonds' road trips both channeled the zeitgeist and directed it. Car sales weren't exploding because automobiles were convenient but because they were liberating. Gallivanting around the countryside had long been the frivolous pastime of Rockefellers and Vanderbilts, not the bank tellers and factory workers they employed. The rise of affordable Model Ts made vacation a widespread rite. Nationalism also propelled the fad, as highway associations and car companies pushed auto touring under the mantra "See America First." So vocal were the See America Firsters that one died of "laryngitis complicated with tuberculosis of the throat" after a vociferous public-relations junket.

And there was so much America to see. For centuries nature had been the unruly enemy of yeoman homesteaders; now the Vagabonds recast it as a manly respite from "feminized" city life. Automobiles, Ford declared, provided "the blessing of hours of pleasure in God's great open spaces," and scenic parkways like the Bronx River and Taconic beckoned to joyriders. Car campers spread across America like ants at a picnic, bearing Airstream trailers and Coleman stoves peddled by the growing outdoors industrial complex. "Camping out used to be done in the wild woods," enthused one reporter, "and now it is done in the wild Fords instead."

At first the wild Fords went wherever they pleased. Early car camping was gloriously anarchic; travelers pitched tents along highways and in the fields of irate farmers. Over time camping formalized into a tamer practice. Towns established "autocamps," the forerunners of motels, and booster groups published maps and guidebooks. With the roadside commons enclosed, many campers descended on national forests, where they commandeered Elers Koch's firefighting roads and antagonized Aldo Leopold. Still more flocked to a novel category of land, one that soon became synonymous with the American road trip: the national park.

To be sure, parks predated the automobile. Yellowstone National Park, America's first, was established in 1872, and Yosemite, Mount Rainier, Crater Lake, and others followed. But most were seldom visited, and then only by wealthy travelers who could afford cross-country train tickets and guided horseback adventures. No formal agency existed to steward parks; instead, they operated under the haphazard aegis of the Interior Department and the U.S. Army, whose charge was to thwart poachers, not kowtow to tourists. Early visitors to Yosemite Valley had to chain their cars to trees and surrender their keys. Allowing automobiles, one visitor opined, would be akin to the serpent "finding lodgment in Eden."

But the serpent was backed by powerful lobbyists. As the motor age flowered, pressure mounted to open parks to cars. Even John Muir predicted that automobiles would soon "mingle their gas-breath with the breath of the pines . . . with but little harm or good." When cars finally wormed their way in, however, park infrastructure was in no condition to accommodate them. Mount Rainier's main road was so treacherous that President William Taft's aides worried that the commander-in-chief wouldn't survive his 1911 visit. (Taft experienced no graver harm than getting stuck in the mud; his car, and his own substantial bulk, had to be hauled out by mules.) A few years later Stephen Mather, a gregarious millionaire who had made his fortune hawking borax, toured Sequoia and Yosemite. Mather was smitten by his surroundings—"Here we get a free shower!" he cried beneath waterfalls—but not the amenities. "Scenery is a hollow enjoyment to a tourist who sets out in the morning after an indigestible breakfast and a fitful sleep in an impossible bed," he carped in a letter to the Interior Department.

Came the reply: "Dear Steve, if you don't like the way the national parks are being run, come on down to Washington and run them yourself."

And so, in 1917, Mather became the first director of an agency tasked with pulling America's best-loved landscapes into a cohesive whole: the National Park Service.

Mather's ward was a bundle of contradictions. Like its older sibling, the Forest Service, the Park Service inherited a murky mission. Its establishing legislation directed it to simultaneously conserve "wild life" and "provide for the enjoyment" of visitors but offered little guidance about how to reconcile those objectives. Mather made his priorities clear. Solving the parks' "roads problem," he declared, would "enable the motorists to have the greater use of these playgrounds which they demand and deserve." Thanks to the Vagabonds, scenery had become a consumer good and the tourism economy a deity that demanded worship. Under Mather's direction the Park Service cranked out films, guidebooks, and other propaganda; cozied up to the American Automobile Association; and finagled millions of dollars from Congress. By 1929 national parks were filigreed by 1,300 road miles, down which wheezed nearly 700,000 cars.

Although Mather was a shrewd operator, his ardor for auto-tourism was genuine. Mather's joviality concealed depression and perhaps bipolar disorder; at one gala a coworker found him in a reception room, "rocking back and forth, alternately crying, [and] moaning." For Mather, scenery offered better treatment than any sanitarium. A visit to the parks, he wrote, "begets contentment" and "contains the antidote for national restlessness." Mather often availed himself of his preferred cure, tooling around in a Packard whose vanity plate read "NPS 1." Park roads, to Mather's tumultuous mind, were egalitarian forces that permitted the old, young, and infirm access to their scenic birthrights—a nationwide prescription for mental health.

Mather wasn't entirely sanguine about his construction boom. A road through southeastern Yellowstone, he cautioned, "would mean the extinction of the moose"; today that area is the farthest you can get from a road in the Lower 48, around twenty miles. Rather than building roads for their own sake, Mather believed in building the *right* roads, monumental feats of engineering—Trail Ridge Road in Rocky Mountain National Park, Skyline Drive in Shenandoah, Generals Highway in Sequoia—as stunning as their environs. Mather's roads were crimped with hairpin turns, punctuated with pullouts, and bedded with local rock, tourist attractions that nonetheless strove to harmonize with their spectacular surroundings. The epitome of his philosophy was Going-to-the-Sun Road, which the Park Service blasted from Glacier's face with nearly a half-million pounds of explosives—a project so hubristic that three workers died during construction. "It is like the route to the fabled Olympus," raved the *Great Falls Tribune* when Going-to-the-Sun opened in 1933. "Each revealing bend in the road is another portal to prodigious splendor."

Yet Going-to-the-Sun was no benign feature. Like other roads of its era, the Park Service designed it to flaunt vistas, to accommodate tourists who consumed nature with eyes, binoculars, and cameras. But it neglected to account for Glacier's nonhuman denizens, many of whom experienced the world primarily through other senses. Mather's grandiose roads showcased landscapes and trashed *sound*scapes, a form of degradation whose consequences would take decades to discover. "They're built to be scenic, they're built to be part of the experience," Kurt Fristrup, a former Park Service bioacoustician, told me of the agency's roads. "Ironically, that also means they're built to project noise about as far as it can possibly be projected."

Most laypeople treat the words *sound* and *noise* as synonymous. To an acoustician, though, they're antonyms. Sound is fundamentally natural, and tickles the ears in even the quietest places: the susurrus of wind, the drone of a bee, the starched snap of a jay's wings. Noise, by contrast, is a human-produced pollutant, and, like its etymological root, *nausea*, it's unpleasant. In the serenest landscapes even an airplane can seem a violation, as sinister as "Flight of the Valkyries" blasting from the helicopters in *Apocalypse Now*.

And noise is still harder on other species. Wild animals inhabit an aural milieu that is sensitive beyond our imagining. Human conversation occurs at around sixty decibels, and sounds that barely register to us—gentle breathing, the rustle of leaves—produce around ten decibels. The most acute predators, meanwhile, can detect negative-twenty decibels. Bats seize upon the crunch of insect feet; foxes triangulate snow-buried voles. Prey is equally perceptive. Scrubwren nestlings freeze at the footsteps of enemy birds. Tungara frogs duck at the flap of bats. Vision is a luxury, hearing a necessity: most animals sleep with their eyes closed, but nearly all awaken at the snap of a twig.

This acoustic arms race evolved in a quiet world, one that roads have vandalized. Although organisms have always contended with loud environments, like blustery ridgelines and crashing waterfalls, cars have made cacophony more rule than exception. For animals that survive by the grace of their hearing, traffic's "masking effect" can be fatal. Ambient road noise drowns out songbirds' alarm calls and prevents owls from detecting rodents. A mere three-decibel increase in background noise halves the "listening area," the space in which an animal can pick up a signal. By disturbing animals, noise also disrupts the ecological processes they catalyze, among them seed dispersal, pollination, and pest control; in Portugal's oak woodlands, for instance, birds like chaffinches and blue tits avoid loud highways, allowing unchecked insects to kill roadside trees. Even the unpaved ocean is overwhelmed with road noise, furrowed with shipping lanes marked by oily rainbows and the grind of freighters. After the September 11 attacks induced the Canadian government to suspend ship traffic, researchers who analyzed right whales' floating feces found that their stress hormones had plunged, as though the cetaceans had taken a deep but temporary sigh of relief.

And traffic doesn't merely interfere with hearing: it makes it hard to be heard. In 1911 a French surgeon named Étienne Lombard observed that when he piped noise into a patient's ear midconversation, the speaker involuntarily raised his voice; subsequent studies found that noise also causes people to raise their pitch and speak more slowly. Since 1972, when researchers documented this "Lombard effect" in Japanese quail, scientists have observed it in dozens of species, from blue-throated hummingbirds to pig-tailed macaques. In Australia male tree frogs croak at higher frequencies near freeways (at risk to their romantic prospects, since females may associate squeakier ribbits with less studly mates). The grey shrike-thrush, too, jacks up the frequency of its whistling to be heard over traffic. Juvenile shrike-thrushes may be more likely to detect and mimic these higher-pitched melodies, leading to the "relatively rapid development of dialects": a regional avian language, its vernacular shaped by roads.

For all of its harms, the pave's blab was historically difficult to study. In 2000, Richard Forman, road ecology's *paterfamilias*, demonstrated that meadowlarks, bobolinks, and other birds gave highways a wide berth, at least two football fields, and hypothesized that traffic noise was the "primary cause for avian community changes." Yet roads were so transformative that it was hard to tease apart their perversions. Noise was likely repelling wildlife, yes—but perhaps animals also detested the sight of cars, or the abrupt forest clearing, or the impaired air quality. "People had been guessing that noise pollution was controlling animal distributions for a long time," Jesse Barber, a sensory ecologist at Boise State University, told me. "But we needed to do an experiment"—to strip out the road's many variables and isolate its sonic impacts.

For his experiment's site Barber chose Lucky Peak, a green shoulder of land that rises above central Idaho's sunburnt scrub. Every fall, rivers of southbound songbirds—lazuli buntings, hermit thrushes, golden-crowned kinglets—alight at Lucky Peak to refuel on berries and insects, like mule deer pausing to browse tender greens as they migrate. The mountain is a superlative stopover owing partly to its paucity of roads; it's a wild oasis in a desert of Boise-area sprawl. In 2012, Barber, a student named Heidi Ware Carlisle, and some colleagues resolved to figure out how Lucky Peak's migrants would respond to a new road: not an asphalt road but an acoustic one. The group mounted fifteen pairs of speakers and amps to tree trunks, wrapped the wires in garden hoses to deter chewing rodents, and shrouded them with shower curtains to keep off the rain. Then they blared a looped recording of traffic from 4:30 a.m. to 9:00 at night—a Phantom Road in a roadless wood.

The Phantom Road was so realistic it fooled even people. "We get hikers and mountain bikers and hunters through there all the time, and at least three times we ran into folks who were like, okay, I thought the highway was *south* of here, not east," Carlisle told me. The birds were equally disoriented. On days when Carlisle played her recordings, bird counts plummeted. Some species, like cedar waxwings and yellow warblers, avoided the Phantom Road altogether.

And the Phantom Road didn't merely drive off birds: it drained those who stayed. When Carlisle captured warblers and examined their tiny bodies, she found they were skinnier after they'd been near the Phantom Road. Songbirds survive by listening ceaselessly for the whir of falcons, the rustle of martens, and the alarm calls of their neighbors: "the chipmunk next door that sees the goshawk before you do," as Carlisle put it. When road noise drowns out sonic cues, birds must look for predators rather than listen for them. This "foraging-vigilance tradeoff" gradually depletes them: every moment you're scanning for hawks is one you're not gobbling beetles. Alertness saves your life, until it starves you.

The Phantom Road was a road-ecology landmark. For the first time, researchers had shown that noise alone could impinge on animals' lives. What made the experiment so compelling, though, wasn't merely its ingenious design: it was where the Phantom Road's noise came from. The traffic that played through Carlisle's speakers hadn't been recorded on an interstate highway or an urban boulevard. Instead, it was from Going-to-the-Sun Road, Glacier National Park's fifty-mile byway.

Going-to-the-Sun was a strategically chosen source for the Phantom Road. If noise tainted even a sanctuary like Glacier, where could birds feel comfortable? America's windshield wilderness had extended acoustic pollution into its best-protected places. A dusky grouse in Rocky Mountain National Park or a cerulean warbler in Shenandoah was safe from hunters and developers, yet vulnerable to the growl of Winnebagos and Harleys—a conundrum not lost on the Park Service itself. "If you went to a freeway and said, 'Hey, these trucks need to go slower because of the birds,' well, there's just absolutely no way that would happen, right?" Barber told me. "The parks are the places where we might actually be able to get some action. They're getting slammed with traffic, they're getting slammed with people, and the managers are willing to try something."

For nearly as long as the National Park Service has existed, its critics have complained about noise. Edward Abbey, the cranky bard of Utah's canyon country, begged tourists "sealed in their metallic shells like molluscs on wheels" to "walk—walk—WALK upon your sweet and blessed land!" *Desert Solitaire*, Abbey's 1968 account of his stint as a park ranger, is both a screed against Motordom and an homage to serenity: the "quiet deer," the "quiet little coyote," the "crystalline quiet" of the Colorado River. Declared Abbey, "We have agreed not to drive our automobiles into cathedrals, concert halls, art museums, legislative assemblies, private bedrooms and the other sanctums of our culture; we should treat our national parks with the same deference, for they, too, are holy places."

When the Park Service awakened to its noise problem, however, it wasn't because of cars at all. In 1986 an Otter airplane and a Bell 206 helicopter collided over Grand Canyon National Park, killing their passengers and painting the canyon walls silver with molten aluminum. The catastrophe

galvanized Congress, which passed a law requiring the Park Service to reg-
ulate the flight tours buzzing around its airspace and acknowledged that
aircraft were having "a significant adverse effect on the natural quiet." The
Park Service, chastened, began to study noise pollution on its lands and in
2000 formed the Natural Sounds Program, a scientific branch tasked with
protecting the parks' aural riches—the bugle of elk, the crash of calving
glaciers, the tremulous dawn chorus of songbirds. The parks' soundscapes
might be subtler than Half Dome or Old Faithful, but they deserved attention as surely as scenery.

What the Sounds Program would do was unclear. Acoustic ecology, like
road ecology, was a nascent field, one that dealt largely with how ship noise
harmed whales. How did that pertain to snowmobiles in Yellowstone or
helicopters over Yosemite? When Kurt Fristrup met with Federal Aviation
Administration officials to discuss limiting flights, they laughed him out of
the room. Scoffed one, "There are no whales in the Grand Canyon."

The Natural Sounds Program's first act, then, was to listen. In the
early 2000s the agency dispatched an army of technicians to install hundreds of "listening stations," rudimentary instrument packages—laptops,
microphones, solar panels—that passively absorbed soundscapes for
weeks at a time. The stations were raked by wind, buried by snow, sabotaged by animals. "We have a lot of measurements where there's the
sound of a bear finding the microphone, and then the sound of their
mouth on the microphone, and then total silence," one biologist told me.
In Great Sand Dunes the instruments detected the clattering antlers of
rutting elk; in the Everglades they recorded the bellowing of alligators. In
one Massachusetts park a skier stumbled upon a microphone and delivered a weather report.

Gradually, the Park Service was discovering that it supported acoustic
treasures equal to its visual wealth. But the recordings also betrayed sonic
pollution. When researchers analyzed nearly 1.5 million audio clips, they
found that noise contaminated more than a third. The causes varied—there
was the train that chugged through Cuyahoga Valley, the cruise ships that
rumbled into Glacier Bay—but roads were a constant. Almost invariably,

the loudest parks were the ones closest to airports and the ones with the densest road networks.

In theory the Park Service was well positioned to address the issue. The agency might strike a wobbly balance between "wild life" and tourism, but it still held conservation in higher esteem than did, say, the Federal Highway Administration. Yet parks weren't easily hushed. When Jesse Barber convinced the overseers of Grand Teton to cut speed limits in 2016, traffic quieted by a couple of decibels and visitors heard more birds. But the birds themselves still avoided the road. A car dawdling along at twenty-five miles per hour might be softer than one cruising at forty-five, but it took longer to pass—a less acute stressor but a more drawn-out one. It wasn't hard to make park roads quieter, but it was exceedingly difficult to make them quiet enough.

"I think all of our work is pointing towards this: the best way to preserve quiet habitat for wildlife is to not build the damn road," Barber told me. "And once you do, you're in big trouble."

Electric vehicles should help to an extent. Unburdened by the rattling gadgetry of internal combustion, their motors are virtually silent. But EVs are no panacea. Above thirty-five miles per hour, tires, not engines, produce most vehicle noise. (The interstate's monotonous drone is a blend of "rhythmic percussion" from rubber on road and "pattern noise" generated by air pockets popping within tread.) The Grand Canyon parking lot will be more pleasant without grumbling RVs, but electrification won't silence the Everglades' main highway, where the speed limit is fifty-five miles per hour.* And while "quiet pavements"—pitted surfaces whose divots dampen tire noise—showed potential when they were trialed in Death Valley National Park, their benefits diminish as grit clogs their pores. It's hard

---

* Although the silence of EVs has allegedly caused a few driveway cat flattenings, I'd wager that it won't lead to more wildlife roadkill since most animal collisions happen at the higher speeds at which tire noise makes electric cars audible.

to envision the Park Service scouring its roads with the "giant vacuum-cleaner-like device" some European cities deploy to maintain quiet streets.

Technology, then, holds few answers. And the Park Service, bound to appease visitors, seems unlikely to overhaul its windshield-wilderness model. But the *number* of windshields—well, that's more malleable. In 2000 the agency banned private cars from much of Utah's Zion National Park and replaced them with public buses. Rangers saw more deer and again heard the burble of canyon wrens echoing off sandstone. Since Stephen Mather's day, the Park Service had bowed to cars, yet in several places it had cast out the serpent, or at least restricted its ingress. I decided to visit one such park myself—not Zion, but a wilder one, with a longer history of anticar warfare.

Alaska is an odd place for a road-curious writer to visit for a simple reason: there aren't many roads. The state is creased by just one paved road mile for every hundred square miles of land, and the roads that exist don't last. There are places in Alaska where temperatures fluctuate between 100 degrees Fahrenheit and 60 below; places blanketed by sixteen feet of snow; places bludgeoned by avalanches and crumpled by melting permafrost. There are two seasons in Alaska, residents crack: winter and construction.

"We're about to experience some land turbulence, so if you've got any beverages you might want to put a lid on 'em," my bus driver intoned over the intercom one June evening. Beneath our wheels the Parks Highway, the road from Anchorage to Fairbanks, dipped and buckled. An elderly couple gripped their armrests. "Remember," the driver added, "it's not my fault, it's not the bus's fault—it's the asphalt."

I offered a courtesy chuckle and watched the horizon. Wizened spruce blanketed the feet of serrated mountains. Rivers churned with glacial flour, milky as cataracts. Fireweed flourished in the churned roadside soils, magenta flowers against gray-green tundra, the definitive palette of Alaskan highways. I disembarked at a turnoff and continued on foot, my thumb out and my least menacing smile plastered on my face. In the past I'd had good luck hitchhiking around parks, but either society had grown less trusting or I'd gotten creepier, and the minutes became an hour. At last a trucker took pity, pulled over, and delivered me to the turnoff for one of America's most iconic byways: the Denali Park Road.

The Denali Park Road epitomizes the tension that tugs at the Park Service's heart. Its curvaceous ninety-two-mile course is smudged with Stephen Mather's fingerprints: in 1924, soon after construction began, Mather's lieutenants urged builders to "avoid long straight lines" and showcase "the best possible views and vistas of the country." From the start the road was a boondoggle. The Alaskan weather and "bottomless" mud frustrated road crews, who earned poverty wages and lived in tent villages. Even so, luxury auto-tourists, inspired by the Vagabonds' escapades, showed up in droves. Caravans of Studebakers ferried sightseers, clad in suits and pearls, thirteen miles down the road to Savage Camp, where they slept in wood-framed tents and waltzed on a polished floor. "You sit down at a table that is covered with snowy linen and napkins—articles scarcely looked for in the wilderness," marveled one reporter.

The Park Service continued to genuflect to cars in the years that followed. In the 1950s the agency launched Mission 66, a massive construction program, cosponsored by the American Automobile Association, that, among many other projects, would have paved and widened the mostly dirt Denali road. The scheme outraged Adolph and Olaus Murie, biologist brothers who feared that a bigger, faster road would ruin the "purity of [the park's] wilderness atmosphere." Denali was a place for deep thought, not cursory sightseeing. "The national park will not serve its purpose if we encourage the visitor to hurry as fast as possible for a mere glimpse of scenery from a car," Olaus admonished. Even as the Lower 48's car culture approached its apogee, the Muries insisted that Denali remain a temple to a forgotten value: slowness.

The brothers won the battle, and the Park Service kept the road primarily gravel and dirt. In 1972 the agency, now squarely on Team Murie, announced an even more drastic policy. The unpaved portion of the road—seventy-eight of its ninety-two miles—would remain off-limits to most private cars, forcing visitors to ride buses that crept along at twenty-five miles per hour. The shift enraged the tourism industry: Denali, inveighed one hotelier, had become "a Washington Monument with no elevator." But the Park Service only piled on more restrictions. In 1986 it declared that just ten thousand or so vehicle trips would be permitted each year. Other parks were bursting at their seams, yet Denali had frozen its traffic.

"Today we're within a thousand vehicles of the number that went out onto the park road in 1986," an ecologist named William Clark told me when, after several hours of hitchhiking and walking, I reached Denali headquarters. "I don't know if there are many roads in the world that can say that."

Clark, whose scruffy beard could have adorned a gold prospector, managed Denali's road-ecology program. He was so laconic that, when we had spoken on the phone earlier, he'd had to reassure me that he was looking forward to our meeting. "Sorry if I don't sound excited," he had said. "I am. But Dave will actually *seem* excited." Dave was Dave Schirokauer, leader of the park's research team, who was as high-spirited as Clark was reserved. The men were tasked with upholding Denali's Vehicle Management Plan, its abstruse prescription for reconciling the needs of wildlife with the demands of tourists. "We can't kill the golden goose," deadpanned Schirokauer when we convened in his office. "We just keep it on life support."

Although the park's vehicle plan is intended to help all wildlife, it's particularly concerned with Dall sheep, snowy-fleeced ruminants that from the road appear mostly as pale pinpricks against distant outcroppings, like grains of salt sprinkled across a tablecloth. ("Rock or sheep?" is the most-uttered phrase on Denali's buses, answerable only through binoculars.) Dall sheep—nimble, far-sighted, capable of surviving on lichen salad—are exquisitely adapted for life among Denali's promontories, where wolves can't touch them. Each spring, though, sheep clamber down the talus to feast on grasses and wildflowers, a migration that the park road disrupts.

In 1985 naturalists watched eight sheep inch cautiously downhill, "oriented toward the road, and displaying attention postures." Over the next hour the jittery animals advanced and retreated, ceding their progress every time a bus rumbled past, like sandpipers scuttling from a wave. Before long, wrote the naturalists, "the sheep had returned to the escape terrain," their journey thwarted. Even minor increases in traffic, from ten vehicles an hour to twenty, can chase sheep from Denali's roadside.

The park's Vehicle Management Plan caters to sheep anxiety. The plan, which runs 428 turgid pages and reliably cures insomnia, dictates how many cars, trucks, and buses can bounce through the park every day, where they can go, and who gets to operate them. In a world whose default is unlimited access, it's a quietly revolutionary document, permitting no more than 160 vehicles every 24 hours. Its most progressive provision is the "sheep gap," a requirement that every hour, at several migration corridors, at least 10 minutes elapse without a single vehicle. "If there's a band of sheep, or any wildlife, that are looking to make their move, they have their opportunity," Clark said. "I describe it as a deep breath in traffic."

Traffic, however, is inclined to hyperventilate. The near-universal goal of transportation officials is to fill service gaps, not create them. The sheep gap, Schirokauer acknowledged, is a "four-letter word" among the concessionaires who run Denali's buses. Like any transit users, the park's sightseers crave a steady stream of buses, a desire that is at odds with the lengthy gaps that sheep require. It's no coincidence that the bus system generally satisfies visitors but falls short of maintaining the sheep gap. Ruminants don't leave one-star TripAdvisor reviews.

"There's incredible pressure on the park from the travel industry to allow more buses," Schirokauer said. "Every time they add a wing to a lodge, there's eighty new pillows. And they're like, okay, that means we need two more buses a day to accommodate these people. The American way is to keep growth alive. People get when a theater is full, but they don't get why a park road would be full. Until they get to a wildlife viewing stop and there are ten buses and all they can see is the back leg of a bear." I thought of all those tourists resenting other tourists for contaminating their solitude and remembered the rush-hour cliché: You don't get stuck in traffic. You *are* traffic.

When the Park Service instituted the Denali shuttle in the 1970s, its goal wasn't to cut noise pollution but to preserve the vague wilderness vibes that the Muries cherished. Yet the bus system has turned Denali into one of America's finest soundscapes, a place that unmasks natural choirs rather than concealing them. In Denali's symphony hall, nature plays strange, unrepeatable compositions—the song of warblers backdropped by a late-spring avalanche, say, like a percussion section pounding behind wood-winds. The road's faint sonic footprint is easily escaped; you can step off the shuttle wherever you please and follow a dry wash into the park's trailless, entropic backcountry. "There's a lot of solitude that can be had just a hop, skip, and a jump away from the main road," one Park Service acoustician told me. "You can get up a few hundred feet and pretty much be out of ear-shot of the thing."

One afternoon I drove the Denali road with Kate Orlofsky, a technician tasked with studying the bus system. We cruised through patchy forests that rolled toward scree. Cloud shadows scurried over the green-gold val-ley, a world without borders; a wolf could walk six hundred miles from here without padding across another road. At Savage River, the end of the line for most cars, Orlofsky checked in with a ranger, who ticked a clipboard. We pressed on down dust and gravel, as the Muries had intended. The road was exquisitely free of visual clutter, unsullied by the official admonitions— Falling Rocks, No Passing Zone, Road Work Ahead—that supposedly make us safer but, like those pesky Deer X-ings, have faded into white noise. A caribou browsed a gravel bar, swinging his velvety chandelier. "Sometimes I'll have the windows down and listen to birdsong," Orlofsky said, and I wondered where else that would be possible.

After a while Orlofsky pulled over at a sheep-gap site and fished out a laptop. To my eyes the mottled tundra looked no different here than any-where else. I scanned the slopes: no sheep. We waited for buses to come so that Orlofsky could record their timing and confirm that the gap was being upheld. Soon a dust plume rose in the distance, a towering column of aerosolized roadbed that spiraled like smoke.* The cloud's creator, a matte green schoolbus, chugged into view. Its passengers gazed out the windows, or fiddled with their cameras, or dozed, cheeks smushed against glass. Orlofsky tapped the laptop, and we got back in the car.

On the drive over, I'd asked Orlofsky about her past gigs. She had trapped wolves, chased pikas, patrolled the Pacific Crest Trail. By comparison, I observed now, counting buses seemed, well, kind of boring. Orlofsky laughed. "It's not the sexiest work. It's not the most active," she admitted. "But I can watch wildlife. I can enjoy the breeze." When she wasn't monitoring sheep gaps, her job entailed riding the bus system herself, a mole among the visitors, covertly recording data—about ridership, rest stops, wildlife, traffic—that would inform future iterations of the park's vehicle plan. She was, in other words, a professional tourist, a luxury she didn't take for granted. "People come from all over the world to do what I get paid to do," she said. Denali has become what social scientists describe as a "near-wilderness," a place where most visitors peer into the wild without immersing themselves in it. But a near-wilderness can still evoke ecstasy. Orlofsky had recently seen a roadside caribou send a tourist into a state of rapture. Even ptarmigan, the pudgy grouse that is Alaska's state bird, got the shutters whirring.

"People get pretty excited about ptarmigan," Orlofsky said. "As they should."

Like the Forest Service's dirt roads, culture-war objects as hotly contested as Confederate statues, park roads are anthropological artifacts—features whose design and function change to reflect their contexts. As our values evolve, roads follow. In the 1920s the Denali road had no higher purpose than injecting humans into parks to ogle mountains, nor did it make

---

* The decision to keep the Denali road unpaved came at a cost. A patina of dust coats roadside plants throughout the park, as though they've been rolled in flour. The Park Service sprays a chemical slurry to tamp the dust, which in turn causes "localized spruce decline."

any allowance for the animals it disrupted. Three decades later the wilderness movement fought to preserve the road's "primitive character" on aesthetic grounds; today, ecologists manipulate its traffic to aid sheep. A road built to display scenery for people is now governed to preserve the acoustic habitat of a ruminant—an acknowledgment that we share our infrastructure with wild users, that we're not alone on the road.

The trouble with parks isn't that they're windshield wildernesses, but that there are too many windshields. When I worked a season for the Park Service in Yellowstone, I was often trapped in miles-long "bison jams" that formed to watch the shaggy beasts lumber over the pavement. I couldn't blame the rubberneckers: Where else could you commune with America's national mammals in all their obstreperous glory? How else could you admire their exquisite detailing—the rippling muscles that control their gargantuan heads, the odd daintiness of their haunches, the ragged fur that sloughs from their shoulders like snakeskin? I would never deny people their bison viewing, but I'd happily shepherd them onto fleets of buses: electric, to avoid engine noise, and slow-moving, to quiet their tires, with ample gaps between them. Even Edward Abbey granted that parks could have shuttles "for those too old and too sickly to mount a bicycle." If buses work for Cactus Ed, they work for me.

Months later, back home, I reread my notes from Denali. As I did, I could hear the transient muttering of an arterial and, below it, the sibilance of the interstate. Its static washed over me like acid rain, elevating my stress levels, shortening my life span. And yet I hardly heard it. We're so awash in what the writer David Haskell called "the fossiliferous racket of industrialized humanity" that we've compensated by deafening ourselves. Paging through my notes, I was appalled by their shallowness. "Birds," I'd written, unhelpfully—and "wind."

Humans, it turns out, *are* good listeners: we just inhabit a world that doesn't reward it. Natural sound is as salubrious as noise pollution is harmful. The crash of waves calms heart-surgery patients; the trill of crickets boosts the cognition of test-takers. Road noise both degrades our bodies and overwhelms the natural sounds on which we, like songbirds, depend. This, then, is the value of parks: as sonic sancta for wildlife and humans, refuges for acoustic experiences that are as endangered as any species. When I spoke to the bioacoustician Kurt Fristrup, he described wandering with a friend one night through Natural Bridges, a national monument in the Utah desert. The hush was absolute, unmarred by traffic, and, after Fristrup sat down, he became aware of his heartbeat thumping softly in the darkness. As he relaxed into silence, he heard another sound, eerie and magical: the heartbeat of the friend sitting beside him, its slow rhythm syncopated with his own, bass drums in the amphitheaters of their chests.

7

LIFE ON THE VERGE

Will the highway's novel ecosystem save America's most beloved butterfly or obliterate it?

Imagine for a moment this scenario: You're cruising a two-lane midwestern highway at the end of a long day—corn rippling out the driver's-side window, westering sun in your eyes—when you realize with a start that your windshield is clean. In your youth, you recall, this same trip left your car as spattered as a Pollock canvas, smeared with the debris of grasshoppers and beetles and spiders ballooning on silken threads, the infinitesimal casualties that Barry Lopez called "aerial plankton." Perhaps, like Lopez, you'd scrubbed at your grille, ashamed of the damn spots that implicated you in mass murder. "I am uneasy carrying so many of the dead," Lopez mourned. "The carnage is so obvious."

Today, however, your conscience is as immaculate as your windshield, unmarred by so much as a flying ant. With growing unease, you wonder, What gives?

In fact, your observation has a name: the windshield phenomenon. Its innocuous moniker belies the horror it describes. We are living through what some entomologists have called an "insect apocalypse," an invertebrate die-off that's all the scarier for being so poorly understood. According to one analysis, insects are going extinct *eight times faster* than mammals, reptiles, and birds; per the same study, the world loses 2.5 percent of its insect mass every year. The apocalypse is both exacerbated by roads—cars kill billions of North American pollinators every year—and revealed by them. Like mobile natural-history museums, windshields once showcased the exuberance of the invertebrate world, its blizzards of moths and midges. Now our cars remain spotless, and gas-station squeegees sit useless in their soapy buckets. Some scientists argue that the windshield phenomenon is an illusion created by modern cars, along whose streamlined bodies insects tumble rather than splat (picture a low-slung Porsche versus an upright Model T). Then again, as one entomologist told *Science*, "I drive a Land Rover, with the aerodynamics of a refrigerator, and these days it stays clean."

Like most animal collapses, the insect apocalypse starts with habitat loss. That's particularly true in the Midwest, where more than 99 percent of historic tallgrass prairie has been sacrificed to agriculture and development. As usual, infrastructure spurred the devastation. The railroads that gashed the Great Plains in the nineteenth century also opened this "inland sea" to farmers, who promptly plowed up the milkweeds, indigos, paintbrushes, and larkspurs that nurtured insect life. The remaining original tallgrass in Illinois, known anachronistically as the Prairie State, could today scarcely cover a half-dozen corn farms.

Yet a funny thing happened en route to the prairie's liquidation: transportation, the handmaiden of its destruction, became its refuge. As row crops overwhelmed the Midwest, the only safe spaces were the rights-of-way that bordered roads and train tracks, the exclusive property of governments and private railroads. Precarious strips of native plants—tickseed and anemone, switchgrass and beargrass, thimbleweed and snakeroot and bluebell and snow-on-the-mountain—clung to highwaysides like stubble on an unshaved cheek, offering succor to six-legged refugees. In Michigan, Amtrak's lines sustained bluestem, rosinweed, and compass plant—among the "last prairie relicts" in the state. In Wisconsin endangered Karner blue butterflies dipped their proboscises into the lupine that lined sandy roads on military bases. Roadsides were eclectic and cosmopolitan habitats, dive bars where precolonial prairie plants like lady's slipper rubbed elbows with

gritty, sun-loving foreigners like dandelion. Above all, roadsides were novel
ecosystems, amalgams of nature and technology that were created by *Homo
sapiens* but had become largely self-governed—Anthropocenic environments that resembled nothing so much as themselves.

I toured Minnesota's roadside ecosystem one July, on a county highway
in the St. Paul outskirts, its drainage ditch abloom with oxeye daisy, bee
balm, trefoil, and milkweed. It was one of those swampy summer mornings
when the whole Midwest feels like a damp armpit. Five students in yellow
vests, members of a University of Minnesota research group called Team
Ditch, flogged the black-eyed Susans and Queen Anne's lace with butterfly
nets. Cars purred past without slowing; Team Ditch could have been gar-
bagemen, or pavers, or line painters, or any of the other unsung personnel
who keep our roads functional. "If you put on a fluorescent vest and carry a
clipboard," a research scientist named Tim Mitchell told me, "nobody ever
asks you anything."

I could forgive drivers their disregard. A road isn't quite a *place*; instead,
it's an interstice between places, a conveyance that we use on our way elsewhere. The verge is the blurry backdrop to our journeys, smudged by speed
into green and gray blotches. At times it literally bores us to death: long-
haul truckers are susceptible to "highway hypnosis," a glazed fugue broken
only when they run off the road.

To billions of nonhumans, however, the roadside isn't half-noticed scenery, but home. Around us, sterile lawns dominated the St. Paul sprawl;
beyond the city, corn and soy fields stretched to the horizon. Roadsides
offered sanctuary from these monocultures, floral confetti splashed against
suburbia's muted green. One student waved her net to capture a cobalt butterfly smaller than a pinkie nail. Another scooped up bumblebees panta-
looned with pollen. A third scythed the flowers and dumped her squirming
harvest into a plastic bag: garden spiders, katydids, parasitoid wasps. "This
is a whole community that we drive past without even noticing," Mitchell
marveled.

Roadside ditches are not the world's most glamorous research sites.
"People who go into biology generally don't say, 'I want to spend my sum-

mers sitting next to a hot, high-traffic roadside, inhaling exhaust and watching semis go by,'" said Emilie Snell-Rood, the ecologist who oversaw Team Ditch's labors. Yet the road's novel ecosystem was as dramatic as a forest or wetland. Snell-Rood stooped to inspect a thistle, atop which a crab spider slurped a bumblebee's innards. "There are bees whose entire home range might be a single roadside," she said.

For all its unlikely benefits, though, the road is still the road: noisy, chemical-laced, perilous, an environment that both nourishes insects and reduces them to goo. As Team Ditch thrashed the flowers, a monarch butterfly sailed into view. She (for she was a she, identifiable by the thick black veins in her orange wings) flew in languid zigzags, seeking a place to lay an egg. There was something of the *flâneuse* about her, the window-shopper drifting between storefronts. We bit our lips as she sailed into traffic. She tumbled like a cork bobber in a rapid, buffeted by turbulence. At last she alit on a milkweed, touched her thorax to its surface, and freckled it with a white speck. We cheered.

Egg deposited, the butterfly took off again, bound for a stand of milkweed across the road. "She just bounced over three cars—kind of impressive," Snell-Rood said. "I'm so worried for her."

"At least she's already replaced herself," Mitchell said.

"Except larval mortality by predation is at least 95 percent," Snell-Rood replied. In other words, for every twenty eggs she laid, nineteen of her caterpillars would be devoured by wasps, spiders, and other foes.

Mitchell shrugged. "Maybe she's already laid a hundred."

We watched the monarch wobble. Every moment I expected to see her hit a windshield. But buoyed by her wings or slipstreams or good fortune, she wafted over traffic and gained the shoulder. There, in the quiet eddy of the ditch, she settled onto a milkweed and laid another egg.

Who are roads for? Well, us, of course. We're accustomed to regarding roads as human spaces, environments inimical to other forms of life. But roads create habitat as well as destroy it. The road, the critic J. B. Jackson noted, is "a disturber of the peace, an instigator of radical change"—it's how the stranger comes to town, the launch point for the unexpected journey. It's likewise an agent of ecological turmoil, a force that births new niches as it snuffs out old ones. In Australia strips of roadside heath sustain pygmy possums; in Norway reindeer retreat into cool highway tunnels during heat waves. House sparrows pluck insects from parked cars. (On winter mornings I've also seen them sipping the melted frost puddled on my sunroof.) Pike and walleye migrate up ditches to spawn; wolves travel faster along gravel roads than through unbroken forest; kangaroo rats dust-bathe on dirt tracks to "keep their pelage from becoming matted and greasy." Female moose in Grand Teton National Park have learned to give birth near roads, exploiting the "human shield" to keep their calves safe from grizzlies.* In the United States the road's novel ecosystem covers at least seventeen million acres, an area eight times the size of Yellowstone. In many states roadsides represent the largest form of public land.

That roads are ecosystems is an insight nearly a century old. Among the first to recognize it was Frank Waugh, a landscape architect who cut his teeth planning Forest Service campgrounds. Like Stephen Mather, Waugh believed that roads should showcase nature. But he went a step further: roads *were* nature. "All along the cleared roadside," Waugh observed in a 1931 article presciently titled "Ecology of the Roadside," "there are dozens, sometimes hundreds, of species of shrubs, vines, and herbs growing which are not to be found in the woods or open fields further back." The road, Waugh wrote, was a world of microhabitats partitioned by variations in soil, sun, and moisture. Tough euphorbias rose from pavement cracks. Goldenrods and dewberries erupted along mowed margins. Jewelweed fringed drainage ditches; sumac formed the "shrubbery border" beyond. While other early road ecologists mourned roadkill, Waugh took a different perspective: roads brought life as well as death.

Waugh's enthusiasm for wild roadsides didn't initially catch on. Thomas

---

* John Muir observed a similar effect in Kenya, where herbivores "learned that the nearer the railroad the safer they are" from lions.

MacDonald's Good Roads crusade had also spawned a parallel movement, one you might call Good Roadsides. Highways, opined a Michigan road commissioner named Jesse Bennett, were the "front yard of the nation," and if homeowners could be expected to keep their property weed-free and well-mown, so could transportation departments. Where Waugh delighted in unkempt roadsides, Bennett espoused micromanagement: "Nature," he declared in 1936, "cannot produce the desired results alone." He urged the replacement of unruly shrubs and vines with turf, which he adored as much as any golf-course groundskeeper. "The necessity and popularity of grass cannot be questioned," he wrote. And, like any lawn, grassy roadsides required "consistent and regular attention to upkeep day after day indefinitely," ideally with mowers and a "number of men equipped with scythes."

Bennett's predilection for manicured monotony influenced a generation of managers. Riotous shrubs and wildflowers might look pretty, averred roadbuilders, but they made "driving hazardous by obscuring warning signs, traffic signals, and the view at curves, crossroads, and driveways." (The proliferation of white-tailed deer offered more rationale for grass, since tasty shrubs and trees lured deer to the roadside and made it harder for drivers to see them.) Engineers waged war against unwanted flora by "mowing, burning, blading, dragging, steaming, hand pulling, hoeing, smothering, and killing with herbicides." The Bureau of Public Roads nuked road margins with cyanide, formaldehyde, and arsenic, and supplanted "useless or troublesome" vegetation with grass, grass, and more grass.

Over time, critics of Bennett's front-yard approach began to rebel. "Keep cow, plow, and mower out of these idle spots," wrote Aldo Leopold of roadsides, "and the full native flora, plus dozens of interesting stowaways from foreign parts, could be part of the normal environment of every citizen." In *Silent Spring*, Rachel Carson saved some invective for the "chemical salesman and the eager contractors who will rid the roadsides of 'brush.'" "There is a steadily growing chorus of outraged protest about the disfigurement of once beautiful roadsides by chemical sprays," Carson wrote, "which substitute a sere expanse of brown, withered vegetation for the beauty of fern and wild flower." Our chemical mastery had imbued us with "a giddy sense of power over nature," Carson added, yet the best thing we could do for roadsides was to stay out of their way.

Over time, benign neglect allowed Carson's and Leopold's visions to flower. The primary drawback of Bennett's front-yard paradigm was that, as any suburbanite can attest, lawns are a hassle to maintain. After the Interstate Highway System spawned another forty thousand miles of roadside, there were too many acres to regularly mow and spray. As highway margins went to seed, they became habitats. In North Dakota, grouse, ducks, partridges, plovers, and other ground-nesters huddled in unmown swaths along I-94. One biologist discovered downy mallard chicks three feet from an on-ramp.

In the 1980s roadsides took another leap forward. Since Bennett's day, transportation departments had subjected highways to "blanket spraying," indiscriminate herbicide applications that contaminated groundwater and destabilized soils. The first state to get fed up was Iowa, which in 1988 swapped out blanket spraying for tactics known as integrated roadside vegetation management. The state resolved to spray and mow sparingly, plant native wildflowers, and burn roadsides to promote fire-adapted plants. "The manicured lawn look is out and the natural look is in," proclaimed the *Des Moines Register*.

As other states adopted Iowa's practices, midwestern roadsides began to rewild. Highways were valuable habitat for the same reason humans prized them: their connectivity. Other remnant prairies were isolated—an unplowed field here, a nature reserve there. But roadsides were inherently networked, an interlocking grid of "long, ribbon-like habitat." When the meadow vole expanded its range in Illinois, it did so by following I-57 and I-74, whose margins and medians blew past towns and cities that might have blocked the rodent's dispersal. *Amblyscirtes celia*, a demure, dime-sized butterfly, so flourished along Texas's highways that it became known as the roadside rambler.

The road's ecosystem was not entirely salutary. Many highways became *ecological traps*, poisoned honeypots that enticed animals, then killed them. Like crafty fishermen, roads present different lures to different quarry.

Warm asphalt is a thermal trap for snakes; hollow signposts are nesting traps for songbirds who enter them and can't escape. (A taxidermist named Homer Dill may have documented the first case of entrapment in 1926, when he found roadkilled woodpeckers engorged with doughnut crumbs and deduced they had been "attracted to the street by waste from the lunch baskets of passing tourists.") As highways became habitat, the trap's jaws snapped shut. After a freak Texas snowstorm in 1978, horrified ornithologists discovered twelve thousand dead Lapland longspurs—handsome songbirds that migrate in huge flocks—along highways near Dallas. The birds, who were "filled with weed seed," had been drawn to the snowless median by an easy meal, then whacked. Other scientists watched cedar waxwings gorge on silverberry in a median, "with several birds being hit with each pass of the flock."

Highway medians, it was clear, could become traps since animals had to cross lanes to reach them. But the benefits of road*sides*—which creatures could potentially access without braving traffic—seemed to outweigh their dangers. After Indiana lined its highways with shrubs, red-winged blackbirds and goldfinches flourished, without suffering any rise in roadkill. The rewilding of American roadsides wasn't merely a subfield of road ecology; it also hinted at a new road geometry. Cars and wildlife tended to travel on perpendicular vectors doomed to intersect, x- and y-axes disastrously colliding at right angles. But animals could also move in parallel with highways—even, in the case of one butterfly, follow them across a continent.

As with all life cycles, it's hard to know where the monarch butterfly's begins. Circles, unlike roads, lack origins. But start here, *in media res*: August in North America, when shorter days and cooler nights spur monarchs into action. All summer long, butterflies, hundreds of millions of them, have drifted over the continent—New York and New Brunswick, North Dakota and North Carolina, Maine and Minnesota—depositing eggs on milkweed and slurping nectar from coneflowers and goldenrod. Now summer is end-ing, and monarchs head south. They fly during the day when the sun warms their scaled wings. They feed so often they gain weight as they travel.

In September the monarchs converge. Streams of butterflies meet in Texas, cross the border, and climb into Mexico's mountains, a rare river that flows uphill. They ascend passes and course up canyons, their flocks thickening as the air thins. Cops in Nuevo León slow traffic to let them pass. At last, ten thousand feet above sea level, they settle in a patchy fir forest and huddle together to spend winter in an energy-saving torpor. When the writer Sue Halpern visited their roost, she found them "so heavy on the branches of the pine trees that the branches bent toward the ground, supplicants to gravity and mass and sheer enthusiasm."

Most migratory animals have straightforward commutes: the same deer who stot into Wyoming's mountains each spring, for instance, ramble back to their winter valleys in November. Monarchs scoff at such simple journeys. In March the butterflies who overwintered in Mexico fly north again, tattered and nearly transparent with age. They make it as far as Texas, lay eggs, and die. Their offspring fly north, spread out, and begin to repopulate the continent; after a few weeks they too breed and die. A couple more rounds of this ensue: breed and die, breed and die, until late summer, when the last crop of butterflies embarks on its own trip to Mexico's mountains. This means, incredibly, that no individual monarch experiences the species' entire life cycle. The butterfly I saw dancing through traffic in Minnesota was likely third generation: her grandparents had gone to Mexico and her offspring would as well, but she would never lay her compound eyes on monarch Valhalla herself. I found something poignant in that—this poor parochial butterfly churning out caterpillars in her corner of the Midwest, her existence brief and utilitarian, living in service to her glamorous, adventurous children.

It seems unfathomable that billions of burnt-orange butterflies could vanish into the folds of the world, yet almost every detail of monarch life—where they go, how they get there, what befalls them along the way—was, until the recent past, shrouded in mystery. Many details still are. Solving these puzzles has been the life's work of dozens of scientists, among them a snow-bearded entomologist named Chip Taylor. In 1992, Taylor launched Monarch Watch, a volunteer-science program that recruited school groups, garden clubs, and retirees to affix numbered tags, each the size of a hanging chad, to butterflies' wings. As tags went out and sightings rolled in, the migration's hazy contours sharpened. Although monarchs appeared to randomly drift hither and yon, Taylor learned they were following two primary flyways. One corridor wandered northeast across a half-dozen lines of longitude, fanning out over the Carolinas, New England, and Canada's coastal provinces. The other flyway shot due north from Mexico, straight as a compass needle, through Texas, Oklahoma, Kansas, Iowa, and Minnesota. These midwestern monarchs had the shorter commute, which meant they were less likely to get blown off course, freeze to death, or smack into a windshield. "Seventy percent of the butterflies that reach Mexico come out of that central corridor," Taylor told me.

But migrating through the Corn Belt had drawbacks. Monarch butterflies lay their eggs exclusively on milkweeds; on hatching, the caterpillars scarf down the leaves, absorb the plant's natural toxins, and become noxious to predators themselves. This relationship survived the twentieth century: even as industrial agriculture plowed up the prairie, enough milkweed persisted between crop rows and along field edges to sustain monarchs. In the late 1990s, though, corn and soy began to overwhelm even these scraps. Farmers planted genetically modified crops that could withstand weed-killer baths, and biofuel subsidies incentivized growers to carpet every last inch with corn. The combination of Roundup and ethanol wiped out thirty million acres of milkweed between 1996 and 2013, an interval that coincided with an 84 percent decline in the monarch population. Come winter, flocks that once blanketed dozens of acres of fir forest shivered on a few trees. The insect apocalypse had come for the monarch butterfly.*

By the time I spoke to Taylor, the rate of milkweed loss had settled at

---

* The monarchs who fly to Mexico are known as the "eastern population." The western population, which overwinters primarily on the California coast, is in even worse shape: since the 1980s it has declined by more than 99 percent.

around two million acres per year. The plant was disappearing everywhere—everywhere, that is, except for roadsides, where 450 million stems endured. Most had popped up on their own, although some had been planted by Monarch Watch and other groups. Roadside milkweeds, Taylor had learned the hard way, were often short-lived. "You can spend thousands of hours creating a habitat, and you say, for God's sake, don't let anybody mow it," Taylor said. "And you know exactly what happens: it's mowed, and it's mowed, and it's mowed, and you go, oh my god, why did we waste all our time and money doing that?" Still, the potential was immense: vast swaths of would-be prairie bordered America's highways, if only transportation departments would stop mowing them down. And there was a theoretical alignment between butterflies and transportation. Highway managers favored the same low-growing prairies as monarchs for the reason that wildflowers, unlike trees, didn't obstruct sightlines—and drivers were less likely to die crashing into milkweed than they were a maple.

As it happened, a disproportionate amount of butterfly habitat fell along one road: Interstate 35, the freeway that runs nearly 1,600 miles from Duluth, Minnesota, to Laredo, Texas. The heartland's metropolises—Minneapolis, Des Moines, Oklahoma City, Dallas—lie along I-35 like rosary beads; between these outposts stretch cornfields, pothole lakes, American Gothic farmhouses, and prairies that slipped through development's clutches. Larry McMurtry called this austerely beautiful corridor the "long and lonesome I-35," but to a monarch it's the most bustling place on earth. By coincidence the interstate parallels the central flyway with uncanny precision. A butterfly could drift from Minnesota to Mexico without leaving I-35's side.

In 2015 that serendipity attracted the attention of the Obama administration, which announced its intent to foster thousands of acres of monarch habitat along I-35. The next year, a handful of midwestern states signed a memorandum to support the so-called Monarch Highway, an agreement to "protect, plant and manage pollinator habitats" along the interstate. The Monarch Highway inverted the concept of a road: it was linear infrastructure whose users were wild, a road designed to support habitat rather than trash it. Although the interstate was the unifying thread, the Monarch Highway's footprint stretched many miles wider, enfolding not only I-35 but the farms, backyards, and other lands around it. In truth, it was less recovery plan than branding exercise—a "symbolic highway," as one conservationist put it to me. While it didn't oblige states to spend money, it did encourage them to post signs featuring its logo: a monarch with dashed lane lines scoring its orange wings, a chimera of infrastructure and animal.

At first the Monarch Highway galvanized some action. Midwestern states planted pollinator gardens at rest stops, like travel plazas stocked with milkweed and nectar. But, like many Obama-era initiatives, it withered when Donald Trump assumed the presidency. "When ideas like that come forward and they don't have support monetarily, there's only so much you can do," Christopher Smith, an ecologist at Minnesota's transportation department, told me.

Smith is the man responsible for making Minnesota's margins habitable for butterflies, and one day I joined him to explore some roadsides near the Twin Cities. We drove south through corn and soy, soy and corn, as he explained the tensions inherent to his job. Roadsides were many things at once—infrastructure, aesthetic spaces, habitat—competing interests that hung in delicate balance. "It's not uncommon for us to get complaints, like, why'd you let your roads get all weedy?" Smith said. "And then we mow it, and the bee people say, why'd you mow it?" Other grievances were harder to redress. After Smith replanted one interchange, geese arrived to eat the grass. Several were hit by cars, to the consternation of an elderly woman who had grown fond of visiting them on her power scooter. "I've had multiple three-hour conversations with this woman," Smith sighed. "She's gotten emotionally attached to these geese."

We parked near a soy field and stretched our legs. The prairie, a strip narrower than the pavement itself, lay between the highway and a railroad track. We nudged past clumps of wild parsnip, whose toxic sap had recently seared a chemical burn into Smith's forearm. Smith was looking for a rare plant called prairie bush clover, which turned out to be a lank, knee-high stalk with slender leaves. I broke away to watch a monarch feed on the mauve flowers of a milkweed, the butterfly unperturbed by the stink billowing off cattle trucks.

It was an auspicious sign. While the Monarch Highway had gone somewhat torpid, the idea that the butterfly's survival depended on roads had become gospel. A year earlier Minnesota had joined an elaborate plan (known in bureaucratese as a Candidate Conservation Agreement with Assurances, or CCAA) to nurture butterflies along its roads. Its name wasn't as pithy as the Monarch Highway's, but the CCAA had some teeth. The transportation departments, railroads, and electric utilities who signed the agreement consented to devote some of their land to supporting butterflies; in exchange, they wouldn't face any new regulations should the feds someday declare the monarch an endangered species. By the time I visited, a dozen transportation agencies had taken the deal, and around eight hundred thousand acres of land were slated for butterfly habitat.

Under the CCAA, Smith's department had agreed to dedicate 8 percent of its roadsides to growing milkweed, and he was confident it would meet the mark. His prescription included being more judicious with herbicides, planting milkweed, and setting controlled fires, practices that broadly resembled the strategies Iowa pioneered back in the 1980s. Mostly, the key was doing less. "The maintenance folks are like, yeah, we're happy to stop mowing some of these spots," Smith said.

But roads rarely stay neglected for long. In Minnesota ranchers habitually mow public roadsides to produce hay for cattle. Although this is illegal without a permit, law enforcement usually turns a blind eye, and many of the highwaysides we drove past looked as freshly shaved as a fairway. When the state tried to crack down, farm lobbyists complained that it was trampling livelihoods. The issue, Smith said, was "a political hot potato." Roadsides were among the monarch's final redoubts, yet they only endured at our discretion.

The chronic mowing had serious implications. Although Minnesota had thousands of miles of quiet two-lane highways, the state had chosen to focus on restoring habitat along its interstates: I-35, I-90, and I-94. Partly this was because the biggest roads had the widest roadsides and partly because they faced the least illicit mowing. (*You* try sneaking a John Deere 8245R Tractor onto I-90.) The problem, from a butterfly's perspective, was that the interstates didn't just have the most milkweed; they had the most everything: cars, noise, turbulence, pollution. "If we eventually find out that the interstates are the absolute worst place we could've put monarch habitat, we'll have to adapt," Smith said as we drove back to St. Paul.

Yet highway habitats had already become public policy. Several months earlier the federal government had declined to rule the monarch an endangered species. The butterfly had floated into a legal purgatory known as "warranted but precluded": it deserved protection, but other plants and animals were ahead in the queue. Besides, the feds added, the CCAA ensured that roadsides and other infrastructure had "a high likelihood" of providing enough habitat. The monarch's keepers had entrusted it to highways—even if no one was certain whether roads would save the butterfly or destroy it.

Fortunately for monarchs, Emilie Snell-Rood has spent more than a decade investigating that question. A couple of days after I'd watched her and Team Ditch wave their nets in the St. Paul suburbs, we rode a rickety elevator together to the greenhouse that tops the University of Minnesota's biology building. I had come to get a closer look at some monarchs.

The greenhouse was hot and close, with a vegetal reek of compost. Rows of butterfly breeding tents stood on steel tables, like a monarch campground. In the largest chamber females perched on nectar-soaked sponges, wings ragged from relentless male attention. "Female monarchs spend a lot of their time experiencing sexual harassment," Snell-Rood said. In other tents J-shaped chrysalises dangled from the ceiling, auguring new butterflies. In still others, caterpillars, clad in pajamas of green, white, and black, munched on milkweed. Their droppings, called frass, littered their enclosures. "It's amazing how gross these guys get," Snell-Rood said.

These butterflies were experimental subjects whose lives Snell-Rood would manipulate to answer a crucial question: what was life like for a monarch on the road? Roadside butterflies inhabit an environment of immense chemical complexity. There's copper from brake linings, zinc from tire wear, nitrogen oxides from exhaust, cadmium from paint, residual lead from leaded gasoline, microplastics from everything. Some pollutants get absorbed into milkweed tissues, while others stick to their leaves. Either way, they enter the bodies of monarchs. "There are all kinds of things combining to influence the chemistry of the roadside," Snell-Rood said. "What does that do for butterfly performance?"

Snell-Rood was the right person to tackle this knotty problem—a "lovable nerd," per her students, with an omnivorous curiosity. She grew up in Virginia, an avid birder who toted binoculars into the woods and recorded sightings on a Dictaphone before catching the school bus. Snell-Rood mapped warblers' territories, dissected doves on the kitchen table, and started a club called "Save Animals and Reduce Roadkill." Her valedictory speech, a harbinger of her future in ecology, was about finding your niche. She illustrated her point by clacking rocks together to simulate the call of the yellow rail.

Thus it was an upset when, after studying birds in college, Snell-Rood switched to butterflies for her PhD. She'd never had any special affinity for insects—"I thought they were bird food"—but she had to admit that butterflies had advantages, namely, that they were easy to raise and study in laboratories. She zeroed in on the neural development of cabbage-white butterflies, entomology's lab rat. "I used to be jealous that monarchs got all the attention," she said.

In 2010, Snell-Rood took a job at the University of Minnesota, and monarchs flitted into her life. She arrived in December, an infelicitous time in the Twin Cities; the week she moved, a blizzard staved in the Metrodome's inflatable roof. Most shocking to Snell-Rood was the road salt. After storms the state salted its highways so heavily that granules obscured the pavement. "The road would be totally white," Snell-Rood recalled. "My husband and I played this game—'Snow or salt?'" Her mind bent toward a new question: what was salt doing to nature in general and butterflies in particular?

As a cheap deicing agent, road salt—the same stuff you shake on your food in coarser, unprocessed form—is peerless. Sodium chloride is a natural compound, mined from the beds of prehistoric seas, that infiltrates the molecular bonds of water and undermines its ability to form ice. The first state to systematically salt its roads was New Hampshire in the 1940s. Salt met a chilly reception: drivers complained that it corroded their cars' undercarriages, and reporters wrote that it caused "Mrs. Housewife to do a bit of storming" when husbands tracked it into the house. Highway departments explored other compounds but dismissed them; Vermont abandoned one because of its "disastrous effect on nylon stockings" and its tendency to provoke "chrome itch." In the end sodium chloride was too useful to ignore. Salt helped roads conquer nature, decoupling winter travel from the elements. As road salt spread, highway departments subscribed to the "bare-pavement concept," the notion that motorists could expect ice-free roads immediately after blizzards. One study, funded by the Salt Institute, suggested that deicing cut accidents by nearly 90 percent. "Some say snow fighters may save more lives than firefighters," the institute's president declared, a self-serving claim that contained a hard, white grain of truth. Today America's road crews apply more than twenty million tons of salt annually, and Minnesota is among the heaviest users. Each winter the state spreads thirty-six tons for every mile of four-lane highway.

Humans have always coveted salt, a commodity so precious that Roman soldiers were paid in it, the etymology of *salary*. Wild animals are likewise drawn to sodium, a "super-stimulant" that their bodies use to contract muscles, transmit nerve impulses, and control blood pressure, among other functions. Large mammals experience such intense cravings that they forge trails to natural mineral deposits, known as salt licks. One traveler in 1765, wandering what is present-day Kentucky, marveled at "a large road which the buffaloes have beaten spacious enough for a wagon to go abreast, and leading straight into the lick." (The Natchez Trace Parkway, the modern highway from Mississippi to Tennessee, follows the trail that bison carved to a lick near Nashville.) Butterflies so adore sodium that many take part

A bighorn sheep satisfies its sodium craving by licking the road salt from a car in Wyoming. *David Swindler*

in "puddling," bacchanals that form around salty mud, dung, carrion, and the tears of basking turtles.

Once, salt licks were rare. Today they line our highways, lodestars as enticing as watering holes. Sodium-deficient moose in New Hampshire elongate their home ranges to lap at roadside salt ponds, while tourists in Alberta's Jasper National Park are greeted by a flashing warning: Do Not Let Moose Lick Your Car. In Elizabeth Bishop's poem "The Moose," the titular animal wanders into the road to sniff at a parked bus, a mystical moment that leaves passengers awash in a "sweet sensation of joy." They might have been less awestruck if they'd known the "grand, otherworldly" creature was merely rectifying a chronic mineral deficiency.

But road salt is an ambivalent gift, the bait in the highway's trap. At one Quebec wildlife preserve, brackish roadside ponds induced so many moose collisions that officials drained the pools, filled them with rocks, and excavated replacement ponds far from the highway. (Elsewhere in Canada, researchers have repelled moose from licks by rototilling human and dog hair into soils.) Salt inflicts even worse damage away from the highway, by contaminating rivers, wetlands, and other fresh waters. Nearly half the lakes in the Midwest and Northeast are experiencing "long-term salinization," and some Canadian creeks have been colonized by marine crabs.

Road salt (principally chloride) slows trout growth, makes frogs more susceptible to viruses, and ferments dead zones. Like greenhouse gases, which will heat our planet long after we've stopped emitting them, sodium chloride is a legacy pollutant that outlives the snowplow driver who applied it. As Snell-Rood put it, "We're turning our roads into potato chips."

As any Pringles junkie knows, too many potato chips can be detrimental to your health. And it was clear to Snell-Rood that Minnesota was creating a lot of chips. The state shared her concern and gave her a grant to find out whether its compulsive road-salting was creating a monarch trap. She raised hundreds of caterpillars in her laboratory, some on a diet of milkweed from a remote prairie, some on sodium-rich plants she collected from a roadside ditch. "I've filled up my entire car with bags of milkweed," Snell-Rood said ruefully. Her results were, well, mixed. Male butterflies reared on roadside milkweed—the potato-chip plants—developed heftier wing muscles, and females gained larger brains and eyes. That was good. But caterpillars of both sexes were less likely to survive to adulthood. That was bad. A pinch of salt helped, Snell-Rood concluded, yet "increasing nutrients beyond some level can be stressful." The dosage made the poison. But what was the proper dosage?

In the years that followed, Snell-Rood and her collaborators tried to figure out how much salt was too much. They gathered milkweed to feed caterpillars, scrubbed out frass-smeared tents, measured tiny eyes under microscopes, and threw away clothes befouled with the sticky latex that gives milkweed its name. At risk of compressing a decade of laborious research into a few soundbites, Snell-Rood found, broadly, that most roadsides weren't too salty for butterflies. Only milkweeds close to the shoulders of busy highways were doused with enough salt to turn potentially toxic. "All of our work suggests that the moderate increases in sodium that you would get on a typical road are fine," Snell-Rood said.

This was reassuring on one level and worrisome on another. A quiet two-lane highway in the middle of nowhere might grow the most nourishing milkweeds, but Minnesota planned to focus its monarch restoration on huge interstates, which received heavy doses of salt, especially near the

Twin Cities. The highways with the most butterfly habitat weren't necessarily the ones with the best.

Moreover, the laboratory was an imperfect substitute for the real world. Scientists prefer their variables isolated, but on the highway no stressor acts alone. Maybe butterflies could handle salt—but could they handle salt *and* copper *and* zinc *and* lead *and and and*? And even if they survived those irritants, did they thrive? Perhaps roadside butterflies were more susceptible to parasites. Perhaps they were too small to complete their arduous migrations. These "sublethal effects"—problems that impaired butterflies without killing them outright—most worried Snell-Rood. "You're producing monarchs on roadsides, but if they're not going to make it to Mexico, should they count?" she asked.

The year before my visit, Snell-Rood and her students had attempted their largest experiment yet. They'd raised three thousand monarch caterpillars in tents on the campus lawn and fed them milkweed that they sprayed with salt in dosages that mimicked various highways. When the caterpillars metamorphosed, they stuck tags to their wings and, one warm summer evening, turned them loose. Then they waited to see whether other scientists spotted them in Mexico. If the saltiest roadsides produced unfit migrants, this would be the way to find out. In the end, though, only seven tags were reported, hardly enough to draw meaningful conclusions. The rest fell into the crevices of the universe, another monarch mystery that might never be solved.

Although we may never know what became of Emilie Snell-Rood's three-thousand-odd monarchs, it's safe to assume that some were struck by cars. Roadkill is a common fate for a butterfly whose migratory routes seem welded to freeways. As one entomologist put it, monarchs "have the disconcerting habit of playing in traffic across about a few thousand miles of North America."

Road ecologists historically paid scant attention to insect collisions. As recently as 2017 the world's bugkill studies could be counted on two hands.

(When engineers concerned themselves with invertebrates, it was to fret about the "cricket menace," katydid swarms that carpeted western highways in such densities that cars skidded on their crushed exoskeletons.) Insect deaths were hard to count and harder to prevent. The inevitability of bugkill was a metaphor for life's propensity to crush us all: as Mark Knopfler sang, "Sometimes you're the windshield, sometimes you're the bug."

If an insect *could* be studied, though, it was monarch butterflies: big as your palm, bright as a flame. In 1999 a couple named Duane and Katherine McKenna undertook one of the first butterfly-kill surveys. The McKennas scoured roads in Illinois and found almost two thousand dead butterflies, among them ninety-nine monarchs. Extrapolating to the entire state, the McKennas calculated that cars may have killed a half-million Illinois monarchs in a single week. In 2019 other entomologists estimated that cars killed around two million butterflies in the southern segment of their migration alone. Even the "wind vortices" generated by trucks, still another team wrote, could shred fragile bodies.

"It's staggering, absolutely staggering," a biologist named Andy Davis told me. "There's a massive chunk of the overall population that's just getting wiped out by cars." On his blog, Monarch Science, Davis has estimated that as many as twenty-five million monarchs are hit every year. To Davis's mind, the data suggest the folly of initiatives like the Monarch Highway and the Candidate Conservation Agreement with Assurances, the federal bargain to create roadside habitat. "Why would you want to draw them closer to this deathtrap?" he demanded. And where was the evidence proving that highways generated more butterflies than they splattered? "You need to be producing tens of millions of monarchs for it to make sense," he said. "Because that's how many are being killed."

Davis has poked other holes in roadside habitat. In one experiment he introduced caterpillars to a "noise room," where he blasted them with eighty decibels of traffic racket played through speakers. (Monarch caterpillars detect sound through a pair of long hairs that jut from their thorax.) Through a microscope Davis peered at their dorsal vessels—the tubes that circulate blood through insects' bodies—and watched their "heartbeats" accelerate. Caterpillars also bit their handlers, which Davis attributed to stress-induced aggression: basically, road rage. The highway's din clearly bothered monarch larvae, perhaps in ways that scarred adult butterflies. "If you want to alter an animal's—or a person's—physiology, you stress them at an early life stage," Davis said.

All of this had earned Davis a reputation as the monarch world's Debbie Downer. A few months before we spoke, he complained on a listserv that transportation agencies were conducting a "gigantic experiment, with the monarchs being the Guinee [sic] pigs." The listserv was irate. "Once again, Dr Davis wants to stick a hot poker in the eye of anyone who wants to do something for the Monarch," rejoined a biologist at the Texas Department of Transportation. But Davis was unmoved. The road to extinction, in his opinion, was paved with good intentions. "We live in a world full of shit," he told me. "There's so much bad stuff happening all the time. And here's this one little butterfly that we can save by planting this one little plant, and people just love it. Everybody high-fives each other and pats themselves on the back. But then here's this mean scientist who says, 'Wait a minute—there's no data.'"

I asked Davis how he would manage roadside habitat if he were appointed Monarch Czar. His answer was swift.

"Mow it all down."

Davis's disdain for roadsides, it's fair to say, puts him in the minority, and not just in the United States. In Ireland, where a third of native bees face extinction, a national Pollinator Plan emphasizes road margins. South Africa has protected networks of "road reserves" to safeguard endemic flora, while Sweden's soft estate supports more than thirty "responsibility species." Britain's roads have been described as its "largest unofficial nature reserve" and "a modern form of wilderness."

Most wildernesses, however, aren't besieged by cars. The idea even seems a little Orwellian: *War is peace. Freedom is slavery. Roads are wilderness.* For Davis is indisputably right that butterflies don't *want* to live along dangerous, polluted, noisy highways; instead, they've been relegated there by agriculture. Time and again, researchers have found that monarchs are less productive along roads than in other habitats. If I had to choose between restoring prairie along a highway or in a cornfield, I'd take the cornfield. But while the quality of roadside habitat may be subpar, its quantity is inarguable. Given the monarch deficit, it's hard to ignore a multimillion-acre milkweed reservoir. "The reality is that, yes, there may be threats on roadsides that you wouldn't find elsewhere," Iris Caldwell, the monarch manager who coordinates the CCAA, told me. "But there aren't many other spots that *are* elsewhere. If you're a monarch trying to cross Iowa, what are your options?"

In Minnesota I asked Emilie Snell-Rood if anyone had proved roadsides produced more butterflies than they killed. She acknowledged that no one had. Then, in her thorough way, she spent that evening crunching the numbers and showed up the next day with a sheet of paper covered in numerical chicken scratch. Her math, she admitted, was rife with assumptions: x milkweeds per mile, on which butterflies laid y eggs, from which z caterpillars reached adulthood, and so on. You could play with the variables all day, but the bottom line was that Minnesota's roadsides produced something like twenty million monarchs. Even if several million became roadkill, the math worked out in favor of highways.

Besides, Snell-Rood pointed out, monarchs will use highwaysides whether we want them to or not. Instead of asking whether roads were "good" or "bad," the better question was how to improve them. Some studies have found that mowing roadsides a couple of weeks before peak egg-laying times could regenerate the young, tender milkweeds that caterpillars favor. Another paradoxical solution may be to make roadsides *more* attractive: if plenty of milkweed blankets the highway's north shoulder, monarchs don't need to dart through traffic to the south. We could also take a cue from Taiwan, which installs several thousand feet of roadside netting to protect purple crow butterflies during migration season and sometimes closes highways altogether. Halting traffic for a bug might seem heretical in

Texas or Illinois, but I'd like to imagine, however naively, that our affection for monarchs transcends our addiction to speed.

As Emilie Snell-Rood and I drove around St. Paul, I watched for monarch roadkill. Along one road we found three dead butterflies in various stages of decrepitude, their brilliant wings dulled to sepia. We also captured an injured butterfly flipping around the median like a windblown bag. I cradled her in my cupped hands, the novice's butterfly-handling technique. (Snell-Rood would have pinched her wings between two fingers.) We brought the monarch back to the lab and laid her in a tent with a nectar-soaked sponge. By the next morning, though, she was dead.

In Snell-Rood's lab every deceased butterfly was an opportunity. "I love dissections," Snell-Rood said. "I could sit here and cut bugs open all day." She placed the butterfly beneath a microscope and opened the abdomen with shears. The monarch's interior appeared on a monitor. Snell-Rood poked around with forceps, and the screen swam with arthropod anatomy, astral as an image from the Hubble Telescope. We saw the silvery threads of the respiratory system, the ghostly tubes of ovaries, the yellowed nebulae of fats. We saw a spermatophore that resembled a garlic clove, proof she'd found a mate.

One feature, though, was conspicuous in its absence. Older females we'd dissected had contained strings of golden eggs, plump as gooseberries. This monarch's eggs, by contrast, were still faint, their yolks undeveloped, the picture of squandered potential. "She hasn't even laid an egg yet," Snell-Rood said. "And she's been hit by a car."

8

THE NECROBIOME

*How ravens, coyotes, vultures, and human scavengers
draw sustenance from the road.*

The space shuttle *Discovery* took off on July 26, 2005—a blue Florida morning, fleecy cumulus over the Atlantic, a lovely day for a launch. The mission was straightforward: the crew would deliver supplies to the International Space Station, conduct a few space walks, and test equipment. But no spaceflight was routine. NASA's last mission, in 2003, had resulted in one of its worst tragedies. A chunk of foam insulation had broken from a fuel tank during takeoff and clipped the shuttle *Columbia*'s left wing. When the craft reentered the atmosphere sixteen days later, the damaged wing tore off and the *Columbia* disintegrated. Seven astronauts died, and the shuttle program was suspended.

Now, twenty-nine months later, nerves at the Kennedy Space Center were frayed. The *Discovery*'s launch was NASA's first since the *Columbia* disaster, and another mishap could spell the indefinite end of American spaceflight. Mike Leinbach, the launch director, watched from his console with clenched jaw. At 10:39 a.m. the boosters fired, and the shuttle lurched skyward in a nimbus of flame and smoke. As it did, three black, ragged shapes flapped around the ascending craft—turkey vultures. Two were instantly incinerated, but a third slammed into the *Discovery*'s fuel tank and tumbled down, limp as a flung sweater.

Fortunately, the vulture didn't hit the shuttle itself, and a fortnight later the *Discovery* landed safely. But the close call terrified Leinbach. The chunk of foam that had doomed the *Columbia* weighed less than two pounds; a vulture tipped the scales at four. Had the bird struck at a different angle, Leinbach told me, "we could have lost that shuttle too." NASA, he realized, had a vulture problem.

Animal conflicts had long vexed the Space Center, which sits within an expanse of salt marsh and pine woods called the Merritt Island National Wildlife Refuge. Woodpeckers drilled holes in shuttle insulation, ospreys nested on signal towers, and feral pigs blundered onto runways. The Space Center scared off those intruders with owl statues and noise cannons. But vultures were not easily deterred. Every day hundreds of NASA personnel—technicians, astronauts, support staff—commuted through the refuge, crushing armadillos, raccoons, alligators, hogs, and other scavenger delicacies. "One of my biggest fears personally was driving down that road at night," said Leinbach, who once hit a pig himself. Vultures had learned to exploit the carcasses and congregated around the Space Center in droves. Each morning they perched, huge and sepulchral, on gates and powerlines—"like hungry diners," wrote one reporter, "waiting for their favorite restaurant to open." Vultures had become intermediaries in a vehicle-strike trophic cascade—their bodies fueled by terrestrial roadkill, then obliterated when they rode thermals into flight paths.*

Leinbach and his team bent their minds to the problem. "I chaired many a bird meeting," he recalled. They developed radar to detect incoming flocks and contemplated, then rejected, the idea of culling vultures. The most effective approach was also the simplest. In April 2006, NASA convened a "roadkill posse," a group of contractors tasked with scouring the Space Center for carcasses and lugging them off to a landfill. Leinbach advertised the program in staff newsletters and bulletins, imploring workers to report

---

* Far more common than shuttle strikes are airplane collisions, and few birds are pulverized more frequently than vultures. Since 1990 over 1,200 vultures in American airspace have become *snarge*, pilot lingo for the smeary remains of a bird.

roadkill sightings and their own collisions. Colleagues giggled when Leinbach mentioned his Roadkill Roundups during briefings. "The chuckles go away when you think about what can happen," he insisted.

The roundup worked wonders. Within three months Leinbach's posse had removed three thousand pounds of meat from the vicinity of the launchpad, and the vultures took notice. At first, one worker said, the birds scattered at his truck's approach. Now they grunted aggressively, as though they recognized him as a competitor. Eventually the frustrated flocks dispersed, and the next launch went off without a hitch. Even rocket scientists must occasionally lower their gaze, from cosmos to pavement.

The parable of the Space Center's vultures tells us much about how roads and nature mingle. Roads, we've seen, are at once bountiful and deadly, novel ecosystems that entrap their members. For butterflies the lure is roadside milkweed; for vultures and other scavengers, it's car-struck flesh, the lives that the road takes and gives. Whereas pollinators have been exiled to roads by development, scavengers seek them out. Vultures— like coyotes, ravens, and other carrion eaters—are *synanthropes*, clever and adaptable creatures who harvest the resources that humanity furnishes. Synanthropes profit from the pizza crusts moldering in dumpsters, the subterranean corridors offered by sewage systems, and the carcasses that line our highways. If the road is an ecosystem, synanthropic scavengers are among its most vital components—plucky, flexible necrophages who court death whenever they approach its shoulder. On the road, there's no such thing as a free lunch.

"For you are dust," Genesis tells us, "and to dust you shall return." Precisely how you melt back into the soil, though, the Bible leaves to the imagination. In truth, no one decomposes alone. Every wild death is also a beginning. Foxes and eagles rend skin and scatter bone, botflies and beetles nibble flesh, plants and microbes absorb the nutrients that remain. Carcasses are self-contained ecosystems that support revolving groups of undertakers as they cycle through phases of decay. Scientists have a name for the living communities that revolve around the dead: the *necrobiome*.

In the necrobiome the car is a keystone species. Roadkill is an unnatural cause of death that simulates a natural process, and the necrobiome's custodial staff operates with remarkable speed and skill. In Florida researchers found that dead chickens and snakes vanished from the road within thirty-six hours, vacuumed up by vultures, raccoons, skunks, and fire ants, which "skeletonized" carcasses. In Brazil scavengers carried off nearly 90 percent of birdkill within a day. It's largely these covert efforts that prevent roads from being paved with corpses. One biologist estimated that Wales's roadkill rates were up to sixteen times higher than surveys suggested, mostly because scavengers so rapidly disposed of the dead. "A blackbird was seen to take a car-damaged slow worm; a blue tit was seen to enter the decayed ribcage of a long dead road casualty sheep to feed on fat and a stoat was seen during daylight to remove an injured juvenile rabbit," he wrote. Just after dawn he watched scavengers carry off 178 toads that had been crushed overnight. "No evidence of the road kill would have been found by a 10 am survey," he added. The necrobiome airbrushes our roadsides, camouflaging a crisis by devouring it.

Every necrobiome is different, but they all feature similar players. Carnivorous birds are one constant. Driving around Mexico's Baja Peninsula, I saw crested caracaras—stout, handsome falcons with blue-and-orange beaks—strutting the roadsides like self-important soldiers. When I visited Britain, the skies teemed with red kites, raptors so renowned for their street-cleaning prowess that they were once protected by royal decree. And in the United States we are graced (NASA might say plagued) with vultures.

Vultures do not, in general, endear themselves to the public. Perhaps it's because they subsist on rotten meat, a foodstuff we associate with illness; Darwin described the turkey vulture as a "disgusting bird, with its bald scarlet head, formed to wallow in putridity." (More complimentary was John James Audubon, who admired the "wonderful quickness and power" with which they projectile-vomit in self-defense.) Or perhaps we dislike them because they portend our own deaths, funereal shadows that smudge the clear skies of our lives. Either way, their dark reputation is unearned. Vultures, one bird rehabber has written, are "the gentle recyclers of the animal kingdom," content to float above the food chain's fray and clean up the messes we strew across the land. Their bald pate is easier to clean than a feathered one, and their gut bacteria and corrosive stomach acid digest anthrax and botulism. Even the turkey vulture's scientific name alludes to its healthful role: ornithologists know Darwin's disgusting bird as *Cathartes aura*, the breezy cleanser.

Our world was once a vulture's paradise. Twenty thousand years ago North America was prowled by lions, cheetahs, saber-toothed cats, and dire wolves, a ferocious suite of predators that littered Pleistocene prairies with dead mastodons, mammoths, camels, and bear-sized armadillos. When some combination of human hunting and climatic change wiped out this megafauna, the continent was briefly carpeted with meat and then bereft, a scavenger's Last Supper. Teratorns, gargantuan vultures with twelve-foot wingspans, went extinct. Black vultures vanished from the West. The California condor, which once floated as far east as Florida, retreated to the Pacific coast to peck at stranded whales. Virtually overnight North America became a carrion desert.

Twenty millennia later, boom times for New World vultures have returned. Modern vultures gorge on gut piles left by hunters, half-eaten hamburgers that spill from landfills, and, most of all, roadkill. Over the last century turkey vultures and their smaller cousins, black vultures, have flourished across the United States and pushed into Canada. That's partly thanks to the banning of DDT and the protection afforded by the Migratory Bird Act. But vultures also owe something to highways, food trails as linear as the breadcrumbs that guided Hansel and Gretel. Roadkill is an unusually salubrious banquet: unlike gut piles, which are often peppered with bullet fragments, car-killed opossums and squirrels come lead-free, Whole Foods for the necrophagous set. Highways are, too, the wind beneath vultures' wings: the warm air that rises off asphalt lifts them skyward to drift in lazy helices, scanning the earth with their acute vision and even keener olfaction. One biologist described cities and roads as a "new, manufactured thermal corridor," both flyway and buffet. The teratorn died out with the dire wolf, but it's some comfort that the turkey vulture thrives alongside the SUV.

Of course, roads aren't just bountiful lands of milk, honey, and carrion. Like butterflies, scavengers are often struck as they feed. One roadkill survey in northeastern Spain, for instance, judged the Griffon vulture the region's most-whacked bird. Vultures, like many birds, are poor judges of speed who flee danger based on its proximity, not its velocity. As ever, cars subvert their victims' evolutionary history and turn their instincts against them.

When ancient rules become maladaptive, roadside animals must learn new ones. The road is a cognitive gantlet that honors the smart and the curious—the synanthrope. Take black bears, fur-clad geniuses who plan for the future, recognize themselves in mirrors, and tutor each other in the dark arts of unlocking supposedly bear-proof containers. Most road-killed black bears are inexperienced young males, suggesting that over time bears wise up and become safer pedestrians. This street-smart behavior may be socially transmitted: while rural bear populations in Connecticut are genetically fragmented by roads, their urban counterparts aren't, per-haps because the city slickers learn to cross safely and teach the trick to their cubs. "If bears can learn to be careful with traffic, they can learn to exploit a lot of the resources that come with human presence," a biologist named Mark Ditmer told me. "They seem to generally be making the right decisions."*

---

* Ursine prudence even shows up in their physiology. When Ditmer implanted heart moni-tors in Minnesota bears, he found that the animals' pulses rose by around a dozen beats per minute near roads—the same symptom of mild but manageable stress that your FitBit might record as you prepare to cross a busy street.

Equally shrewd are corvids, the brilliant bird family whose members include crows, jays, ravens, and magpies. Examples of corvid road intelligence abound. Florida scrub jays get better at avoiding cars over their lifetimes, either by escaping close calls or by watching comrades get smacked. British scientists have observed the "sensible crow family" slurping up earthworms driven from roadside soils by vehicle vibrations. Japan's carrion crows famously drop walnuts onto roads to break open their shells. Some crows also *place* their booty at intersections to let cars do the nutcracking for them, human conveyances ingeniously appropriated as avian tools. The birds lay down their nuts at a red light, retreat when it turns green, then swoop down at the next red to collect their prize. If much time elapses without their nut being crushed, they reposition it. In the city of Sendai the behavior originated near a driving school, whose slow, light traffic—all those student drivers puttering cautiously along—gives crows plenty of time to snack between cars. Novice motorists and bold birds learn the rules of the road together.

Corvids dine on roadkill with gusto. Particularly avid are common ravens, which the naturalist Bernd Heinrich described as "arguably the most important carcass consumers in the Northern Hemisphere," apt to devour "all sorts of roadkill, from raccoons to porcupines." Ravens are eager morticians and plucky synanthropes who commandeer transmission towers as nesting platforms and highways as dining halls. As their populations have exploded, ravens have become crucial operatives within the road's ecosystem. By caching morsels of roadkill in tree hollows and duff, they disperse carrion to fishers, weasels, flying squirrels, and other road avoiders, gifting the highway's bounty to their shier neighbors.

Ravens play another vital role in the necrobiome: advance scout. Beginning in the 1990s a biologist named Roy Lopez and his colleague, Teryl Grubb, launched a messy, years-long investigation into the complex interactions that coalesced around carrion along Arizona's I-17. They scoured the interstate for dead elk, loaded carcasses into trucks, and dragged them into the forest. Then they watched through binoculars to see who would come eat. "Colleagues would rib us about being the ghoul patrol," Lopez told me. He once strapped a half-mummified elk to the truck by its leg, only to watch the limb detach and maggots gush from its socket.

Carcasses, Lopez and Grubb discovered, attracted a rotating cast of diners. First to arrive were ravens—bold, wide-ranging, sociable. They picked at the elk but mostly waited, kept at bay by the thick hide. A day later bald eagles showed up, America's national symbols come to avail themselves of its run-amok car culture. "Eagles can get a little bit more of the flesh, especially soft tissue: eyes, lips, anal area," Lopez said. But even their beaks couldn't penetrate hide. The impasse broke when all the avian activity attracted coyotes, who tore into the elk and exposed its innards to the birds. Scuffles sometimes broke out: Lopez once watched an eagle attack a feeding coyote, who "lunged upward snapping with open jaws." Usually, though, the three species coexisted, drawn by roadkill into partnership: the birds discovered the carcasses, and the coyotes broke them open.

This symbiosis isn't unique to highways; biologists have also observed ravens following wolves to kills in Yellowstone. This, perhaps, is what makes scavengers such effective roadside synanthropes: they're accustomed to exploiting the largesse of big, dangerous rivals. Coyotes skulk behind wolf packs, black bears gnaw at grizzly-slain elk, ravens steal bites wherever they can. Carrion eaters learn to detect and interpret the signs of a kill and negotiate the social contracts that spring up around carcasses. Scavenging demands intelligence, bravery, and circumspection, qualities that the highway also rewards. To scavenge is to seek the alms of an apex predator, be it wolf, bear, or tractor-trailer.

Along the highway, however, scavenging opportunities are rarer than they might be. A vast army of human maintenance personnel trolls America's roads, scooping up or dragging away the bodies along their shoulders—a necrobiome in its own right, layers of federal, state, county, and private workers tasked with sanitizing the violence of automobility. A handful of states compost carcasses, while others bequeath them to animal shelters and zoos. In most places, though, workers simply cart the dead to landfills, incinerators, or dumpsites, thus stymieing vultures and coyotes. Our "aseptic" approach to carcass management has short-circuited processes, like scavenging and decomposition, that have buttressed ecosystems since the dawn of life. Because objects interred in landfills don't readily break down, many dumps have become tombs for the unprocessed corpses of deer and raccoons, as eerily preserved as pharaohs in their pyramids.

Rather than reflexively discarding carrion, we might do better to treat it as a resource. Consider the plight of the golden eagle, mottled skylord of the West's mountains and high deserts, possessed of a seven-foot wingspan and a hooked, flesh-rending beak. Like many raptors, golden eagles are drawn into the highway's trap by roadkill. In summer, when eagles hunt rabbits and ground squirrels, they're rarely struck. In winter, though, their prey burrows under the snow, and they resort to scavenging. Once, eagles tore into deer and pronghorn killed by starvation and disease; today they congregate around highways to tragic effect. An eagle weighed down by a bellyful of venison takes about as long to achieve liftoff as a 747. "They're just not very agile compared to a magpie or a raven," Steve Slater, the director of a nonprofit called HawkWatch International, told me.

In 2016, Slater launched a systematic investigation of eagle-kill. He dispatched crews around Oregon, Utah, and Wyoming to scour roadsides for dead eagles and set up cameras to observe live ones. Slater didn't know what his teams would uncover. Eagle carcasses get scooped up quickly by black-market feather salesmen, making collision data scarce. The Hawk-Watch surveyors, however, found dozens of dead goldens, suggesting that hundreds, possibly thousands, were dying annually—no small crisis for a raptor that takes five years to reach sexual maturity.

Just because a road is dangerous doesn't make it a trap. For an eagle, the nutritional benefits of eating roadkill could theoretically outweigh the risk of becoming it. But here, too, roads cause trouble. Ordinarily, a golden eagle perched atop a carcass is utterly content; Slater once watched eagles pick at a pronghorn in a field for hours, nursing their repast like day-drinkers in a pub. Put that pronghorn beside a highway, though, and the birds get jittery. Slater's team found that golden eagles spend an average of just nine minutes on a roadside carcass at a time, and "flush"—birdspeak for fly away—every fifth car, a tendency that leads them into the paths of passing trucks. Eagles also burn precious calories every time they flush, a serious problem in winter, when many teeter on the edge of starvation. Instead of a feast, the road provides only an unsatisfying snack.

While eagles' taste for roadkill is presently a liability, some careful intervention could transform carcasses into cornucopias. By Slater's math, dragging deer just forty feet from the highway would allow eagles to gorge themselves in peace. This would admittedly complicate the lives of road workers, who tend to either haul carcasses to landfills or nudge them just off the shoulder. "If they move the deer two feet off the edge of the pavement, that's a job well done as far as human safety goes," Slater said. "But that's still a high eagle risk." Slater envisions a sort of eagle-credit system, whereby wind farms whose turbines kill raptors expiate their sins by hiring workers to lug dead ungulates out of the danger zone. "We could increase survivorship of overwintering eagles by making this plentiful food source safe to them," Slater said.

Novel ecosystems are by definition self-willed: humans inadvertently set them in motion, then step back and let their inhabitants interact. Yet Slater's research suggests that the necrobiome can also be actively governed—and that, under the right circumstances, carcasses can do more good on landscapes than in landfills. Schlepping a rigor mortis–stricken deer forty feet is no pleasant task (I've tried), but it's the least we can do for our most majestic raptor. Like Dr. Frankenstein, we're responsible for managing our monster.

Alongside ravens and golden eagles, *Homo sapiens* is also a member of the road's ecosystem. We occupy every trophic level: we are the road's top predators and too often its prey. And sometimes we're scavengers ourselves, necrophages who eat the dead as eagerly as any crow.

Like tax codes and firearm regulations, roadkill harvest laws differ between states. New Jersey residents are allowed to claim only deer, while enterprising Wyomingites can also scoop up elk, pronghorn, moose, bison, and turkeys. West Virginia is a free-for-all, where anything goes aside from fawns, bear cubs, and some protected birds. Roadkill consumption is not without its detractors: groups like the Humane Society fear that legalization gives drivers a perverse incentive to hit animals, and food safety advocates object that wild meat could be contaminated with salmonella, *E. coli*, or the like.* (To my knowledge there are no documented cases of roadkill-to-human disease transmission.) Since the 1990s, however, the United States has trended toward the liberalization of salvage; today more states permit it than proscribe it. Per one philosopher, roadkill offers the annual equivalent of eighty thousand cows or eight million chickens—no factory farm required.

American roadkill cuisine has a long and storied history. In his mordant 1973 essay "Travels in Georgia," the writer John McPhee embedded with naturalists who took macabre delight in cooking what they called DORs, shorthand for Dead on Road. As a hunk of meat sizzled over an open flame one night, the chef asked McPhee how he liked his weasel. Rejoined the writer: "Extremely well done." While McPhee gamely masticated his mustelid, other scribes have been fussier. Snarked the humorist Dave Barry, commenting on the consumption of possibly diseased squirrels, "Doesn't a person who eats road-kill rodent organs pretty much deserve to die?"

It's hard not to detect a note of classism in Barry's disgust. Roadkill offers food security for thousands of hungry families who couldn't otherwise afford free-range, organic meat. America's foremost roadkill service was once a nonprofit called the Alaska Moose Federation, which, between 2010 and 2020, distributed four thousand car-struck moose to poor, elderly, disabled, and Native households. More than a quarter were dispensed by Laurie Speakman, a stay-at-home mom who began driving for the federation on a lark and became locally famous as Laurie the Moose Lady. For years Speakman slept by her phone, waiting for state troopers to call her with carcass alerts. When they did, she drove down

---

* Intentional roadkill is rare but not unheard of. In 2013 a Wisconsin man pled guilty to illegal hunting after security camera footage revealed he'd accelerated to run down a whitetail. He received a thousand-dollar fine and a ninety-day jail sentence.

to the crash site, winched the thousand-pound body onto her flatbed in the subzero Alaskan night, and drove off to personally deliver it. "I don't know how many times I pulled up to somebody's house, and they were like, 'We just put our last meal on the table, we didn't know what we were gonna do,'" the Moose Lady told me. After the federation shut down due to budgetary issues, the closure sent ripples throughout the forty-ninth state. "I actually learned to sleep again," Speakman said. "But people miss it tremendously."

Some of America's best roadkill salvage conditions prevail in the Methow Valley, a rural enclave in northern Washington, and one March, I drove there to look for a carcass myself. The valley is a long, winding defile through which the Methow River and Highway 20 twine like mating snakes. In winter, mule deer trek down from the mountains to amass in cottonwood galleries along the valley floor, crosshatching the foothills with their trails. As I entered the valley near sundown, I passed deer at every curve, ribs showing through their hides. They paced the roadside anxiously, nearly invisible against patchy snow. Brake lights flared in front of me as a couple of skinny does clacked over the pavement.

At first light the next morning I set out with a friend to search for an overnight kill. We found no shortage of meat, most of it fit only for nonhuman scavengers. The towering snowbanks that bordered Highway 20 had begun to melt, disgorging deer they'd entombed over the winter, like receding glaciers coughing up mummified mammoths. Ravens and bald eagles swarmed the putrid offerings, ripping at glassy eyes and nasal cavities. We drove on, stopping at every deer we saw: two, four, a half-dozen. One day-old doe was borderline, her cold flank firm as river mud. My friend, a hunting guide, advised against it.

As we debated whether to claim the doe, a highway worker rolled up in a frontloader, a deer already slumped in its bucket. We let him scoop up our carcass and asked him where he would put it. "Wherever I feel like," he said. (Later, my friend took me to one popular dumpsite, an embankment strewn with spines and skulls, like a buffalo jump.) One recent Sunday, the workman said, he'd helped a mother and two kids load twin fawns into

In Washington's Methow Valley, highwaysides have become reliable but dangerous food sources for bald eagles. *Ryan Bell*

their minivan. The family had been on its way to church; with the decompositional clock ticking, they bailed on services and went home to butcher.

The Methow Valley is a salvage hotspot not merely for the quantity of its deerkill but for its quality. Most of Highway 20's traffic is composed of vacationing Seattleites rather than commercial trucks, and the speed limit tops out at fifty-five miles per hour. Deer are often killed but rarely obliterated. Although harvesting roadkill was long illegal in Washington, pesky laws didn't stop anyone from nabbing the occasional carcass. Then, in 2016, the state legalized deer and elk salvage, and the *sub rosa* roadkill community emerged from the shadows. "Before salvage was legal, you would see dead deer daily on the highway," Jason Day, an enforcement officer for Washington's wildlife department, told me. "And now I can think of just two in the past couple of weeks. They get scooped up pretty quick."

Day was pro-legalization; he'd even tried to claim an injured deer himself. (The animal, alas, fled into the woods, presumably to die in lonely agony.) But the law had caused him difficulty by introducing ambiguity to the formerly clear-cut job of busting poachers. "Once upon a time, if we found deer parts in the woods in February"—outside of hunting season—"we'd say, well, gee whiz, I don't think a cougar boned out that deer with a knife," he said. "That would have obviously been poaching. Now we don't know." Day believed that poachers were laundering illegal deer as salvaged roadkill. Even a bullet hole could be passed off as the handiwork of a compassionate Samaritan who had euthanized a suffering animal.

And performing salvage is even more complex than regulating it. There is, first, the question of whether a carcass can even be salvaged, or whether the vehicle strike liquefied the organs and tainted the meat. "I give it the ol' physician's pat-down," Nils Knudsen, one of the valley's many salvagers, told me. "If you feel broken ribs, don't open up the animal—the guts and stomach might be rolling around." A deer struck broadside may have one unharmed flank; ruptured organs may spare the choice backstraps along the spine. Roadkill, like any food, is seasonal: a hundred pounds of raw meat doesn't last long under the August sun. "I wouldn't bother salvaging a deer in summer unless I saw the collision happen," Knudsen said.

I spoke to Knudsen outside his home along Highway 20. He was angular and frank, a cabinetmaker by trade, a homesteader by lifestyle. He'd begun salvaging one night in 2013, when his wife, Sarah, hit a mule deer down the road. They were vegans and distraught. Salvaging the meat, Knudsen thought, would make them both feel better. He woke up their neighbor, a moose hunter, who walked them through the process of butchery. It was the first meat they had eaten in three years. Knudsen wasn't sure how many deer he'd salvaged since—eight, maybe, or ten. Some he dumped in a gully after hacking away a few slabs; others he buried in the garden. "Things seem to grow really well a couple years later," he said.

If roadkill represents a wanton loss of life, salvage, to Knudsen, was a form of gustatory accountability—the senselessness of death redeemed by oven or skillet. Harvest fit his ethos of self-sufficiency and locavorism. "Getting a pig from a farmer in Alabama is like hiring a hitman to do your dirty work," he said. He showed me around the property: the carport built from pine logged on his land, the chickens he and Sarah bred for their sky-blue eggs. Many summer nights they sourced dinner entirely from their garden and the highway beyond. Knudsen was partial to *bigo*, a hearty Hungarian stew. "Once, midway through the meal, I mentioned the venison came off the highway," Knudsen said. "Dad was a little disturbed."

Knudsen's father isn't alone in his revulsion. Today scavenging is déclassé: we obtain hamburgers through opaque supply chains rather than cutting them ourselves from bloated cows. Yet scavenging is a noble, timeless pursuit integral to our evolution. Our ancestors were hobbitish vegans who spent their days foraging fruit and being eviscerated by enormous cats on the African savanna. Two and a half million years ago, though, our lives changed: we learned to scavenge. At first, like ravens, we were "passive" scavengers, opportunists who nibbled dead antelopes after leopards had eaten their fill. Over time we upgraded into "power" or "confrontational" scavengers, banding together to run lions off their kills. What we lacked in claws and fangs, we made up in team chemistry.

Our predilection for scavenging changed us forever. It forced us to grow: once we were competing with big cats, it behooved us to be big ourselves. It drove us to innovate: lacking carnassial teeth, we dismembered warthogs and zebras with stone tools. It made us social: the cooperation required to chase off lions also disposed us to live in clans. And it transformed our neurology. As our diets shifted away from fibrous plants and toward easy-to-digest meat, our bodies reallocated energy to large, powerful brains. Scavenging made puny apes a planetary force.

Like a vulture or coyote, Nils Knudsen had learned the modern trick of exploiting the highway's novel ecosystem. Yet he'd also slipped back into humankind's ancient role, one that intimately bound him to other necrophages: eagles, bobcats, ravens. One winter a deer was struck in front of Knudsen's house and stumbled uphill to die. That night snow fell, preserving the body in a coverlet of frost. Knudsen discovered it a few days later and, finding its neck meat and backstraps still good, carved out some cuts for the stewpot. By the time he returned later that week, a cougar had cached the leftovers beneath leaves—a hominid and a felid, uneasily sharing carrion once more, the necrobiome reunited by the road.

9

THE LOST FRONTIER

Where roads and rivers cross, salmon cease to migrate.

Perhaps because they don't create game trails or visit backyard feeders, fish don't get much credit as migrants. Yet they're remarkably peripatetic. Our planet's coastal waterways teem with diadromous species, fish that move between rivers and oceans, from Atlantic herring to Pacific lamprey. Many strictly freshwater fish migrate, too. Every spring spawning suckers nose up tributaries of the Great Lakes in such numbers that their feces and eggs nourish the streams themselves, like glittering bison fertilizing a liquid prairie. Like wild bison, too, fish migrations have become perilously rare. Since 1970 migratory fish populations have collapsed by three-quarters, an extinction crisis as troubling as any on land. Mega-dams, overfishing, and pollution have taken their toll. And so, improbably, have roads.

Think of waterways and road networks as vast, overlapping meshes: one carved over millions of years by glaciers and gravity, the other superimposed by humans. Everywhere the meshes intersect is a potential calamity. Take Skokomish Valley Road, a rural two-laner that traces its namesake river through western Washington. The Skokomish River is, bar none, the state's most flood-prone, partly because it's clogged with sediment shed by logging roads. The chum salmon that surge upriver to spawn every November follow the engorged Skokomish wherever it leads—through dairy farms, beneath fences, and across the road. They line up in ditches like sprinters in the starting blocks, then hydroplane over the blacktop, the olive prows of their backs cutting wakes through ankle-deep water. Pickups skid to a halt, hubcap-deep. The media treats the spectacle as comedy: "Why did the salmon cross the road?" is a headline that's been done, and done again. Most of the migrants wind up stranded when the water recedes, inglorious coyote fodder.

Roads rarely doom fish so flagrantly, though. Instead, disaster comes most often in the form of culverts, the hidden matrix of conduits through which rivers and streams flow beneath roadways. Culverts take many shapes: they can be corrugated steel pipes, rectangular concrete tunnels, semicircular arches with gravel-lined beds—any channel that conveys water beneath infrastructure. (Engineers in Alabama historically funneled creeks through the hollowed-out trunks of gum trees.) Whatever their design, culverts are constantly subjected to the erosive force of water, which destroys them in endlessly creative ways. They get plugged by sticks and silt; they wash out; they crumple like tin cans. For a salmon or herring fighting upstream, a defective culvert is a barrier as imposing as the Grand Coulee Dam.

Culverts keep a low profile by design, which makes it tough to say how many are out there. Around the Great Lakes, perhaps a quarter-million culverts frustrate pike, whitefish, and brook trout. In New England another quarter-million curtail the movements of eels, alewives, and shad. Early colonists described rivers that "ran silver" with spawners; at fords in Virginia, wrote one nineteenth-century traveler, herring were so multitudinous "that it is almost impossible to ride through, without treading on them." Now it's our planet's highways that gleam with chrome travelers, not our waterways. To say that roads kill fish by a thousand cuts understates the case: more than ten thousand crossings mutilate some Amazonian tributaries. Australia's Murray cod, a massive dagger-toothed predator, is so powerful that, according to Aboriginal cosmology, the beating of its tail sculpted the bends in its home river. Today impassable culverts are among the reasons that it's classified as vulnerable. The Murray cod can recontour a river, but it can't cross a road.

No creatures, however, are as harmed by culverts as Pacific salmon, the aquatic world's most famous commuters, a group of fish that race up coastal rivers from California to Alaska. Born in fresh water, juvenile salmon spend anywhere from a few days to two years flitting around their home streams, slurping midges and dodging kingfishers, before their stripes fade, their flanks brighten, and they drift to the ocean to grow fat on krill, squid, and small fish. Some species while away a couple of years at sea; others linger for seven. Eventually they return to their natal rivers, attuned to magnetic fields and familiar smells, and power upstream. They cease to eat; their organs deteriorate. Some swim hundreds of miles inland and thousands of feet upslope, halting at last in high deserts fragrant with sage. With vigorous tail-sweeps, females excavate gravel nests and deposit sticky orange eggs. Males joust for primacy and contribute milt with spasmodic shivers. Mission accomplished, they die. Their carcasses nurture eagles, bears, otters, mice; the nitrogen and phosphorus inhered in their flesh nourish the insects that will comprise the matter of their fry. Their life cycle is a circular economy, a perpetual reincarnation. A salmon never dies—she's merely recycled.

For as long as roads have crossed water, they've short-circuited this loop. When a logjam clotted a culvert in British Columbia in 1893, for instance, impatient fish spilled onto a wagon track and piled up several feet deep. (Workers hauled them off in horse-drawn carts.) The problem worsened as the Northwest developed. Engineers designed culverts to usher streams beneath roads as expeditiously as possible; as a result, many sloped steeply downhill, forming "velocity barriers" of rushing water. Thousands more were built narrower than the channels that ran through them, cinching streams into torrents that blasted fish away like fire hoses. The force of this discharge scoured out pools and turned culverts into waterfalls—"perched" culverts, engineers called them. Fish hurled their bodies toward pipes like desperate half-court shots and clanked off the rim.

Just as interstate highways blocked elk from crucial winter range, faulty culverts prevented salmon from reaching their upstream habitats. Unable to access their entire riverine domain, adult salmon often dug their nests where other fish had already laid eggs, excavating the life's work of their dying comrades. The fast flows within culverts tumbled baby fish like Cessnas in a hurricane, denying them productive refuges like sloughs and beaver ponds. Some biologists retrofitted culverts with baffles, rungs that created stepwise chains of gentle pools; others tried angled ramps known as fishways. But these were only partial fixes. Five distinct species of Pacific salmon—not to mention sea-run rainbow trout, called steelhead, and sea-run cutthroat trout, called bluebacks—swam northwestern rivers, and no two salmonids experienced culverts the same way. A pipe that mighty Chinook salmon navigated with ease might slow down weaker chum salmon a few months later, and completely block juvenile coho salmon a few months after that. It wasn't enough for a culvert to be passable for some fish some of the time. A culvert had to work for a stream's every species, in every life stage, on every day of the year.

Few of Washington's culverts met that standard. By 1949 the state acknowledged that a "lost frontier" of erstwhile salmon habitat lay beyond its pipes and tunnels. If the Northwest was, as the writer Timothy Egan put it, "anywhere a salmon can get to," culverts had shrunk the region itself.

- - - - - - - - -

Among the people who felt the loss most acutely was Charlene Krise. The daughter of Native parents—Colville on her mother's side, Squaxin Island Tribe on her father's—Krise grew up in Tacoma, the scrappy port south of Seattle. She and her brothers were often hungry. One day, sometime in the 1960s, Krise attended a party at a relative's house. Her aunts spread newspapers as tablecloths, and from the kitchen emerged a bubbling pot of stewed salmon heads. Laughter and talk filled the warm room. Krise picked tender meat from the fish's cheeks. It was the best food she'd ever tasted. She glanced at her father, expecting to see joy on his face. Instead, he looked angry—angry, she realized, at having only the heads.

"Really," he fumed, "we should have the whole salmon."

His fury was justified. For thousands of years Krise's Squaxin Island ancestors, the People of the Water, had plied the southern end of Puget

Sound, the glacier-chiseled arm of the Pacific that elbows into western Washington. Theirs was a world of tidal flats and brackish rivers, one that collapsed the line between land and sea. They hunted elk and gathered clams, but salmon were their staple, a food as sacred as it was nourishing. The People of the Water harpooned salmon with obsidian points, endured winter on their dried flesh, honored their seasonal return with First Food ceremonies. They carved salmon petroglyphs, inscribing their devotion upon the earth.

That all began to change on December 26, 1854, the day that Isaac Stevens, the scurrilous governor of the Washington Territory, strong-armed the People of the Water into signing the Treaty of Medicine Creek. The treaty was a diplomatic travesty: the negotiations were conducted in a pidgin language ill-suited for a complex land transaction, and the dismal weather kept away many delegates. By day's end the tribes of southern Puget Sound had ceded four thousand square miles in exchange for a four-mile-long reservation that would become known as Squaxin Island, though they did retain the right to fish in their "usual and accustomed" places. Other northwestern tribes—the Tulalip, the Lummi, the Snoqualmie, the Makah, and many more—were cajoled into similarly inequitable deals.

With the tribes exiled, colonists poured into western Washington. As the region industrialized, salmon dwindled. Canneries swallowed runs; hydroelectric dams throttled rivers. By the middle of the twentieth century, salmon landings had fallen to less than a fifth of their peak. Native people bore the brunt, barred by the state from catching salmon in the misguided name of conservation. Many exercised their right to fish anyway, risking harassment, arrest, and gunfire. Once, during Krise's youth, her father led her down a streambank off a dirt road. "I'm going to get us a salmon," he declared. A few minutes later her brothers, who had been keeping watch, ran back to warn them of an approaching game warden. Krise had no idea what a warden was and, picturing some sort of bogeyman, hid behind her father's back. The warden turned out to be a white man with red hair. He was polite but firm: under no circumstances could Mr. Krise take salmon from the stream. The Krises trudged away, empty-handed and -stomached.

Native people refused to tolerate the abuse. In the 1960s tribal fisher-men staged a series of widely publicized protests, known as fish-ins, on the Nisqually River, setting nets and daring officials to stop them. The response was violence. In September 1970 police raided a fishing camp, deploying tear gas and clubs, an attack that turned public opinion deci-sively in favor of the tribes. The federal government sued the state of Washington on the tribes' behalf, and in 1974 a judge affirmed that Native people had the right to fish at their traditional grounds—and harvest half the total catch.

With the Fish Wars finally ended, scattered Squaxin Islanders trickled home from California, Montana, and Alaska. "We all had to reacquaint ourselves and find out, 'Oh, you're my cousin,'" Krise recalled to me. She and other tribal members took to Puget Sound in aluminum skiffs, hand-pulling gillnets that came in laden with salmon they ate and sold. When the fishing slackened, they tied their boats into flotillas and talked. They swapped stories, explored familial connections, rebuilt ancestral knowl-edge. Krise designated herself a "rememberer," an archivist who preserved tribal history before it washed away like wave-swept sand. She spent hours chatting with elders, jotting notes on the backs of clamming permits.

Fishing was a physical life. Krise leapt onto rocks, chucked anchors, and hauled nets; she tore tendons and threw out her back. But the inju-ries didn't faze her. She loved Puget Sound in all its moods, even its cold-est and wettest. And she revered the salmon themselves, her daily bread: salmon eggs for breakfast, salmon sandwiches for lunch, smoked salmon with potatoes for dinner.

In 1990, though, the salmon runs collapsed again, this time worse than before. When the Squaxin Island armada convened now, it was to discuss who had sought help from food banks, who'd had their car repossessed, and whose electricity had been shut off. Some tribes, unable to catch enough salmon for their ceremonies, resorted to buying fish from sympathetic First Nations in Canada. "It was like a loved elder had died," Krise said.

Krise felt angry, as her father had. She channeled her rage into political action, joining watershed committees and serving as Squaxin Island's representative in conclaves with state agencies. Often she showed up fresh off the water, in sweatshirts stained with salmon blood. "They were probably wondering, who *is* this vagabond?" she said. She attended meetings about water quality, logging patterns, suburbanization, harvest rates. And at every one, she asked the same questions: what had become of the salmon, and what would bring them back?

Puget Sound's salmon, Krise knew, faced myriad stressors. Deforestation hastened erosion and smothered spawning gravels in silt. Highways cut off rivers from their floodplains, eliminating the marshes in which young salmon sheltered. And then there was stormwater. The rain that pummeled Washington's concrete skin—its roads, driveways, parking lots—rushed downhill to the Sound, carrying with it a toxic brew of motor oil, transmission fluid, gasoline, copper, and other automotive tinctures. Deadliest of all, though no one knew it then, were the particulates shed by tires. Scientists would eventually pin decades of coho salmon die-offs on 6PPD-quinone, a chemical that manufacturers apply to tires to protect them from ozone. We'd paved the earth for cars, then used them to poison it.

Those threats, though, were diffuse and overwhelming; even the most powerful tribe would be hard-pressed to rein in the growth of Seattle's suburbs. Culverts, by contrast, were easy targets. Each one foreclosed salmon from a discrete amount of potential habitat, allowing the harms to be easily quantified. (State biologists later estimated that repairing the region's culverts could produce an additional two hundred thousand adult fish.) And Washington knew its culverts were a problem. In the early 1990s the state's Department of Transportation had launched a program to replace fish-blocking culverts on state highways, which, it acknowledged, were among "the most recurrent and correctable obstacles" to salmon runs. The department slowly began to tear out narrow and derelict culverts, installing bigger ones—or, even better, bridges—that allowed streams to flow unrestricted. But it was a daunting task. The state had installed its culverts decades earlier, when it first built its roads; in the intervening years, hundreds had been concealed by sprawl. For every culvert the state replaced, it seemed to find two more beneath shopping malls, schools, and subdivisions. At the

Billy Frank Jr. led the 1960s "fish-ins" that resulted in Washington's Native tribes being allocated half the state's salmon catch, and paved the way for the replacement of hundreds of culverts. *Tom Thompson, courtesy of Northwest Indian Fisheries Commission*

rate Washington was going, it wouldn't solve the problem until the twenty-second century.

That wasn't nearly fast enough for the Squaxin Islanders and other Native people. In January 2001, twenty-one Washington tribes sued the state over the sluggish pace of its culvert repairs. Their argument was a strong one. The treaties that Isaac Stevens had foisted on them in the 1850s guaranteed them the right to fish in their customary places, but what good were rights to a destitute fishery? The treaties did not merely promise fishermen access to salmon, the tribes argued: they also guaranteed the salmon access to habitat. By reducing the number of salmon that Washington's streams churned out, culverts were preventing tribal members from practicing their culture and earning their livelihood, thus violating the rights Native people had fought so hard to reaffirm. "We need to start fixing them right now," said Billy Frank Jr., a longtime Nisqually leader who'd spearheaded the 1960s fish-ins. "That's all we're asking—fix the culverts."

The rate of repairs continued to lag, though, and years of depositions, evidence gathering, and negotiations slowly passed. In October 2009 the Culverts Case finally went to trial. Biologists testified about the swimming aptitudes of various salmonids; engineers debated arcane points of culvert design; budgetary officials tallied the cost of repairs. The most affecting testimony came from Charlene Krise, by then a librarian in the tribal museum. She'd always been a rememberer by disposition; now she did it for a living. On the stand she dug into her tribe's store of memories. She waxed poetic about the People of the Water, the pleasure she took in fishing, and the heartbreak of watching Squaxin Island's children go without new shoes. "Whatever happens to that salmon will happen to us as humans," she cautioned.

Four years later a judge finally decreed that Washington had to fix its worst-offending culverts by 2030. Krise's people seemed to have triumphed. But the state protested: it was already repairing its stream crossings, albeit at its own pace. Appeals followed, and in 2018 the Culverts Case ascended all the way to the U.S. Supreme Court. Road ecology would have its day in the country's loftiest courthouse.

That spring Krise went to Washington, D.C., to observe the proceedings— among the "highlights of my whole life," she said. The tribal delegation arrived before dawn on a frigid April morning to secure their seats. Nearly twenty years of struggle came down to an hour of argument. Samuel Alito grappled with the semantic difference between a "substantial degradation" of salmon and a "large decline"; Stephen Breyer expounded on the history of fish-passage law; Neil Gorsuch and Elena Kagan made common cause on esoteric points of treaty interpretation. It was over by noon. Afterward, Krise and her delegation went to lunch. At first they debriefed about the case. And then the talk turned, as usual, back to fishing.

Two months later the Supreme Court ruled. The justices had deadlocked, four to four. (Anthony Kennedy, who was once involved in a related case, recused himself.) The split decision meant that the lower court's decision stood: the tribes had won. The culverts would come out.

One fall morning, a little more than a year after the Supreme Court decision, I drove to Olympia, Washington's capital, to see what all the wrangling had accomplished. There I met Paul Wagner, the manager of the Washington State Department of Transportation's biology branch. Wagner had shaggy white hair, a white beard, and a laidback mien, as if Jeff Bridges had been cast as Santa Claus. He'd worked for the department since 1990 and had prepared some of its earliest culvert reports. I had heard several people describe him as one of road ecology's most esteemed doyens. "It's really interesting to meet someone in college who says, 'I'm a road ecologist,'" he told me as we escaped Olympia's traffic for the Puget Sound hinterlands. "And it's like, huh, I remember when we made that term up."

Wagner's deep background made him an appropriate person to helm the state's fish-passage program, which, owing to the Culverts Case, was perhaps America's single largest road-ecology initiative. The case had created a knotty situation, and Wagner attempted to untangle it as we drove. Although the ruling required Washington to replace around a thousand culverts, it didn't have to fix them all at once. Instead, it had until 2030 to reopen 90 percent of the salmon habitat that its roads had blocked, a complex formula that reflected a simple truth: not all culverts were created equal. Some streams boasted miles of prime spawning grounds, just waiting for a new bridge to open them to fish. Others fizzled out in a parking lot after a hundred yards. The Culverts Case had impelled the state to prioritize the streams with the most potential habitat, which, logically enough, meant larger creeks and more expensive projects. In 1991, Wagner's branch had been allocated a few hundred thousand dollars for culvert upgrades; now the state was shelling out more than $15 million on some individual streams. "Back then I couldn't have conceived of hundreds of millions of dollars being committed to fish passage," Wagner said.

And yet it still wasn't enough to fulfill the case's mandates. One analysis had suggested that new culverts could cost north of $3 billion, and, despite its obligation to the tribes, the state legislature seemed hesitant to pony up. In part, that was because culverts posed a collective-action problem: while hundreds of state-owned culverts had been earmarked for replacement, twenty thousand more lay on county, town, and private roads. What was the point of repairing a culvert on a state highway, some politicians objected, when salmon would bump into one on a county road just upstream? As one researcher told me later, "The fish don't care who owns the road."

Still, the tribes' victory had produced undeniable gains. Half an hour outside of Olympia, Wagner pulled into a casino owned by the Squaxin Island Tribe. A screen of spruce and alder bordered the parking lot; somewhere beyond trickled Little Skookum Creek. We shouldered through blackberry bushes into a gully. Wagner tilted his head and sniffed. "I can smell them," he said. I could, too: a briny, maritime funk, like a seafood counter during a power outage.

We peered into the creek, and there was the source: a male chum salmon, long as my arm and as deep-keeled as a battleship, finning in the shallows. His belly ground against gravel; the green hillock of his back gleamed. He'd been at sea for around four years and had begun decomposing the moment he entered fresh water, his digestive tract and muscles yielding proteins to the production of sperm. He was crosshatched with scars, his fins white-edged with decay, his flanks barred in purplish stripes. He was so trans-fixing that it took me a minute to notice the trio of fish undulating in a downstream pool. One broke away and thrashed to the next run, odiferous with incipient death.

Just upstream of the salmon loomed, yes, a culvert, above which a state road crossed Little Skookum Creek. At a cost of $2.7 million, the state had torn out the old, narrow pipe and installed a new stream crossing, twice as wide and forty feet shorter, more bridge than tunnel. Daylight shone on the other side. Wagner described it as a "stream simulation" model, the apogee of culvert design, meant to resemble a natural creek in all particulars—slope, width, substrate. Like a deer underpass, it was a structure that animals would never notice, a facsimile of nature created through the artifice of engineering.

We climbed out of the creek, crossed the road, and scrambled down the other side, now upstream of the crossing. Here, too, we found fish, both living salmon and spawned-out carcasses. I knelt to one ragged corpse. The fish's kype, the hooked snout with which he battled rivals, was studded with canine teeth, his jaw frozen by death in a fierce leer. Beside him floated a can of Miller High Life, the profane bobbing past the holy. A UPS truck rumbled by; in the casino retirees pulled the slots. A person could live many lifetimes without seeing a cougar trot across an overpass, yet here a wild migration flowed through one of the country's most densely colonized corners in broad daylight.

I had become accustomed to thinking of wildlife crossings as a form of ecological empathy, an expression of concern for the creatures who used them. When I spoke to Charlene Krise, she pointed out that roads had also severed reciprocal *human* relationships. Replacing culverts was not merely ecological restoration; it was environmental justice—a means of remedying, or beginning to remedy, more than 150 years of state-sponsored inequity. "Tribal people use salmon as an analogy of our life—you know, you go through a lot of struggle in your lifetime," Krise said. "And sometimes that means throwing yourself up onto some jagged rock or getting tangled up in some roots and getting yourself free. You have to be like that salmon. You have to know what your destination is."

Like the roads they underlaid, culverts, I was learning, were symbols as well as structures. In Washington each humble pipe carried the weight of colonialism, civil disobedience, and perseverance against deferred justice. Elsewhere, culverts connoted the deterioration of our rural infrastructure and its vulnerability to climate change. A road was never just a road; a culvert was never just a culvert.

I hit the road again in pursuit of salmon the following summer, bound this time for Tillamook County, a wedge of seaside farms and fishing villages tucked along Oregon's North Coast. Tillamook, which is scored by tidal rivers and pressed by the Pacific, has a defiant relationship with water. In fall and winter the county reliably floods as storms drench Oregon and its swollen rivers—the Nestucca, the Kilchis, the Nehalem—rocket seaward. Famous floods mark time like mileposts; kids milk cows while balanced on beams above submerged barn floors. "There've been more presidentially declared disasters in Tillamook County than in any other county on the West Coast," Chris Laity, Tillamook's director of public works, told me with rueful pride.

Tillamook is as cash-strapped as it is disaster-prone. The county's tax base has dwindled as the timber industry has withered, and its small population gets short shrift in state budgets. As a result, much of its basic infrastructure is in a state of disrepair, particularly its culverts, which pose a simmering public-safety debacle. When pipes clog with wood and mud, streams tend to flow over roads rather than beneath them, washing out their surfaces and rendering them impassable. A few blown-out culverts and a couple of mudslides during a big storm could shut down emergency response along the entire Oregon coast.

It's not often that public works directors and fish biologists share interests, but Tillamook's culverts make for a useful common enemy. Around 2010 a handful of scientists began to contemplate how to aid wheeled and finned commuters alike. Fish-friendly culverts were being installed across the Northwest, but repairs followed what a biologist named Jim Capurso described to me as a "shotgun-like pattern"—that is to say, no pattern at all. Better to pool resources and heal one region, Capurso figured, than to pick away at many. He and some colleagues alit on Oregon's North Coast, whose rivers hadn't been girdled by the mega-dams that stifled salmon elsewhere. Replacing fewer than a hundred culverts, Capurso's team calculated, would grant salmon unfettered access to nearly all their spawning grounds.

Their scheme needed a name. Capurso and his collaborators hoped to scrounge up millions in donations and grants, and the Focus Basin Aquatic Organism Passage Partnership didn't seem likely to inspire much philanthropy. The biologists hired an advertising firm to lead the rebrand, and a few months later their staff showed up to present their ideas. It was an odd meeting. The marketing gurus were young and Portlandy—"tattoos, tight pants, that kind of garb"—while the scientists skewed old, white, and male. "There was a lot of closed-off body language, crossed arms and legs," Capurso admitted. But limbs began to unfurl when the execs unveiled their proposal: the Salmon SuperHighway.

By the time I visited Tillamook County, the Salmon SuperHighway had become an unqualified success. The initiative had by then replaced almost thirty culverts, at once reclaiming eighty miles of fish habitat and effectively flood-proofing many major roads. The SuperHighway drew its funding from federal conservation grants, nonprofits like Trout Unlimited, and private donations, thus leveraging salmon-recovery dollars to repair crumbling rural infrastructure. According to Laity, the concept had sparked none of the acrimony that usually attends environmental projects. Instead, it had broadly allied loggers and farmers with the "tree huggers and fish squeezers"—a community of interest gathered around the niche cause of culverts.

I'd timed my visit to coincide with the late-summer "fish window," the ten-week interval when Oregon's salmon are mostly at sea and construction crews can replace culverts without disturbing spawning grounds. The SuperHighway's contractors were scattered across Tillamook County, all in a rush to finish their work before the window slammed shut. To see some of the action, I drove south through hazy cattle pastures to Jenck Road, a meandering track overrun with free-range chickens. I inched through the flock and came upon a stream, excavators and backhoes squealing and beeping along its banks. Steel I-beams lay stacked like cordwood; pumps slurped away water. This was Clear Creek, a lazy watercourse that, like so many others, once gurgled through a too-tight culvert. Now the Super-Highway was installing a bridge that would allow it to flow beneath Jenck Road unencumbered, freeing it at last from concrete bondage.

A burly man in his sixties sidled over. He wore suspenders and an unbuttoned striped shirt, revealing a wedge of sunburnt chest, like a train engineer kicking back after a long haul. He introduced himself as Mike Trent, the owner of the poultry operation up the road. Every Thursday he trucked eggs to restaurants in Portland, which whipped them into Benedicts for the brunch set. Almost by default, that made him a passionate supporter of the SuperHighway: "If the roads are out, I can't get up there," he said.

The descendant of nineteenth-century homesteaders, Trent had grown up fishing Clear Creek with his grandfather, a dairy farmer. "The stream was alive with salmon," he said. "You could walk from the mouth of the creek all the way up and see coho and chum." That halcyon era had ended in the 1970s, yet Clear Creek was not entirely bereft. Just downstream of the construction site, cutthroat trout lurked in willow shade. A dipper bounced on her yellow legs. It was as though the stream was holding its breath and would soon exhale life into the channel. "There'll be fish under that bridge the second they're finished," Trent predicted.

There would be other beneficiaries, too. At one Salmon Superhighway site, a biologist told me he'd seen a Pacific giant salamander paddling upstream fifteen minutes after a culvert replacement. "They were literally still hauling away the rubble," he said. Streams concentrate animals as readily as they collect water, and it's a rare otter or mink who won't slip through a properly sized culvert as she stalks a creek. A remote camera in California once captured a coyote leaping and bowing before a dry culvert, playful as a labradoodle. A moment later, up waddled a badger, his unlikely hunting companion. (Badgers dig up ground squirrels, coyotes chase down any who flee.) The odd couple trotted side by side into the culvert's depths, a highway-spanning symbiosis.

That odd, charming interaction pointed at a big idea: if culverts are a vulnerability, they're also an asset. There are perhaps a couple of thousand dedicated wildlife crossings in the United States, which might sound substantial but is virtually nothing in the context of our four-million-mile road network. By contrast, there are something like two million culverts just waiting for a few cheap tweaks to convert them into crossings. Adding an elevated concrete ledge turns a half-flooded box culvert into a boulevard for bobcats. Replacing rocky abutments with forgiving dirt can coax elk beneath a bridge. In that sense, the Culverts Case is a grand opportunity: as the state of Washington has upgraded its stream crossings, it has also attempted to make them amenable to terrestrial animals, thus averting collisions and recouping some costs. Beneath one capacious new bridge that Paul Wagner showed me near Olympia, deer tracks scored the floodplain, and a cherry-red coho salmon lingered in a pool—mammals and fish, moving together through a world newly able to accommodate them.

One particularly fine example of culvert innovation occurred on Highway 93 in Montana, that exemplar of ecological ingenuity. In the early 2000s the state installed long metal shelves, like catwalks, along three culverts to allow small mammals to tiptoe down them during the rainy season, when roadside wetlands submerged the pipes' bottoms. "We had no idea whether even a single animal would use these things," a biologist named Kerry Foresman told me. He needn't have worried: his cameras captured thousands of crossings, including raccoons, skunks, and weasels, in the three years after the shelves' installation. Yet one animal remained conspicuous in its absence: the meadow vole, a prolific rodent that's prey for practically every carnivore in Montana. To such a furtive and delectable creature, Foresman realized, scurrying across the exposed surface of a catwalk must have seemed tantamount to suicide. He headed to Home Depot, where he bought 180 feet of plastic rain gutter and taped it into a cylinder. Then he returned to the culvert to hang his jerry-rigged "vole tube" beneath the metal shelf. To test his makeshift tunnel, Foresman dusted a sheet of aluminum with soot from an acetyline torch and surrounded it with sticky paper, ensuring that any animal passing through it would leave telltale tracks.

The next morning Foresman returned to the culvert, aflame with curiosity. He shuffled into the darkness and shone his flashlight through a small door he'd cut in the gutter. There, pressed in black soot onto white paper, perfect as fingerprints, were hundreds of tiny, exquisite tracks laid by *Microtus pennsylvanicus*—the meadow vole.

The curse of culverts—like dams and power lines, gas mains and fiber-optic cables—is that they only cross our minds when they fail. In the public imagination, infrastructure is practically synonymous with the dull nitty-gritty of governance; we perceive politicians fixated on potholes as lamely pragmatic figures, more small-minded than inspirational. Today, however, infrastructure no longer has the luxury of invisibility. Climate change has destabilized systems that were never stable to begin with. Storm surges claw at levees and seawalls. Wildfires erupt from power lines. Floods overwhelm sewage plants. As Alaska's permafrost melts, "frozen debris lobes" of ice, rock, and soil inch toward highways like bizarro glaciers. Officials in the Florida Keys have discussed ceding causeways to the rising Atlantic, like sacrifices laid on the altar of an implacable god. Fossil fuels consume the roads on which we burn them, a global-warming ouroboros.

The same climatic changes that sabotage our roads also bedevil the creatures that cross them. Our landscapes are changing in a cruel and ironic way: they're becoming less permeable to wildlife at the precise moment animals need to move most. As our climate warms, beings large and small will be forced to relocate, either north or upslope, to remain within their "thermal envelope," the band of temperatures in which they evolved. (Moose, for one, are already pushing onto the tundra to feast on emerging willows, even as ticks are literally bleeding them dry in southern New England, where winters no longer last long enough to kill off parasites.) This displacement compels animals to cross roads more frequently and in new places. The migration pathway that exists today may shift northward tomorrow; the wetland across the highway may become a refuge from megafires the day after that. Wildlife passages would be valuable on any planet, but on one that's heating up, they're indispensable.

Few animals are as susceptible to climate change as salmon and trout, a fish family that depends on cold water as closely as polar bears need ice. Global warming has stressed their fragile bodies, exposed them to disease, and devoured the snowpack that feeds their streams. Some biologists have proposed protecting a network of icy creeks, known as a "climate shield," as a hedge against warming. Yet it's precisely those chilly mountain streams, narrow and steep, that are most susceptible to being cut off by crummy culverts, especially as climate change whips up more intense deluges and bigger floods. Tens of thousands of stream crossings are becoming more susceptible to failure, even as aquatic connectivity becomes more urgent. "The more culvert work we do, the more options fish will have," one biologist told me.

One afternoon in Oregon, I stopped by a tea-colored stream called Peterson Creek, where the Salmon SuperHighway had hired a construction company to replace an old culvert with a bridge. All that remained now was to pave the new, wider span. A steamroller stood by, waiting to finish the job. "This is gonna be the nicest stretch along the whole road when we're done," Dave Walczak, the company's owner, vowed. He had three culvert jobs to finish in two weeks; after that, salmon would pour into Oregon's rivers, and the fish window would close. The SuperHighway, Walczak added, had turned fish passage into a reliable, lucrative business that buttressed an ecosystem of labor, including his own forty-odd employees and a legion of subcontractors. "There's a lot of local support for these projects," he said.

Conservationists and construction companies made for axiomatic enemies. Yet the imperative of climate adaptation had forced environmentalists into an unfamiliar position—advocates of building things, not just of stopping them from being built. Tree-sits were out, residential solar panels in. Even the Sierra Club had recently put forth a $6 trillion infrastructure proposal. The Salmon SuperHighway was a manifestation of that trend: a climate-sensitive conservation plan that embraced infrastructure rather than fighting it, a stimulus for people and planet.

The next morning I went down to the Nestucca River to commune with the fish who would soon swim the SuperHighway. Mist hung over the current, as pearlescent as abalone. I assembled my fly rod and flicked a minnow imitation into a riffle. After a few minutes I hooked a cutthroat trout—a species I'd caught in countless Rocky Mountain streams but that here, so close to the ocean, had adopted a salmon's maritime lifestyle. The fish came to the net seabright and freckled, a flawless creature.

I hiked downstream a hundred yards and cast into a pool. A step deeper, cast again; another step, another cast—and then the hole erupted, as though someone had dropped a paving stone into the river. The sparkling back of a Chinook salmon broke the surface, as incongruous in the pool as a porpoise in a bathtub. He hadn't taken my fly, was no longer feeding at all. I watched him gather himself for the run, his last, beneath bridges and up rapids, toward the cool heart of the Northwest, to release his milt in a swift shudder and relax into death's embrace. It seemed an injustice that such a wondrous fish could be denied fulfillment by an object as banal as a culvert. Nothing could be more mundane than a road, or more sacred than the beings that swim beneath it.

PART III

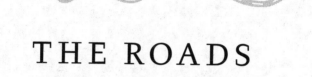

THE ROADS
AHEAD

10

THE GRACIOUSNESS AT THE HEART OF CREATION

In the Roadkill Capital of the World, wildlife carers confront the highway's trauma.

Ask a road ecologist why we should build wildlife crossings, and she'll reel off a litany of perfectly valid reasons: the prevention of dangerous crashes, the cost savings, the conservation of rare species. Ask a layperson why it's worth averting roadkill, by contrast, and the first answer typically pertains to animal welfare. Motorized transportation inflicts suffering on an almost unimaginable scale; there may be nothing humans do that causes more misery to more wild animals than driving. In some places nearly three car-struck ungulates stagger off to die for every one who immediately expires on the highway. This has profound repercussions for ecologists: if collisions are more frequent than most roadkill surveys suggest, it stands to reason that we should build more wildlife crossings. Yet this knowledge—that vehicular death, far from being instantaneous, may unfold over minutes, hours, days—has ethical implications as well as infrastructural ones. It's horrific to contemplate the pain, fear, and confusion of a whacked animal's final moments: broken-limbed, hemorrhaging, scrambled and shattered and shocked. The writer Leath Tonino, following an injured deer into the New Jersey woods, found her "curled on the ground, looking at me, shaking"—the protracted ebbing of a sentient life, a torture appalling and commonplace.

Many ecologists, I've found, are not particularly comfortable discussing animal suffering. In part, this is because acknowledging it can veer toward anthropomorphization: who can comprehend the subjective experience of a badger, a gopher snake, a barn owl? And in part, it's because nature is rife with torment, and thinking too hard about it can lead you to preposterous places: to argue, as some academics have, that we should wipe out lions to stop them from hurting zebras. Death in the wild is brutal; what's the car but another predator?

As the writer Emma Marris has pointed out, though, humanity's planetary dominance has obviated "the wild." Few corners of the planet are free from our cities, pollutants, and roads. Given our power over animals, aren't we responsible for the agony that we inflict in them? The idea, Marris admits, engenders a "kind of intellectual vertigo": do we humans conceivably have ethical obligations to "all the untold millions of animals on Earth, to every sparrow and ground squirrel and city rat and white-tailed deer?" Some scientists have concluded that the answer is yes, or at least maybe. The Wild Animal Initiative, a group dedicated to wildlife welfare, has funded research on birth control for pigeons and light pollution's effects on owls—not for the sake of managing their populations but for making their lives more pleasant.

If we're duty-bound to reduce pain, it follows that we must deal with our roads. Historically, animal-rights advocates have focused on "negative rights," like the right not to be killed, tortured, or confined: basically, the things we shouldn't do to animals. But what about our positive obligations, the actions we should take on animals' behalf? To some scholars those may include "design[ing] our buildings, roads, and neighborhoods in a way that takes into account animals' needs." The practice of road ecology is not merely a set of engineering principles but a moral mandate.

When our roads do inflict injuries, we may likewise be bound to redress animal pain. Often, amends come in the form of wildlife rehabilitation, the treatment and eventual release of injured creatures. Granted, nursing raccoons and squirrels back to health doesn't generally interest conservation biologists, who, reasonably enough, devote their limited resources to species that would decline or go extinct without their intervention. Every dollar that's lavished on the care of a common critter, the argument goes, is one that isn't available to protect the habitat of a rare one; as one biologist has put it, most rehab "wouldn't survive a proper cost-benefit analysis." But wildlife crossings and wildlife rehabilitation strike me as justified moral responses to the same problem encountered at different points: the former proactive, the latter reactive. In the journal *Pacific Conservation Biology*, Daniel Lunney, an Australian ecologist, bemoaned the "mutual isolation of these intellectual communities"—the population-focused ecologists on one hand, the animal welfarists on the other. "Yet the subject of concern is, for all intents and purposes, identical," Lunney wrote. Whether you're worried about the perpetuation of rare species or the lived experience of common ones, you can't escape the road.

Wildlife rehabilitation is particularly vital in Australia, for an odd reason with deep evolutionary roots: it's populated with two-hundred-odd species of marsupial, a class of animals who complete their development in a maternal pouch. For a young North American placental mammal, losing a mother to roadkill usually means gradual and hidden death; an orphaned bear cub or fox kit concealed in a den is almost certain to starve, unnoticed by any human who might care for her. But baby marsupials—kangaroos and wallabies, koalas and quolls, phascogales and bandicoots—often survive car collisions within their mothers' broken bodies, the ultimate parental sacrifice. According to one study, Australia's traffic orphans a half-million marsupial babies, called joeys, every year. Animal-loving Aussies have thus been trained to check the pouches of fresh roadkill, extract living joeys, and deliver them to one of the country's fifteen thousand rehabbers, known as carers. The way Americans take in stray cats, carers raise road orphans.

Carers are especially prominent in Tasmania, the island that hangs off Australia's southern coast like a shark's tooth. Tasmania is a dubious car-strike Mecca, a morbid reminder of road ecology's importance. Outraged visitors deplore the "bloody carpet" that mars their holidays. Biologists have suggested, only half-joking, that the state offer tours in a glass-bottomed bus. In 2014 a newspaper columnist named Donald Knowler saddled Tasmania with an ugly appellation: Roadkill Capital of the World.

The more I read about the Roadkill Capital, the more resolved I became to visit it. I wanted badly to explore its road ecology: what made the state so collision-prone, and what was it doing about the problem? But I also hoped to plumb road *psychology*. Roadkill, Knowler wrote, was "something [Tasmanians] grow used to, inured to, and it even forms part of the local lexicon, to say nothing of the subject of jokes." (A platypus crossing between wetlands became, yes, a splatypus.) What was it like to devote yourself to alleviating the suffering that your neighbors ignored and inflicted—to routinely face the death, grief, and trauma that the road incurred? "People just don't care about wildlife here," the Australian philosopher Glenn Albrecht warned me. "And then there are people who care so much that it almost kills them."

Fable came into the world climbing. She emerged blind and furless from her monthlong gestation, an embryonic jellybean endowed with little more than grasping claws and an instinctual map of her mother's topography. She clambered from cloaca to pouch, found a teat in the warm dark, and sealed her lips around the spigot, conjoined to her mother like a limb. Milk flowed into her. After a month she'd attained the mass of a quarter; after three, a deck of cards. Her ears unfurled like spring leaves. Her eyes opened. She sprouted whiskers. She grew a layer of silver velvet. She attained the weight of a human heart.

And then, on a rural road in Tasmania's Central Highland one December, Fable's mother met the fate that befalls thousands of her fellow wombats every year: she was killed by a car. That evening a good Samaritan examined the still-warm body, extracted Fable from her pouch, and delivered her—cold, weak, hungry, alive—to a carer named Cory Young.

When, twenty months later, I met Fable at Young's home outside Hobart,

Cory Young shares a tender moment with Fable the wombat, one of the many orphaned marsupials that Tasmanians rescue every year. *Ben Goldfarb*

Tasmania's capital, she'd grown into a powerful chunk of muscle with the squat frame of an English bulldog. Young reached into Fable's pool-sized pen, hoisted her out, and cradled her in his arms. She rested against his chest, claws splayed to reveal her belly. Her fur was gray, but complexly so, a weave of black and white hairs that, when the late-morning sun struck them at the right angle, glowed mahogany. She looked like she could run through a wall.

Young rearranged Fable's bulk and encouraged me to tap her hindquarters. I rapped with my knuckles, expecting soft flesh, but instead met a hard surface, as though she wore Kevlar beneath her fur. "Wombats have this cartilage plate in their bums," Young explained. "It's basically the door to their burrow. What they do is run down and cork the entrance with their bottom and stop any dogs or dingoes getting down there." The plate also functioned as a rear bumper that buffered joeys against car strikes. "Babies can be perfectly fine in the pouch, mum's so big and robust," Young said.

He set Fable down. She nuzzled his leg and gummed him. "Bitey-bitey, are we?" chided Josh Gourlay, Young's fiancé. Gourlay pushed against her square head with his palm to rile her up, then darted away. Fable followed, nipping at his high-tops.

We watched the playful chase. Eucalyptus trees, bark peeling in ropy strips, nodded in the salt wind off the Tasman Sea. Wombats, Young said, spend the first two years of their lives with their mother, then turn aggressive toward their own flesh and blood. "They become horrible," he said affectionately. "They try to drive mum out of her territory and take it over." Fable's combativeness suggested she was nearing her Mr. Hyde moment, at which time Young and Gourlay would release her into the wild. "There's some sadness," Young admitted. He had bottle-fed Fable every four hours throughout her infancy. He'd wiped her bum, cleaned the cloth pouch in which she slept, and driven her to his office, where she napped on an electric heating pad. He and Gourlay had taken one vacation in two years. Who could blame him for feeling attached? "But the emotion goes out the door," he added, "because she wants to kill you."

We left Fable to her murderous devices and went back to the house, whose minimalist décor consisted mostly of taxidermy: a sea eagle electrocuted by power lines, an aquatic rodent known as a rakali. For as long as Young had driven cars, he said, he'd cared for the animals damaged by them. When he was sixteen, he found a dead pademelon—a short, chubby member of the kangaroo family—plucked a joey from her pouch, and adopted the orphan. He soon got his hands on a second paddy and bottle-fed them before and after school. (He was fairly certain he'd named the pair after Disney characters, though he couldn't remember which ones.) He had since become one of the island's most accomplished carers, graduating to trickier species like the eastern quoll, a polka-dotted predator now extinct on the Australian mainland.

I asked Young and Gourlay how many animals they raised annually. Gourlay shot his fiancé a look. "We were meant to be having a break last year," he said, "and we did forty-three."

I'd caught them in a slow phase. Besides Fable, they had only two orphans on hand: another wombat, named Archie, and a still-nameless bettong. When I asked what a bettong was, Young disappeared into a back room and emerged with what appeared to be a dingy Christmas stocking. The stocking wriggled, and I realized I was looking at a surrogate pouch. "There's a ladies' knitting group that makes the pouches for us," Young said. "Other carers make their own, but I can't sew to save myself."

Young rolled back the pouch's lip and extracted the bettong, a handsome rat-like creature with twitchy ears and a white-tipped tail. Young dandled the bettong on his thigh and inserted the rubber nipple of a bottle between the animal's lips. Marsupials are lactose intolerant and require special formula; a few die every year when novice carers feed them cow's milk. The bettong drained his bottle. Young snatched up a paper towel and dabbed away greenish turds in mimicry of the bettong's mother, who would have licked her baby clean. "You've had a bit of diarrhea lately, which is not ideal," he said with a grimace. *Carer*, it occurred to me, was a beautiful word, one that captured the tender reality of animal work—the bottle-feeding, the bum-wiping, the relationship-building—better than *rehabilitator*, with its connotations of incarceration and addiction.

After he'd fed and toileted the bettong, Young returned the animal to a heating pad in the back room, a converted clinic whose cupboards overflowed with medications and syringes. Young slid open a drawer to reveal a tray of mealworms, a bettong delicacy. "I'm very lazy with my mealworm farming," he said apologetically. Out the back door stood a freezer packed with ice-rimed meat set aside for his next litter of Tasmanian devils, voracious scavengers capable of consuming nearly half their body weight in a sitting. The meat was roadkill, salvaged from neighborhood streets.

"Raising carnivores is definitely not for everybody," Gourlay said. "It's a pretty grisly job."

"One minute you're bottle-feeding a pademelon," Young added, "the next you're cutting one up."

Despite the free meat, Young estimated he and Gourlay spend around $10,000 each year on their wards. (They receive assistance from Young's family, who finds his habit charming if perplexing.) That outlay wasn't atypical. Australian carers spend an average of $5,300 annually; some have shelled out a half-million dollars over their caring careers. And time, that most precious resource, is an even greater expenditure. Carers report tending their furry children for 32 hours a week, nearly a full-time job; collectively they give 186 million hours every year, around $6 billion worth of time. Volunteer rehabbers, one carer has written, "are a national asset that requires strategic nurturing with empathy, understanding, financial and psychological support." The road not only taxes Tasmania's animals; it drains the people who nurture them.

"We overload ourselves, every one of us," Young said. He knew carers who had devoted so much time to their animals their spouses had threatened divorce, carers so overwhelmed they'd considered killing themselves. "A lot of people within the community are too proud. They feel that not taking on an animal makes them a failure. You need to know when to say, 'I'm at my limit, I need to stop, I need help.'"

In the annals of conservation, Tasmania is infamous for what it's lost. When British colonists arrived in 1803, they found the island inhabited by Aboriginal Tasmanians, who had beaten them there by forty thousand years. They also found a "species perfectly distinct from any of the animal creation hitherto known": a lithe, carnivorous marsupial with powerful jaws, a stiff tail, and black stripes. The invaders called it the dog-headed opossum, the striped wolf, and the Tasmanian tiger. Scientists called it the thylacine. Whatever its name, colonists considered the beast a sheep-eating abomination and hyperbolically accused it of perpetrating "dreadful havoc among the flocks." Livestock companies launched a ruthless bounty program, a version of which remained in place until 1912, by which time there were no tigers left to kill. (Genocidal officials simultaneously dispatched soldiers to imprison and kill Aboriginal Tasmanians.) On September 7, 1936, the last known tiger, Benjamin, died at the Hobart Zoo. Today the species likely persists only in the imaginations of an obsessive klatch of thylacine truthers who comb the bush for evidence of its survival—and in the hearts of wildlife carers, for whom the tiger serves as sobering testament to humanity's capacity for cruelty. "They're my reminder of everything we're fighting for, is that Tassie tiger," one rehabber told me.

Nearly a century after the thylacine's demise, an island known for extinction has, in an odd twist, become an ark. The Bass Strait, the stretch of ocean that divides Tasmania from the Australian mainland, also insulates its wildlife. Dingoes, which spread across Australia four thousand years ago, never reached Tasmania, nor have invasive foxes. Even agriculture benefited some native fauna. Visitors to Tasmania often remark on its Britishness—the sheep-speckled hills, the gray-green damp—a resemblance enhanced by early colonists, who tried to terraform this "antipodean England" to match their home countryside's idyll. Native marsupials exploited the same pastures and watering holes that colonists cultivated for livestock. "Populations of things like wallabies and brushtail possums are probably elevated compared to what they were historically," a biologist named Alistair Hobday told me.

Hobday, as much as anyone, deserves credit (or, if you're a tourism minister, blame) for Tasmania's reputation as the Roadkill Capital. He's a lifelong roadkill obsessive; as a child, he entertained himself by taxidermying birds and possums. "I had very tolerant parents," he said when we met near the Hobart waterfront. In the early 2000s he and a colleague drove more than 150 roadkill-counting forays around Tasmania and logged over 5,500 dead animals—Dayton and Lillian Stoner with a laptop, a Garmin GPS, and a Toyota Land Cruiser. All told, they spotted one animal every 2.7 kilometers, among the highest roadkill densities ever recorded. Extrapolating, they estimated that 300,000 creatures were being flattened each year, an astonishing count for an island smaller than South Carolina.

Why is Tasmania's roadkill so profuse? Some carers blame the state's narrow, tortuous highways; others lament the carelessness of its drivers. According to Hobday, the most probable explanation is also the simplest: Tasmania has a lot of animals to kill. Just as gardens and dumpsters have transformed American suburbs into havens for synanthropic deer and raccoons, the cultivation of antipodean England bolstered wallabies, pademelons, and possums. "In most cases, the species here can probably handle that amount of roadkill," Hobday said.* Far from being a conservation cri-sis, roadkill could even be perceived as a perverse blessing—a "trade-off for living in a beautiful place with lots of animals," as Hobday put it.

To outsiders like me, the pervasiveness of Tasmanian roadkill is glar-ing. Driving around the island, I seemed to pass a slumped wombat or wallaby every mile, many marked with spray-paint to indicate their pouches had already been cleared of joeys. I stopped to examine every unsprayed carcass, parking on precarious bends and dashing into the road, expecting to plunge my fingers into a still-warm marsupium. To my frustration and relief, every animal turned out to be male, his sex revealed by a pinkish penis extruded in morose salute. Granted, I'd come to the island precisely because of its roadkill; still, the carnage shocked even tourists who weren't looking for it. At a café one morning, I over-heard a couple from Adelaide discussing the topic in scandalized tones. "In one spot," the woman told me when I butted in, "we saw two kanga-roos lying together, their noses almost touching, like mirror images." She seemed near tears.

To many Tasmanians, by contrast, roadkill is anesthetizing: a problem so prevalent that it no longer looks like a problem. Surviving other kinds of car crashes has been shown to make drivers more worried about future accidents, yet prolific roadkill may produce the opposite effect—a popu-lace desensitized to tragedy. "I've been a passenger when the driver hits an animal and doesn't even pause the conversation," one Tasmanian biologist told me, "or says something like, 'bloody things.'" When motorists aren't blind to wildlife, they deliberately target it. Farmers have been known to instruct their children to run down animals during driving lessons so that they'll get used to killing without swerving off the road. Another biologist told me that drunk drag-racers compete to flatten wallabies between bars, a murderous pub crawl.

---

* The notable exception is the Tasmanian devil, up to 6 percent of which, Hobday found, were being roadkilled every year—a devastating loss to a species already ravaged by an epidemic of contagious facial tumors.

The abundance of Tasmania's wildlife may be a gift, but it also provides an implicit license to kill: if animal *populations* remain healthy, *individual* creatures must be expendable. Species are devalued by their own commonness. (Would you be more upset at crushing a squirrel or a lynx?) The most frequent roadkill victim in Hobday's surveys, the brushtailed possum, is considered a nuisance for its habit of raiding gardens and crapping in attics. The Bennett's wallaby, another high ranker, is reviled by farmers who deem it a crop-thieving pest. One night in Hobart, I asked my cab driver if he'd ever struck an animal. He laughed.

"Oh, yeah, I've hit a few things in my time," he said.

"Like what?"

"Wallabies, mostly," he said. "Got plenty of those things. No problem if you squish a few."

I'd come to understand road ecology as the purview of government agencies, scientists, and nonprofit workers—in short, salaried professionals. And yet the Tasmanians who most directly confronted the road's horrors seemed not to be ecologists at all. Whereas American wildlife rehabbers must typically pass an examination or work as veterinarians, Tasmania's 250 or so carers formed a scattered, underresourced army of volunteers.* To care was to open your heart and to harden it, at once and every day. Some carers told me they kept scalpels in their glove compartments in case they encountered an orphan so tightly fused to their mother that they had to slice off the dead animal's teat.

One afternoon in Tasmania, I met a carer who'd embraced the ethical responsibilities that roads incur. I found Teena Marie Hanslow at the end of a Hobart cul-de-sac, in the home from which she operates a carer network called Wildlife Bush Babies and Snake Rescue. The first time I knocked on her door, no one answered. As often happens, a neighbor had called her

---

* By comparison my home county at the time had roughly the same population as Tasmania but just one licensed rehabber, a veterinary hospital.

away to handle a car-struck wallaby. Hanslow knew from the moment she beheld the wallaby's broken leg that he wouldn't survive. Hanslow wrapped the animal, docile with shock, in a nest of blankets, bundled him into her backseat, and drove him to a field. Then she laid the wallaby on the grass, leveled her .22 Ruger, and put a bullet in his brain. She stayed with him, as she always did, until his heart stopped.

Hanslow described this scene to me a few hours later in her living room, where evening light filtered through floral drapes. The discordant aromas of barnyard and hospital mingled in the close air. Between us sat a laundry basket brimming with pastel fabrics. The basket stirred, and a baby walla-by's pink muzzle popped out. His head seemed several sizes too large for his scrawny torso. Hanslow pulled back the pouch to reveal his feet, which, with their furless toes and hooked black claws, looked like they belonged on a bird of prey.

"This one's had a broken toe—you can probably see how thick that one's there," Hanslow said. I nodded, reluctant to betray my ignorance of wal-laby podiatry. "If it's just a joint in the toe, they quite often can still cope. But if it was a bone in the leg, it means that animal will always be vulner-able. Then I euthanize." Most rescuers dispatch injured animals with blunt force; Cory Young keeps a sledgehammer in his truck for the purpose. Hanslow prefers the precision of a gun. On an end table, bullets stood in a neat row, noses up, like soldiers at attention. Hanslow favored hollow points that mushroomed upon entrance. "That way it's even more instan-taneous," she said.

Hanslow had grown up detesting firearms. Her father, a former Austra-lian Army sniper, participated in "roo shoots," brutal kangaroo culls. Then, in her twenties, Hanslow started a job whose commute took her along a dangerous stretch of highway. "I'd be pulling these animals off the road that were still alive, but they would be paralyzed," she said as she bottle-nursed the wallaby. Their anguish became her own. "It got to a stage where I wanted to have control over the pain the animals were in. So I thought, well, gotta get your firearm license."

Although Hanslow had no great love for guns, in her hands they were instruments of mercy, delivery mechanisms for what the poet Robin-son Jeffers called the "lead gift." Through anatomical study and practice, Hanslow taught herself the finer points of euthanasia. Wombats had to be shot through the back of the head, so their thick skull wouldn't deflect the bullet. The brains of macropods—the family that includes kangaroos, wal-labies, and pademelons—sat just behind the ear. Hanslow caressed their furred jawlines to find the best entry point, a gentle touch at the end of a horrific ordeal.

The beauty of caring is that it sanctifies life; its ugly underbelly is that it requires dealing death. Every rehabber I met in Tasmania remembered ani-mals they couldn't save. "You take it as a devastating blow when something does pass away, and then you get no sympathy from family or friends," Josh Gourlay told me. Most carers have experienced *compassion fatigue*—a condition, popularized in the 1990s by a psychologist named Charles Fig-ley, that describes the suffering of therapists, nurses, social workers, and others exposed to their patients' pain. "Animal caregivers are among the most susceptible to compassion fatigue because of the toll that performing euthanasia takes on their psyche," Figley wrote. Studies have found that half of animal care workers combat depression or hopelessness and that employees of American dog pounds commit suicide more often than the national average. Empathy is both a gift and a vulnerability.

Several months before my visit, Cory Young had organized a compassion fatigue workshop for Tasmania's carers, led by a counselor from the main-land. Rehabbers discussed anxiety, burnout, and depression. "People say, oh, it's just a possum, just a wallaby, there are shitloads of 'em—why do you care if one dies?" Young said. "Even your own family doesn't appreciate how hard you might take an animal's death." Yet wildlife rehabilitation provides comfort as well as pain. Young once contemplated suicide himself, a crisis that caring helped him survive. "There were three or four other lives that were dependent on my life," he said. Another rescuer told me she'd grappled with feelings of emptiness and that saving animals filled the void. "I would give all my time if I could," she said. "I just feel like it's given me a purpose."

Teena Hanslow had weathered her own trials. In her twenties she'd been diagnosed with lupus, an autoimmune disease. Her vision deteriorated; the tissue around her heart grew inflamed; she suffered a stroke. She spent her nights in a wheelchair, awake and in agony. "The animals kept my mind off the pain," Hanslow recalled. "Back then I had bandicoots, devils, pademelons, potoroos—you name it, I had it. They gave me the will to keep going. I feel like I owe it to them now for the rest of my life. They keep me going and vice versa."

Hanslow is healthy enough today to work as a home aide, sometimes for terminally ill patients. Her joeys accompany her on house calls like palliative props. "I can turn up and pop them in bed with people," Hanslow said as she warmed the wallaby against her chest. "They look forward to seeing me. It's always, you know, 'Have you got any babies at the moment?' It gives them a glimmer."

For all its therapeutic benefits, the ostensible purpose of caring isn't to soothe its practitioners but to help the animals themselves. Implicit in wildlife rehab is that wildlife is being rehabilitated *for* something, namely, a return to the wild. Yet carers know little about how their wards fare after release. Less than 5 percent of cared-for Tasmanian creatures receive identifying tags, microchips, or collars. In the Roadkill Capital of the World, their prospects seem murky at best. "The paradox here," a carer named Bruce Englefield has lamented, "is that animals are being rescued, raised for release and returned to the site of the traumatic event that brought them into care." Caring, an avocation devoted to the diminution of suffering, may perpetuate it.

The uncertainty of caring presents an existential threat to the field. In 2017, Englefield distributed surveys to carer networks to collect data on their demographics, expenses, and techniques. The findings disturbed him. More than a quarter of Tasmania's carers had experienced financial hardship, and an equal number suffered moderate to severe grief, in part because they lacked proof that their labors were effective. "Unless we do something in the immediate future, the outlook is pretty grim," Englefield told me. To improve the welfare of both rescuers and wildlife, Englefield has called for an overhaul of Tasmania's animal rehabilitation system, including mental health services for carers and subsidized food banks for joeys. Most radically, he believes that hand-reared orphans shouldn't be released directly into the wild, where food is scarce and predators abundant. "It's not a bloody rest home out there," he said. Instead, he advocates loosing orphans into safe, artfully managed enclosures in which they can learn the ropes of wildness without experiencing undue risk—a marsupial halfway house.

Halfway houses and food banks, however, would require significant support from governments that don't hold caring in high regard. Many Australian states cull kangaroos and wallabies at the behest of farmers, who accuse them of depleting crops and livestock fodder. The work of carers undercuts these control efforts. In 2016 the state of New South Wales issued permits to kill 136 wombats, 308 swamp wallabies, and an unfathomable 170,290 eastern grey kangaroos; meanwhile, its carers rehabilitated more than 14,000 animals, many of them the same species being culled. In 2018 the state of Victoria floated a ban on rescuing animals deemed "overabundant," a proposal that critics decried as a "strategy to silence" carers.

"Their attitude is, well, you don't have to be a carer," Englefield said. "Nobody asked you to do it. It'd be much simpler if we just shot or euthanized all the animals. But who's going to pay for that? The carers provide one hell of a service for free."

The best solution, of course, would be for road ecology to eliminate the need for carers altogether. The Australian mainland abounds with bizarre fauna and with ecological innovation to match. There are bridges for koalas, rope ladders for squirrel gliders, "love tunnels" for pygmy possums. "We are a lucky country in terms of funding and political will," one Australian ecologist told me. Yet Tasmania doesn't share that fortune. The state's small population generates paltry tax revenue, and its timber-based economy often sputters. "We've got hospitals that are in disarray here," one carer fumed to me. "There's no proper drug rehab centers. We're twenty years behind the rest of Australia." It's hard to justify shelling out hundreds of thousands of dollars for possum tunnels when your human populace is wracked by poverty.

Tasmania thus finds itself in a frustrating predicament. The state is aware of its roadkill problem, but it's stuck fiddling around the margins—installing speed bumps, clearing brush, posting signs—rather than retooling its infrastructure. With expensive fixes off the table, Tasmania has put its eggs in cheaper, less certain baskets. In 2014 a company called Wildlife Safety Solutions installed the island's first "virtual fence," a headlight-triggered warning system that purportedly chases animals from roadsides with flashing strobe lights and high-pitched tones, like high-tech deer reflectors. The system seemed to work wonders: one widely publicized study found that it halved roadkill, and the manufacturer rolled out virtual fences around the island. But skeptics soon poked holes in the study's methodology, and Englefield's own research found that virtual fences didn't actually reduce roadkill. The concept, Englefield said, overlooked the likelihood that animals would acclimate to the alarms and cease to flee. Just about all novelty wears off, whether you're a person or a pademelon.

To some scientists the virtual fences were a familiar disappointment—yet another instance of a government casting about for One Weird Trick to Solve Roadkill rather than deploying physical fences and wildlife crossings. "Mitigation isn't impossible, it's just not a priority," one road ecologist who'd worked for the Tasmanian government told me. When she had recommended the state install an underpass for devils and quolls, she said, "It didn't even factor."

From that perspective, wildlife caring looked less heartwarming. In the United States studies have shown that rehabbing and releasing injured turtles and bats can prevent those species from sliding toward extinction—but only if rehabilitation is paired with "threat mitigation" to protect animals from harm in the first place. On Tasmania's roads, however, the threat remained unmitigated. In some ways caring reminded me of those sappy viral stories in which a person makes a dramatic personal sacrifice to patch a hole in the social safety net: the son who opens a lemonade stand to pay for his mother's chemotherapy, the diabetic who crowdfunds his insulin. We force individuals to bear the weight of societal failure, then celebrate their resolve. It's wonderful that wildlife carers exist. It would be better if infrastructure rendered them obsolete.

- - - - - - - -

A few months after I returned home, Australia's carers, who had toiled for so long in obscurity, became celebrities. Wildfires, kindled by years of drought, charred more than twelve million acres in the country. International news coverage mourned not only human deaths but the toll on wildlife. Scientists estimated that more than a billion animals died; the *New York Times* ran an apocalyptic photo of a kangaroo backdropped by flames. Carers were everywhere: knitting mittens for koalas with burnt hands, evacuating carloads of possums, bottle-feeding fruit bats. As pleased as I was to see carers receive recognition, the accolades seemed both overdue and beside the point. Rain eventually quenched the fires, but roadkill never rests. Carers were heroic not merely because they responded valiantly to a finite disaster, but because they rose daily to confront a crisis they knew would never end.

Around the time that Australia ignited, I encountered an essay by the writer Tom Junod that seemed also to reveal something essential about caring. Junod had written about his friendship with Fred Rogers: that's Mister Rogers to you and me. Their correspondence had been far-reaching and free-form, encompassing family, faith, love. In one letter Junod told Rogers that he'd recently seen five people usher a snapping turtle off an exit ramp. Rogers suggested he write about it. Junod, a hard-boiled author of celebrity hatchet jobs, was nonplussed: What made that a story? Replied the world's kindest man, "Because whenever people come together to help either another person or another creature, something has *happened*, and everyone wants to know about it—because we all long to know that there's a graciousness at the heart of creation."

We may yearn for grace, but we don't extend it to wild animals. Instead they're cannon fodder: displaced by development, hunted for sport, run down on roads. The very word *roadkill* degrades animals by combining many deaths in a single category, discrete lives lumped into a squashed collective. Wildlife care opens our eyes to animal individuality and reminds us that roadkill is an ethical crisis as well as a biological one. Rehabilitation, one American carer has written, is a way of "extending social justice and reparations" to the beings we harm—an admittance that we are morally bound to redress our infrastructure's wrongs. A hand-reared wombat or two won't much affect the species' numbers, it's true; yet the act of caring affirms their intrinsic worth. Caring is a practice motivated less by conservation than by contrition, a practice that measures the graciousness in our own hearts.

11

SENTINEL ROADS

How citizen scientists and self-driving cars are
shaping the future of road ecology.

One of the road's quirks is that it brings us into contact with the animals it destroys. Most conservation problems are distant in space and time, their causes displaced from their effects: I'll never meet the orangutan evicted from her tropical forest by the palm-oil plantation that produced my margarine, nor the polar bear whose ice floe my transatlantic flight melted. Roadkill, by contrast, is an intimate crisis, one that reveals the physicality of wild creatures; I'd warrant that more Americans have seen a pancaked opossum than a live one. I think of my father, who would pluck the quills from dead porcupines and arrange their shafts like bouquets in the cupholder. I think of the exquisite weaponry of the great horned owl lying just off I-90, a species I'd previously glimpsed only as a distant silhouette: the scimitars of his talons, the dagger of his beak, the soft cloak of his wings. I think of the farmer who caresses a nightjar with Barry Lopez along an Idaho highway, gobsmacked by anatomy: "He runs a finger down the smooth arc of the belly and remarks on the small whiskered bill," Lopez writes. "He pulls one long wing out straight, but not roughly. . . . He asks if I would mind—as though I owned it—if he took the bird up to the house to show his wife."

This, too, is a form of reparations: attending to, and learning from, the lives that our highways squander. Roads open a window upon our wild brethren. Roadkill can tell us what animals live where, how many persist, and the direction in which their numbers are trending. It can tell us when a disease strikes a population or when a nonnative organism infiltrates an ecosystem. It can tell us when a species evolves, as Charles Brown's short-winged swallows demonstrated. It is, ideally, self-defeating—revealing in its presence and patterns how to reconfigure our infrastructure to prevent it. Per one 2020 study, roadkill is nothing short of "the most useful single wildlife observation and sampling approach available for ecology."

Just as all drivers contribute to the road's harms, we're all capable of contributing to road ecology. Counting dead animals, unlike accelerating sub-atomic particles or sequencing DNA, requires no extraordinary expertise. Nor does it require a laboratory—or, rather, our road network itself is the lab, a vast cauldron of unintentional experimentation. The world's drivers are an untapped legion of field assistants, blithely cruising past millions of potential data points, just waiting to be pressed into scientific duty. This may be the future of road ecology: as one of the first truly democratic disci-plines, a field as dispersed and globalized as the highways it studies. With a bit of prompting, we are all road ecologists.

For nearly as long as scientists have tabulated the road's dead, they've enlisted the public's help. In 1960, Terry Gompertz, a BBC producer who became a bird lover after pigeons nested in her flat, organized the Road Deaths Enquiry, which solicited volunteers through newspaper and radio ads to count bird strikes—"by foot, cycle, car etc."—on British roads. Alto-gether, 176 volunteers tallied more than five thousand birds, from house sparrows to song thrushes. Around the same time, the Humane Society of the United States roped in road-trippers to count the dead animals they saw each Fourth of July. From these surveys came a rough estimate of daily American roadkill that's still commonly quoted: one million vertebrate ani-mals struck each day.

In the decades since the Humane Society's surveys, technology has

made participatory science, known also as citizen science, far more powerful.* In the 1930s roadkill collectors had to worry about postmen throwing
away mailed skunks; modern citizen scientists can snap an iPhone photo of
a skunk and upload it to an app. Fertilized by smartphones, participatory
road-ecology projects have sprung up on six continents. British Columbians
can turn to Roadwatch BC, Belgians to Animals Under Wheels, and Bra-
zilians to Sistema Uburu, the Vulture System. After Israel added a roadkill
button to the traffic app Waze, drivers reported twelve thousand animals
within six months.

For years the United Kingdom's largest road-ecology project was called,
with grisly élan, Project Splatter. One spring I tracked down Sarah Perkins, an ecologist at Cardiff University in Wales, where Project Splatter was
based. Perkins had offered to show me some Welsh roadkill hotspots, but
my vision of a bucolic drive through sheep-speckled countryside quickly
evaporated. No sooner had she picked me up at the bus station than tor-
rential rain began pounding the windshield. It was a day made for nursing
a warm, flat beer, not crouching over dead foxes. Perkins's mind was on
roadkill nonetheless. "Days like this, the rivers flood, and the otters are
driven out and cross the road," she said over the whump of wipers. "Quite
bad news for otters, this."

After a pub lunch that, in true Welsh fashion, consisted mostly of leeks,
we retreated to the dry comfort of Perkins's lab. Project Splatter, Perkins said,
was the brainchild of one of her students, who set up a Twitter account in 2013
and encouraged the public to tweet their sightings. The media latched on, and
the idea blew up. By the time I visited Wales, Project Splatter—a very loose
acronym that stood for Social Media Platform for Estimating Roadkill—had
garnered more than fifty thousand observations via social media, its website,

---

* Ecologists increasingly prefer "participatory science," "volunteer science," or "community science" to "citizen science," in recognition of data collectors who may not be documented citizens.
Mary Ellen Hannibal, author of the book *Citizen Scientist*, rejoins that the term separates the
word "citizen" from its immigration context and reclaims it in the sense that Aldo Leopold used
it: to describe humans as humble citizens of the land rather than conquerors of it. To Hannibal,
citizenship is an ethical category, not a political one. I use the various terms interchangeably.

and an app. Many contributors turn on their phone's audio recorder every time they get behind the wheel, describe the roadkill they see, and then transcribe and submit their recordings. A paramedic once logged a dead bat while driving her ambulance. Granted, not everyone appreciated the morbid humor of Project Splatter's moniker. A few years after my trip, a brigade of incensed cat-lovers bombarded Perkins with "exceptionally fruity language" and compelled her to change its name to the disappointingly anodyne "Road Lab." (Happily, Project Splatter lives on as the title of the data set.)

The United Kingdom is well suited to participatory road ecology. It's a nation of creature lovers, with a fair claim at hosting the world's first wildlife crossing: in the 1950s a naturalist named Jane Ratcliffe pestered the government into including a "large culvert with side walks" to usher badgers beneath the M53 motorway. Today the British ardor for animals extends to roadkill in occasionally odd ways. In 2008 the BBC aired a program called *The Man Who Eats Badgers*, which chronicled the exploits of Arthur Boyt, a mild-mannered consumer of rabbits, owls, and other roadkill. Poor Boyt was promptly hounded by angry crank callers claiming to be the ghosts of his meals. Certain British roadsides do seem haunted. In 2020, Dorsetians began installing "ghost hedgehogs": pale cutouts, like the white bikes that memorialize cyclists, to commemorate animals squished on country lanes.

"I'd literally drive off a cliff to avoid hitting a mouse," Perkins said. She'd once asked a policeman how many crashes could be attributed to the same impulse. "He said, 'If I had a fiver for every time someone told me they swerved for a fox, I'd be a millionaire, love.'"

Much as tight-knit groups coalesce online to discuss sports, music, and politics, Project Splatter's Facebook page had become a rendezvous point for roadkill aficionados. Perkins's weekly Splatter Reports often sparked acrimony. "If they see a polecat"—not actually a cat but a sort of European ferret—"and then I don't mention it in the report, it's like, where's my polecat?" Perkins said. Although Project Splatter doesn't encourage public photographs, some users can't resist. "There's one guy who's always posting pictures of roadkill on Facebook, and I have to wonder if he's getting some kind of gruesome kick out of it," Perkins said. She opened Facebook on her phone. "He hasn't posted anything today." She swiped, looked again. "Oh no, he has. It's a hedgehog. There it is." She sat back with what I can only describe as a mirthful sigh. "Sometimes I worry about him."

On her laptop Perkins pulled up a map displaying the project's thousands of roadkill observations. I'd expected isolated clusters—a few badgers here, a handful of hedgehogs there. Instead, the sightings formed a well-defined outline of Britain's entire road network. In such a crowded country, roadkill wasn't confined to discrete danger zones. It was everywhere. "The U.K. is horrendous. It's just one massive road," Perkins said. "How can you mitigate against that?"

More significant than where British roadkill occurs, then, is when it occurs. Just as roadkill accumulates in hotspots, it can also occur at "hot moments," time periods marked by heightened risk. Hot moments can last a season—like summer in Maine, when female turtles trundle to nesting sites—or a single evening. Kangaroo kills spike during the full moon, whose illumination makes them more active. Deer collisions have historically surged after Daylight Saving Time ends each autumn: when clocks "fall back," dusk arrives at around 5:00 p.m., rush hour for both commuters and whitetails. Enacting permanent DST, researchers have calculated, would prevent enough nighttime driving to spare the lives of thirty-three drivers and some thirty-six thousand deer every year.

Project Splatter's data has likewise unveiled animal behavior. Polecats perish in a "bimodal distribution," a pattern that on a graph resembles a two-humped camel: one spike in spring, when males search for mates, and another in autumn, when naive young disperse. Otters die in winter, when their twilight movements coincide with peak traffic. Pheasant roadkill also climbs in winter, perhaps because hunting clubs stop feeding the birds and force them to strike out on their own. Hedgehogs are vulnerable in autumn, as they scurry around fattening up on worms and slugs. Rat roadkill spikes in fall, too, as farmers plow their fields and the grain-starved rodents scatter, then splatter.

"They're considered a pest species—no one studies them, no one cares," Perkins said. "It's this insight into their behavior, their ecology, that we never would've picked up without citizen science data." Unlike biologists, cars

don't discriminate; they capture endangered species and common rodents
with equal gusto. Some scientists have compared roads to motion-activated
trail cameras, perpetually documenting whatever crosses their path.

For as long as road ecology has existed, it's had a data problem. The issue
stems in part from the atomized nature of our road system. Every transportation agency has a different process for quantifying the carcasses that
its personnel scrape off the pavement. Some states instruct workers to log
roadkill in custom apps; some ask them to scribble down notes in pencil;
some have them lug bodies off to landfills without registering them at all.
Data management is pocked with pitfalls. In Arizona a computational error
once changed hundreds of deer, elk, and pronghorn to beavers, ruining two
years of records. One highway contractor in British Columbia, tasked with
submitting carcass locations to the province's roadkill app, got in the habit
of uploading his reports over coffee rather than from the field. Biologists
figured that out when the app showed a roadkill hotspot at Tim Hortons.

Volunteer road ecology would seem to offer a corrective, by marshal-
ing an immense force of data collectors who log what maintenance workers miss. But participatory science is still a loosely organized discipline
rife with methodological inconsistencies. Every project uses a different app
or web portal, collects different data, asks its members to obey different
protocols. There are "opportunistic" programs whose volunteers jot down
roadkill wherever they stumble upon it and "targeted" ones that canvass
specific roads at specific times. There are projects that only log roadkill and
others that collect "zero values"—places where volunteers don't encounter
animals, which is helpful information in its own right. Some ecologists
have suggested standardizing methods, while others insist that each initiative be tailored to its context. This diversity of approaches has caused some
agencies to dismiss the validity of participatory science altogether. "If you
bring data forward about a problem, they may have to address it and it may
cost money," Fraser Shilling, the founder of a participatory road-ecology
program called the California Roadkill Observation Network, told me. "It's

easier just to say, well, it's publicly collected data, we can't really trust it"—
like clapping your hands over your ears to avoid unpleasant news.

From its inception Project Splatter had opted for an opportunistic approach
rather than a rigid one. To Sarah Perkins's mind, that informality was a
strength, as it lowered barriers to entry and generated more data. "As soon
as you make it hard, people drop out," she said. But the ever-shifting group of
volunteers also introduced potential bias, since not everyone is equally adept
at spotting roadkill. If one year's cohort was especially sharp-eyed, their
reports could create the illusion that collisions had increased when in reality
the new participants were simply better at noticing it. "Early in the project,
people were like, well, this is a bit aimless, isn't it?" Perkins admitted.

Yet Project Splatter's casual approach has been vindicated elsewhere.
When Shilling analyzed nearly four thousand roadkill sightings submitted
to the California database, he discovered that volunteers had accurately
identified 97 percent of the animals. In South Africa ecologists likewise
found that data collected by trained patrollers aligned with submissions
from the hoi polloi. Unsurprisingly, the skilled surveyors had a keener eye
for the small stuff, like rabbits, while the rank and file tended to notice conspicuous carnivores like the aardwolf, a stripy hyena cousin. Both groups,
though, noticed the same zones of carnage—proof that volunteers could
identify roadkill hotspots "in a reliable and robust manner."

If roadkill observations are valuable, then physical carcasses are gold.
Project Splatter shares an office with the Otter Project, another Cardiff
University initiative that makes use of the dead. After I'd spoken with Perkins, a student named Alice Hill whisked me down the hall to see some
dismembered otters. She pried open a chest freezer to display chunks of
fur, bone, and organs packed in Ziplocs like hamburger. Most samples had
come from roadkill specimens found by the public and shipped frozen in
cold-storage crates. "It's always risky," Hill said as she held up a jar con-
taining a putrescent uterus. "We had one that was supposed to come on
a Thursday and didn't turn up until Monday. It had already started the
decomposition process." Assuming an otter reaches Cardiff intact, it under-
goes a workup as intensive as any autopsy. Researchers check its size, age,

and sex; inspect it for injuries; and collect ticks, fleas, and other parasites. The unfortunate mustelid is then sliced open, throat to genitals, and its liver, kidneys, adrenal glands, and other organs pried out for analysis. "It's much less cute and cuddly than I would have liked," Hill admitted. "But ethically, it's quite nice working with dead otters."

The resultant tissue archive, which dates to 1992, has revealed much about otters and their milieu. Like rolling snowballs, environmental contaminants accumulate as they pass through the tissues of insects, fish, and apex aquatic predators. Otters have the misfortune of serving as sponges, absorbing and amplifying whatever we dump into their habitat. In 2014 scientists in Illinois found that the livers of car-struck otters contained alarming concentrations of long-banned pesticides, including some linked to Parkinson's and Alzheimer's diseases. The tissues collected by the Otter Project have likewise demonstrated the frightening tenacity of toxic flame retardants called PBDEs. Happily, they've also shown that tougher laws have reduced lead pollution.

"There's this constant injection of new chemicals into the system, as well as things that are quite persistent," Elizabeth Chadwick, the biologist who leads the Otter Project, told me over a pint that evening. "The chemical signatures that we're detecting in otters are the same chemicals that we're drinking." And, since otters are protected, their tissues wouldn't be available without roadkill. "How else would you sample the environment if you didn't have those otters?"

Otters, Chadwick said, are a *sentinel species*, an organism whose health indicates much about its environment. You might likewise consider roadkill a sentinel *phenomenon*, a process that reveals transitions over time. In Florida the abrupt absence of roadkilled raccoons, opossums, and rabbits showed that invasive pythons had plundered Everglades National Park; in Washington a dead wolf proved that the carnivores, long extirpated, had returned to the state. The changing seasonality of green whip snake roadkill in Italy may be an artifact of climate change, which causes the snakes to wake up earlier in the year and return to hibernation later. Roadkilled ferret-badgers showed that rabies had reached Taiwan. Eighty percent of struck raccoons in California carry a roundworm that can infect human brain tissue. Roadkill has allowed scientists to track the spread of pinworms in black howler monkeys, chronic wasting disease in white-tailed deer, and leprosy in nine-banded armadillos. "Be sure to cook your 'dillo thoroughly," advised one Missouri newspaper.

For all their grimness, roadsides are cabinets of curiosities. Australian scientists rediscovered the pygmy blue-tongue lizard, a skink presumed extinct, in the form of half-digested remains within a flattened snake—a Lazarus species, resurrected by roadkill. Researchers in Ethiopia's Nechisar National Park made an even more stunning discovery: jouncing along a dirt road one evening, the crew spotted a nightjar carcass and, from it, salvaged a reddish wing with a distinctive white patch. Experts confirmed the limb belonged to a new species: the Nechisar nightjar, *Caprimulgus solala*—solala for "only a wing."

The Otter Project has been revelatory in its own right. As the project has grown, so has its carcass supply, to nearly three hundred annually. In part, that reflects its elevated profile and in part the otter's recovery from a brush with extinction. But it's also conceivable that otters have become more collision-prone. Around a quarter of British otters carry *Toxoplasmosis gondii*, a parasite that makes its hosts bolder, sometimes too bold. Infected mice lose their fear of cats; humans get in more car crashes. Car-killed pademelons are nearly three times likelier to be infected by *T. gondii* than those that die from other causes. Roadkill reveals the conditions that beget roadkill, a mirror that reflects itself.

More foolhardy otters bumbling across roads at the command of their parasitic overlords would mean more specimens for the Otter Project—not that Chadwick would be pleased. "You become an ecologist because you want wildlife to survive," she said. "If we could stop all road mortality, my job would cease to exist, and that would be fine. But I don't think that's likely to happen." All she could do was to obey the moral obligation that roadkill imposes. Let no otter die in vain.

- - - - - - - -

As volunteer road ecologists, we humans have one significant flaw: we're imperceptive. The speed of cars both inflicts violence and conceals it.

Behind the wheel you're moving too fast, seated too high, and your field of vision is too restricted to notice any but the largest carcasses. Even trained biologists undercount. When Brazilian researchers scanned one highway from within a car, they spotted just twelve animals; retracing their route on foot, they registered more than two hundred, mostly frogs. You can't study what you don't notice.

We're blind to roadkill, but our vehicles may not always be. Most transportation experts consider it inevitable that self-driving cars will someday take over our roads, though the timeline may be better measured in decades than in years. Autonomous vehicles, or AVs, detect their surroundings via cameras, radar, and a laser-guided system known as Lidar, an array of instrumentation vastly keener than human eyes. It isn't outlandish to suggest that AVs will become the most ingenious road-ecology tools ever devised, a roving squadron of automated scientists perpetually uploading terabytes of real-time data about the animals, dead and alive, whose paths they cross. As ecologists speculated in *Nature*, their sensors could "monitor pond-breeding amphibians during migrations, reptiles thermoregulating on warm asphalt, butterflies and birds flying over the road, and small mammals crossing roads or moving in surrounding areas." And they should dramatically reduce large-animal roadkill: the moose that looms out of the darkness before our feeble focal vision can process it will be easily spotted by an AV's sensors. Autonomy's masters could program their creations to drive slower at dusk, when animals are active, or avoid hotspots. In a 2017 article titled "The End of Roadkill," one *New York Times* writer envisioned panthers "spread[ing] out of the cramped confines of Florida's southern tip" to repopulate the Southeast, safe to recover beneath autonomy's wing.

As promising as that sounds, AVs remain buggier than a windshield in a locust swarm. They struggle to navigate traffic, identify lanes, and cope with rain and snow. And they have an inglorious history of being flummoxed by animals. Consider the saga of Volvo's Large Animal Detection system, a radar-and-camera array designed to tap the brakes for wildlife. Volvo developed the technology in Sweden, where moose pose the primary threat. The company's engineers, boasted one executive, "put a lot of effort in seeing how animals moved and teaching the computer to look for that movement," a process that involved filming moose and deer and digitally simulating their gait. The trouble arose when Volvo's Australian arm began testing the system. Moose had been an easy pilot species: huge, solid, about as hard to detect as a brick wall. But Australia didn't have moose. It had kangaroos. Rather than ambling like an ungulate, kangaroos hop, a form of locomotion that resembles a bouncing pogo stick. Understandably, the engineers back in Sweden hadn't accounted for such saltatory fauna and used the ground as a reference point to calculate the distance between animal and vehicle. By refusing to remain earthbound, kangaroos defied the algorithmic rules that moose established. "When it's in the air, it actually looks like it's further away, then it lands and it looks closer," lamented a project manager. Kangaroos were an effervescent anomaly, as inscrutable to machine intelligence as fear or joy.

After months of trials Volvo resolved its 'roo snafu. But the episode revealed a worrisome truth: it's hard to predict how self-driving cars and wildlife will interact. While it's comforting to contemplate roadkill's end, AVs will do little to protect most species. One Nissan executive told me the company was developing technology that would detect animals "down to the size of a small dog," which might be cause for celebration in the coyote community but won't help lizards or chipmunks. Surrendering control to computers could even lead to *more* roadkill: you might brake for snakes, but your robotic chauffeur won't.

And autonomy will transform more than who's at the controls: it will overhaul where we live, how we receive our goods, and how our societies are configured. In 1994 an Italian physicist described a rule known as Marchetti's constant, or his "wall": the precept that, from Greek villages to American cities, humans have arranged their lives to travel for little more than an hour per day. We sprawl, but only as far as our limited tolerance for commuting allows. Autonomy could break that hoary rule. When your self-driving car or ride-hailed taxibot doubles as a mobile workstation or entertainment center, travel feels like less of a chore. Autonomous vehicles should also iron out traffic (robots don't rubberneck), which will put more

exurbs within reach of urban centers. Autonomy—in tandem with electrification, which will make driving cheaper and thus encourage more of it—
might turn out to be a suburban carcinogen that triggers metastatic sprawl.

Moreover, focusing on human travel is a red herring. Most AVs won't shuttle people; instead, they'll move products. No longer do we drive places to
acquire things: we summon the things to us. Amazon warehouses have supplanted malls, restaurants have pivoted to takeout, and groceries are a button-
push from our doorstep. Our delivery fetish will shape development patterns
as surely as the interstates did. Factories and warehouses will press into the
hinterlands; after all, you can afford to ship goods longer distances when you
don't have truckers on the payroll. The resultant civilization, writes the futur-
ist Anthony Townsend, may look dystopian indeed: "A fractal web of farms,
factories, and towns will spread without limit across the land."[*] Liberated
from drivers, autonomous trucks will have no compunctions about prowling
the highways at night, and darkness, once a reprieve for wildlife, will become
as dangerous as daylight. Far from ushering in a halcyon, roadkill-less era,
AVs could be the gravest challenge to road ecology since, well, roads.

This nightmare feels disconcertingly probable, but it isn't inevitable. As
a pair of Australian scientists opined, self-driving cars have the potential
to dramatically reduce animal collisions—"but only if conservation biologists play a role in the development and implementation of these vehicles
on the road." To that end some scientists are shouldering their way into
the driverless world. Among them is Cara Lacey, a Nature Conservancy
planner who interviewed several AV manufacturers. The conversations dismayed her. "The industry isn't really thinking about the environment," she
told me. Sprawl, Lacey realized, was one likely outcome of autonomy, and
open space, already a dwindling commodity, could become even scarcer.
"Could car companies invest in land protection, so that we're creating a

---

[*] Our skies, too, will get more cluttered. Amazon's Prime Air already delivers parcels by
drone, and tech-bros dream of flying cars. Three hundred million birds are whacked by
vehicles every year; wait until hawks and hummingbirds have to contend with flocks of self-
piloting package-copters and sky taxis.

buffer against how far sprawl can actually go?" she wondered aloud. Per-haps AV developers could contribute to a conservation fund for every mile their vehicles traveled or pay for the construction of new wildlife cross-ings. "People think, 'Oh, autonomous vehicles are just a fad, or it's not like they're getting any closer,'" Lacey said. "No, this is *happening*. They're on the road right now. Why would we wait to get involved?"

Some road ecologists already are. In the mid-2010s a Portuguese research team began developing an automated roadkill mapping system. The sys-tem's first iteration entailed towing a clunky industrial computer and a gas-powered generator around on a trailer, more steampunk than space age. Subsequent versions, however, streamlined the tech to a couple of sleek, bumper-mounted laser cameras and a laptop running a roadkill-identification algorithm developed via machine learning. In trials the system captured nearly 80 percent of dead amphibians and birds, the same animals that human observers overlook. The program struck me as a leap toward road ecology's prospective future, in which our cars will function as near-infallible surveyors. I imagined millions of Teslas and Waymos canvassing highways, reaping near-infinite data about the world they were destroying.

One October morning, six months after I visited Project Splatter, I found myself straddling a double-yellow centerline in Montana's Bitterroot Valley, camera in hand and helmet atop head, speculating aloud about the identity of a matted lump of roadkill.

Heather Hicks, the volunteer surveyor whom I'd joined for the day, snapped an iPhone photo. "Think it's a skunk?"

I prodded the bristly wad with my toe. Traffic flowed around us like a river around boulders. "Could be a badger."

Hicks nodded thoughtfully. The fur was charcoal and accented by whit-ish highlights. On one hand, it looked more black than gray—point for skunk. On the other, it didn't smell. Point for badger.

Hicks settled it. "I'm gonna go with 'unknown,'" she declared, and tapped her phone.

We waited for a break in traffic, then scampered back to the bicycles leaning against the guardrail. "That was pretty hairy," Hicks said. She eyed the fluorescent pinney I'd donned over my jacket, a style best described as crossing-guard chic. "Good thing you're wearing that dorky vest."

Inspired by Project Splatter, I'd decided to try my hand at citizen science. My first call was to Adventure Scientists, a nonprofit that trains outdoor-loving laypeople to gather data. Ultramarathoners have scoured mountains for wolverine tracks; kayakers have collected samples from frothing rivers. Adventure Scientists' latest initiative, the Wildlife Connectivity Project, was more my speed, as it required no special skills beyond the ability to stay upright on a bicycle. Teams of cyclists would fan out around Montana and ride fifty miles—twenty-five out, twenty-five back—on their chosen road segment. Along the way they'd photograph every creature they saw, alive or dead, and input its species, location, and other information into an app. The plan was to repeat the ride four times a year for three years: a structured approach to citizen science in contrast to Project Splatter's opportunistic one. Granted, Montana wasn't about to overhaul its roads on the basis of some volunteer observations, but Adventure Scientists might add another layer to the state's deepening foundation of maintenance records, crash reports, elk satellite collars, and other data.*

By coincidence I'd been assigned to survey U.S. 93, the same highway where, years earlier, I'd walked up my first wildlife overpass on Confederated Salish and Kootenai land. While that famous bridge lay north of Missoula, the Adventure Scientists ride began south, in the Bitterroot, where wildlife-friendly infrastructure was comparatively scant. I'd been assigned to support Hicks, a family physician and cyclist who often photographed dead deer during her rides. When she received an email from a cycling club about the roadkill ride, she immediately enrolled. "I was like, oh, I already do some of that," she said as we tugged at our water bottles. "But I was never in the mind-set that you could do anything to change roadkill."

---

* After funding ran out in 2020, Adventure Scientists discontinued the project early, though its staff told me Montana is still using its data.

Hicks—a conspicuous figure with her hot-pink windbreaker, yellow GT mountain bike, and billowing strawberry-blond curls—had chosen a glorious day for a ride. The air was clear and cold as a headwater stream. To the west the snow-brushed Bitterroot Mountains filled the sky; the Sapphire Range's brown shoulders rose to the east. The Bitterroot River curled toward the road and away, exhaling mist. White roadside crosses bloomed above the mullein. I thought of the old bumper sticker: "Pray for me, I drive 93."

If the road was dangerous for people, it was positively murderous to wild animals. Our bicycles, slow and low, unveiled the highway's hidden violence. Had we been cruising at seventy miles per hour, we'd never have noticed the magpie whose breast was glazed with a patina of hoarfrost, nor the pair of song sparrows tucked beneath the guardrail like suicidal lovers. Nor would we have seen the fawn pretzeled against a modular concrete Jersey barrier—a highway feature, named for the state in which it was invented, that engineers love and road ecologists loathe. Jersey barriers reduce head-on collisions by separating traffic lanes but also trap animals on the road, and it was awful to ponder this fawn's chaotic, terrifying end.

The morning wore on and we rode north, watching for ravens as fishermen scan for gulls. We settled into a rhythm: at each kill we'd drop our bikes and dart into the highway, me keeping watch while Hicks snapped a photo. Then we'd retreat to answer the app's queries. What species had we seen; where had we seen it; what natural features were nearby? The road was a crime scene, we its clumsy forensic detectives. Was that brownish smear a slick of blood or motor oil? Gopher or ground squirrel? "Is this part of *that* dead thing, or its own dead thing?" Hicks asked, pointing to what appeared to be a hoof. I could only shrug.

Although one morning on twenty-five miles of highway was an impossibly small sample, we couldn't resist seeking patterns. Whenever a stream flowed beneath the highway, a deer died nearby, suggesting that animals followed creeks. Where the road ran through clusters of industry—a gravel pit, a martial arts studio, a casino–gas station hybrid called Lucky Lil's—we found only a housecat. Roadkill began to seem not senseless but significant.

Mostly, though, we tried to avoid becoming roadkill ourselves. Like pebbles tossed into the cogs of some great machine, our bikes clotted the flow of traffic. Drivers gesticulated angrily from within their soundproof bubbles, flung into road rage by us nonmotorized anarchists. The ride was somehow both leisurely and exhausting. After a few hours, surveying road-kill came to feel like watching undercover videos from animal feedlots, a grinding reminder of the brutality that lurks beneath society's systems. There was something nice about the idea of outsourcing this grim work to self-driving cars, the dispassionate robotic eyes that will scan highways without grief or anger.

Yet much, too, will be lost in this transition—when it's our vehicles who practice science rather than we plain citizens of the land. The car already cleaves us from nature; the self-driving car will only make our separation worse. (Imagine a windshield that displays TikTok instead of scenery.) Volunteer road ecology forces us to confront modernity and converts humans, real ones, into canny ambassadors for an oft-ignored crisis. In Alberta a cohort of volunteers, operating under the program Collision Count, spent years tallying dead elk, deer, moose, and sheep along a tortuous mountain route called Highway 3. They conducted their surveys on foot, both along the highway and in the nearby brush; as a result, they spotted not only the roadside dead but the many more animals who had stumbled off to die and gone unnoticed by maintenance workers. Ecologists had suspected that Highway 3 was a collision hotspot, a hunch the citizen study both con-firmed and publicized. No longer were academics and environmentalists the only ones carping about roadkill; now friends and neighbors muttered about it in grocery lines. The mounting pressure prodded Alberta's govern-ment to erect fencing and an underpass. "It's a really powerful shift in how information moves," Tracy Lee, the biologist who developed Road Watch, told me. "We can have all the science in the world, but if we don't have the community pushing politicians to invest in this infrastructure, then we aren't going to see it happen."

Like wildlife rehabilitation, participatory road ecology is a form of both reparations and resistance, a collective acknowledgment of automobility's destructiveness. Science, Robin Wall Kimmerer has suggested, can be "a powerful act of reciprocity with the more-than-human world"; noticing the dead honors all that animals give us in life. If attention is the most basic form of love, Hicks and I could have been said to love roadkill that day, or at least to love the animals that had become roadkill. I also newly empathized with them. Cycling along U.S. 93 was to experience a highway from the outside, the wrong side, to feel the wind and the kicked-up gravel and the impossible mass of a logging truck chewing up a straightaway, to understand the vulnerability of a soft and living body in a world made for machines. It was to feel a lot like prey.

12

THE TSUNAMI

*What does Brazil's growing road-ecology movement
reveal about the future of our paved planet?*

When, back in 2013, I first began learning about road ecology, I conceived of it as a one-way cultural export, spread with missionary zeal by scientists from thoroughly developed nations like my own. This belief, I see now, reflected a sort of scientific imperialism: that "they" would learn from "us," that the great feats of western ecological engineering—the Banff overpasses, the Dutch ecoducts—would guide similar structures in, say, Myanmar. The reality, I learned, was far more multidirectional. In the United States protecting animals from roads tends to be a rear-guard action: we retrofit, tinker, and slap up sporadic wildlife crossings, forever trying to undo intractable, decades-old errors. By contrast, less intensively built nations aren't saddled with an ancient, sclerotic highway system; instead, they're still shaping themselves, giving ecologists the chance to intervene before their road networks are calcified. When the Indian government constructed a new highway through a tiger reserve in 2019, for instance, it elevated the road on concrete pillars, permitting animals to wander the forest floor unperturbed. "We think we're such leaders," one American ecologist told me. "And then you go to India and you're, like, wait—they're just going to *lift the highway*?"

What's more, different countries are home to different animals who

interact with roads in distinctive ways. Rainforests, for example, are rife
with primates—howler monkeys, lemurs, lorises—who spend their lives
clambering through the canopy, seldom descending to earth. Tropical habitats are thus three-dimensional, requiring conservationists to heal the
road's gash in the treetops as well as at ground level. Road ecologists have
responded with "canopy bridges," flimsy-looking rope courses that have
reconnected forest fragments for colobus monkeys in Kenya, gibbons in Taiwan, and capuchins in Brazil. Although North America's porcupines, martens, and flying squirrels could surely use a few bridges of their own, the
United States remains a laggard in the field of canopy connectivity, with
much to learn from its creative equatorial peers.

The developing world's penchant for ecological innovation is welcome,
for new highways are coming, and coming, and coming. Ours, the ecologist William Laurance has written, is "the most explosive era of infrastructure expansion in human history," and while power lines, fiber-optic
cables, and railroads are all extending their tendrils, no form of development is growing faster than roads. (In the Maldives, indeed, the word for
development is *thara'gee*, a neologism perhaps derived from *tarmac*, con-
cretization bending both land and language.) Some highways are spawned
by noble goals, like access to schools and hospitals; some by cynical ones,
like funding from rapacious foreign investors eyeing farmland and min-
erals. Most are justified by some combination of motives. As the infrastructure tsunami thunders ashore, it promises to wash away biodiversity.
The Trans-Papua Highway portends palm plantations in New Guinea, the
Dawei Road threatens to shatter elephant habitat in Myanmar, and the
Cross River Superhighway imperils gorillas in Nigeria. Name a species of
charismatic fauna, and infrastructure likely poses the primary threat to its
survival. Near one Kenyan park, so many cheetahs have been hit by cars
that researchers have given up on studying them; across Asia more than
fifteen thousand miles of proposed roads threaten to splinter tiger habitat.
Highways and other infrastructure are largely responsible for endangering
65 percent of the world's primates—creating, as scientists tragicomically
put it, "No Planet for Apes."

Among the places the infrastructure tsunami will crash hardest is Brazil, a country that exemplifies the clash between highways and nature. Brazil's biodiversity and its roads both invite superlatives: it hosts the world's most amphibians, the most freshwater fish, the second-most mammals, the third-most reptiles, and the fourth-longest road network. (It trails only the United States, China, and India.) One Brazilian biologist has estimated that cars strike more than four hundred million animals there each year and that future development stands to "cause the [additional] loss of half a billion vertebrates annually." The country is at once an established powerhouse and an emerging nation, with a shoddily built road system that's forecast to grow by 20 percent in coming years. "The expansion of roads in Brazil is important for the quality of life of the people," one ecologist told me. "The thing is, how will they do it? Is there enough legislation to protect wildlife? I'm not sure."

Brazil's expansionism, fortunately, is matched by a concomitant growth in road ecology. No tropical country is home to as many road ecologists, and projects spring up constantly: underpasses for pumas and tapirs, a new bridge for a resplendently furred monkey called the golden lion tamarin. And Brazil's road-ecology movement hasn't merely replicated the Northern Hemisphere's successes: it's built on them, embedding the discipline into the country's legal fabric. Brazilian law requires highway operators to gather data about the roadkill that occurs within their jurisdiction, transport injured animals to veterinary clinics, and reimburse drivers for damages incurred in collisions. (Imagine if you could sue your state transportation department every time you whacked a deer.) To my American eyes Brazil seemed to sit at road ecology's forefront—a country squarely in the infrastructure tsunami's path, but one where ecologists were frantically throwing up wave breaks. So I got on a plane.

"Today," Mario Alves announced, "we will catch Evelyn."

I mumbled my enthusiasm through a mouthful of *chipas*, the cheesy, resilient dough that was my breakfast. We were rolling east across Mato Grosso do Sul, a southern Brazilian state whose flat terrain, broad skies, and abundant cows recalled Texas. The endless gray thread of the BR-262 highway unspooled into dawn's shimmer. Alves, a veterinarian who sported tattoos depicting owls, horses, and pumas, rested one hand on the wheel of our Mitsubishi Triton. In the backseat perched two other vets, his colleagues. Around us the *cerrado*, the sere savanna that blankets a fifth of Brazil, stretched to the horizon like a beige sea. And somewhere out there was Evelyn.

Evelyn was a giant anteater, a creature that's on the short list for the title of world's strangest mammal. Massive front claws force anteaters to walk on their knuckles, endowing them with the shuffling gait of a gorilla holding a fistful of steak knives. Body-length tails billow behind them like pennants in a breeze, supplying their Portuguese name, *tamanduá bandeira*, "flag anteater." Oddest of all is what passes for an anteater's mouth. Entirely toothless, *Myrmecophaga tridactyla* possesses a tongue so prodigious that it anchors to the sternum and furls, Fruit Roll-Up–style, into its owner's palate. Anteaters deploy their tongues to probe anthills and termite mounds, lapping up prey with a sticky lacquer of saliva. These sieges are intense—an anteater can flick its tongue 160 times per minute—but brief, curtailed when the ants flee into the earth or bite back. Giant anteaters are thus rotational grazers, endlessly circuiting insect-filled pastures. A few termites here, a few there, and by day's end they've slurped down thirty thousand bugs.

The giant anteater's Brazilian range, alas, is bisected by BR-262, a highway that crosses Mato Grosso do Sul as it winds from the Bolivian border to the Atlantic Ocean. Eucalyptus, iron, cattle, and cocaine pulse through this infrastructural aorta in obliterative trucks. Driving BR-262 earlier in the week, we'd come across the broken armor of an armadillo, the mangled remains of a capybara, the scattered plating of a caiman, the torn plumage of a raptor whose too-apt common name was the roadside hawk, and a lump of meat Alves could only categorize as "indeterminate mammal." We'd also found an anteater, six feet from snout to tail, in a state of advanced decomposition, fur sloughing from sun-blackened skin. When Alves hauled the carcass into the brush, it left behind a damp impression of grease and fur, like the footprint of a descending whale. I'd asked Alves if studying roadkill upset him. "I'm sad about the situation," he replied, "not about each animal that I see. Or I would be"—he glanced skyward, casting for an English phrase—"under depression."

I'd joined Alves to observe the operations of Anteaters and Highways, a multipronged study of how Brazil's exploding road network affects its oddest mammal. From its inception in 2017, Anteaters and Highways had been one of road ecology's most comprehensive ventures, an inspiring testament to Brazil's commitment to the field. Every two weeks its scientists tabulated roadkill along more than 1,300 kilometers of highway; by the time I visited, Alves and his colleagues had driven far enough to circle the earth twice. They'd affixed satellite collars to more than forty living anteaters, documented hundreds of highway crossings with camera traps, and necropsied dozens of roadkill victims, collecting everything from lymph nodes to skin parasites to determine whether living near highways impaired anteater health. Evelyn was an involuntary conscript in this epic endeavor. Nearly a year earlier, Alves had wrangled her in a cattle pasture and outfitted her with a tracking collar. Now the time had come to recapture her, download the last year of her life, and figure out how she'd survived so close to Brazil's deadliest highway.

As we cruised through the savanna, my companions regaled me with anteater lore. "Some people say that if they cross the house or the farm, somebody will die," one of the vets whispered.

"They are solitary, they are dark, they are nocturnal, they are weird," Alves said. "People say if you see a giant anteater, you won't be successful when you go fishing. Or your wife will cheat on you. To avoid that, you need to spank the animal."

"Spank them?" I asked. "Like, with a board?"

Alves nodded. Anti-anteater sentiment, he added, made highways a still graver threat, since superstitious truck drivers were rumored to target the animals. Anteaters and Highways had employed a social scientist to interview more than two hundred truckers, only one of whom copped to intentionally whacking an anteater. Still, myths abounded.

"Another story is that they are only female, and they reproduce with their tongues," Alves said.

"Another one is that they have no bones," a vet added.

"No bones?"

"Yes! My mom is from the countryside, a very tiny little town, and she said, 'Oh, but what's the problem with car crashes? They have no bones!' I said, 'Okay, mom, I'm a biologist—we need to talk about this.'"

The morning was cloudless, a world away from the dim, damp Amazon. The *cerrado* is the most biodiverse savanna on earth, home to ten thousand plant species, and its austere wonders lined the roadside. We passed gnarled trees with corky bark; rust-colored termite mounds; shrubs whose seeds, smeared between our fingers, formed a waxy vermilion paste. The most conspicuous organisms, though, were human commodities. Soybean seeds spilled from an eighteen-wheeler, pinging off our windshield like hailstones. We saw herds of bone-white cattle, walls of nonnative eucalyptus, and lime-green tracts of sugarcane, the latter destined to become biofuels to power vehicles that would, somewhere on the planet, drive on more roads.

The commercial exploitation of the *cerrado* is a relatively recent development. Brazil's savanna was once considered worthless, underlaid by soils so acidic that the region's very name means "closed." But the *cerrado* wouldn't stay closed for long. In the 1970s scientists learned how to "correct" its soil with lime and fertilizer, triggering a farming boom that never abated. Around half the savanna's original extent—about eight hundred thousand square miles—has been lost, and just 8 percent is blessed with any official conservation status. All that land-clearing was not only abetted by roads; it also birthed them. The construction of new *cerrado* highways and the paving of dirt ones had made it easier to bring soybeans and other products to market, driving up land values and enticing agribusiness. Highways encourage commodity crops, which justify highways, which encourage commodities, ad infinitum.

After a time Alves pulled onto a dirt road, then into a bumpy field. He nudged the Mitsubishi through cattle whose dewlaps rippled as they stepped aside. We parked near a thicket of spiky bromeliads, and Alves unfolded an antenna to listen for Evelyn's collar. We heard a metronomic clicking. Satisfied his quarry was nearby, Alves drew a cocktail of sedatives from a tacklebox overflowing with syringes, creams, and swabs. He turned to me.

"What do you want to do?" he asked. "Do you want to take pictures, or do something . . . adventurous?"

"Adventurous," I said cautiously.

Alves extracted two long-handled nets from the truck and handed me one. "You will help me to catch Evelyn."

I examined the flimsy net and recalled a paper I'd read titled "Human Death Caused by a Giant Anteater (*Myrmecophaga tridactyla*) in Brazil." Anteaters are docile creatures who, in captivity, affectionately hug their keepers. In fairness to the homicidal individual whose claws had perforated the femoral artery of a forty-seven-year-old worker at the Boa Fé rubber plantation, she'd been provoked by hunting dogs and menaced with a knife. Still, the phrase "severe bleeding in the left inguinal region" has a way of sticking with you.

"I'm, uh, not very fast," I said.

Alves wouldn't let me off. "You don't need to be fast. You must be like a hobbit. Very quiet," he said. "She will not attack. But when she's in the net, don't approach her." With that, he was gone.

Minutes passed. Trucks droned down BR-262. I crept through the undergrowth like Elmer Fudd on a wabbit hunt, fearing and hoping I'd see Evelyn's kielbasa-like snout poking above the scrub. I needn't have worried: a crash in the bushes indicated that Alves, the trained professional, had made first contact. By the time I caught up, Alves was doubled over, hands on thighs, net on the ground. And beside him, straining against the mesh, was Evelyn.

So alien are anteaters, so cobbled together from components with no real analog in Animalia, that it was hard to discern where Evelyn began. I experienced her as an amalgamation of parts: tubular, velvety snout; curved claws, as large and hard as gardening tools; fanned bustle of a tail, coarse as broomstraw; stunning swatch of black fur, trimmed in white, that ran down her flank like a racing stripe. The sight of her decaying compatriot had no more prepared me for her beauty than an art-history class readies you for the Louvre.

After Alves administered a tranquilizer and Evelyn's thrashing ceased, I clumsily scooped up her ungainly body. I cradled her elongated head with one hand, slipped the other beneath her bristly hindquarters. She emanated the not-unpleasant smell of a horse stable in need of mucking. In her tracking collar, which was more like a harness, she recalled a sled dog separated from her pack. I laid her on a tarp and stepped back. Alves and his colleagues worked with swift surety: duct-taping her formidable claws, cutting off her collar, shielding her snout with a sock. Unfurled, her purplish proboscis perfectly explained her taxonomic suborder: Vermilingua, worm tongue.

The next thirty minutes passed in a blizzard of data collection. There were heart rates to monitor, orifices to swab, blood samples to draw, feces to gather. Had she reproduced before? Her teats said yes. Was she pregnant? Her belly said no. Were Evelyn male, Alves would have harvested semen with a contraption dubbed the electro-ejaculator. "We would collect their souls if we could," he deadpanned.

The most revealing information, however, wasn't stored in Evelyn's body but in her collar, which had recorded her coordinates every twenty minutes for a year. By tracking anteaters' whereabouts, Anteaters and Highways could in theory pinpoint locations for wildlife crossings, as deer collars had helped American biologists site underpasses. But anteaters, vexatious to the last, frustrated facile analysis. "There are species that are easier to predict where roadkill will happen," Alves said. "For example, tapirs and capybaras are associated with water. But giant anteaters, when you see the map that we are generating, it seems that someone just put arbitrary points everywhere. They live in forests, they live in eucalyptus plantations, they live together with livestock."

If collaring anteaters has revealed anything, it's that bravery is deadly. More than 80 percent of anteaters who live near roads regularly cross them, and many meet predictable fates. Christoffer, the project's most audacious anteater, averaged two crossings every three days before he was struck. Pequi, a male named for a common *cerrado* fruit, was so acclimated to highways that he slept by their shoulders. "I knew he would be roadkill," Alves said. "Not Evelyn—she approaches the road but never crosses. Pequi wanted more from life." Like a tube-snouted Icarus, Pequi had refused confinement, only to be undone by his own exuberant questing.

Giant anteaters are conspicuous animals—big, bold, fond of loitering on treeless savanna. Their visibility creates the impression of abundance; one farmer told me he saw them daily. Yet giant anteaters are in rapid decline, and roadkill is among the main reasons. Although Anteaters and Highways had only been operating for two years at the time of my visit, its surveyors had found nearly six hundred dead anteaters, no doubt a severe underestimate of the real total. Their sluggish life history and wholehearted commitment to child-rearing only worsen matters. A single anteater pup gestates for six months, then clings for another six to his mother, splayed across her back like a saddle. The deaths of a few precious females and juveniles can slow the growth of an anteater population; in concert with other threats—habitat loss, pesticide poisoning, superstitious ranchers—they can extinguish one. "When you add other impacts, that can cause local extinction," Arnaud Desbiez, the biologist who founded Anteaters and Highways, told me later. "Roadkill is kind of the last straw."

Yet timidity is as dangerous as boldness. Anteaters, no less than mule deer and mountain lions, must roam to survive, and habitat fragmentation may pose a graver threat than roadkill itself. Giant anteaters thus find themselves in a bind: either cross roads and face death, like the impetuous Pequi, or restrict their movements to avoid them, like the prudent Evelyn. When she finally trotted off at the end of her ordeal, grunting in irritation, I couldn't help but pity her—still alive, but her world circumscribed by pavement.

And more pavement was coming. Less than a year before my visit, Brazil had elected as president Jair Bolsonaro, a far-right development fanatic

Female giant anteaters spend months rearing a single pup, which makes roadkill all the more devastating. *Alessandra Bertassoni*

whose platform included new Amazonian highways. No sooner had "Captain Chainsaw" assumed office than his infrastructure minister, whose own ominous sobriquet was "Paver General of the Republic," announced his intent to commission $27 billion in highway contracts. One bill introduced during my trip proposed eliminating environmental review on highway upgrades. Another reopened closed roads in national parks.*

Everywhere I went in Brazil, new highways were being built or existing ones paved and expanded. One afternoon I visited Flavio Gomes da Silva

---

* In 2022 Bolsonaro lost his reelection bid to Luiz Inácio Lula da Silva, Brazil's former president, who vowed to restore many of the environmental protections that his opponent had gutted. Yet even Lula wasn't averse to development. "We can't turn Amazonas state into a nature reserve," he said during his campaign. ". . . It's totally possible to address climate issues and to construct good roads."

Filho, the transportation secretary of Aquidauana, a city of fifty thousand souls at the *cerrado*'s edge. Caimans and anacondas swim the city's ponds, and anteaters occasionally break into its hotels. Silva Filho, jowly and voluble, struck me as a sincere advocate for Aquidauana's nature. In his office he flicked through cell-phone photos from "Yellow Month," a citywide campaign dedicated to driver safety. In one gruesome shot an actor rested his gashed forehead against a spiderwebbed windshield, the imagined result of veering to avoid an anteater at an unsafe speed. "We see the importance of both human lives and wildlife," Silva Filho insisted.

Despite his concern for animals, Silva Filho was cautiously excited about the new highways that Bolsonaro was planning for the *cerrado*. Chief among them was an extension that would connect Aquidauana with the Interoceanic Highway, a 1,600-mile artery through the Amazon. Silva Filho hoped the road would accommodate nature; on his desk, where another civil servant might display photos of his kids, he'd framed a picture of a wildlife overpass. Regardless, there was no denying that the new linkage would expedite travel, connect farmers to markets, and foster development. All told, Silva Filho said, it was "pretty good news for the city."

For the *cerrado* it was far grimmer. In the Amazon the Interoceanic Highway—dubbed "the most corrupt highway in the world" for the bribes that a construction company disbursed to secure the contract—had exacerbated deforestation from the day it was completed. Cornfields, papaya farms, and the murky wastewater ponds spewed by gold mines now clustered around it. And the Interoceanic was only the latest in a litany of injurious Amazonian roads. The story of the rainforest's unraveling is that of its highways, the "lines of penetration" through which logs poured out and agriculture flooded in. Today 160,000 road miles gouge the world's greatest tropical forest, most unauthorized or illegal.

For all of human history, roads have been instruments of conquest, at once ostentatious displays of imperial power and whips that bring unruly subjects to heel. King Leopold III subjugated the Congo by rousting riverside villages and relocating them along roads; the Dutch marshal Herman Willem Daendels drove the Great Mail Road across Java, killing thousands of Indonesian laborers, to ward off the British. The Amazon's highways, too, were meant to extend governmental control into a region perceived as lawless. In the 1960s Brazil's military dictatorship began to punch highways into the rainforest, the vanguard of a campaign to open it to development and subdue the Indigenous tribes who'd gardened it for millennia. Early Amazonian highways were malleable symbols, capable of shape-shifting to fit whichever goals, real or rhetorical, politicians needed them for. One moment they were weapons for fighting the Communists allegedly hiding within the forest, the next release valves that funneled slum-dwellers, poor farmers, and other "surplus population" into the hinterlands. The Amazon, the government declared, was "a land without people for a people without land"—and a grand new road, the Trans-Amazonian Highway, would make settlement possible. *Ocupar para não entregar*, went the military's mantra: occupy so as not to surrender.

The Amazon, however, wasn't easily occupied. Road crews floundered in its "green hell," and unpaved proto-highways devolved into impassable wallows. Trucks bogged down; bus passengers died in their seats. Aspirant farmers found thin soils and dysentery. The Trans-Amazonian came to be known as the Trans-Misery. "The government distributed rice seeds that failed," Philip Fearnside, an American biologist who moved to the highway's bleeding edge in 1973, told me. "The plagues of biting blackflies were just incredible." As he interviewed destitute farmers and analyzed infertile soils, Fearnside realized that the government schemers touting the Amazon's potential suffered from the "illusion of endlessness," the misapprehension that the forest was inexhaustible. In fact, it was withering fast. "In the beginning it sounded like you had howler monkeys right there in the house," Fearnside recalled wistfully. "They got fainter and fainter until they disappeared."

A question about roads, much debated by economists, is whether they cause deforestation or follow in its wake. Like the chicken-or-egg dilemma, the answer would seem circular: there's no sense clearing the forest without a highway to bring crops to market, and there's no sense building a highway until there's demand for transportation. But the Amazon is a clear point in favor of the road's primacy. Thousands of settlers poured into the forest along the Trans-Amazonian and other highways, then claimed land through the Lockean doctrine of "right of possession," which allowed colonists to gain title by lopping down trees, releasing cattle, and otherwise putting the rainforest to commercial use. By connecting remote forest tracts to international markets, highways ensured their claims would pay off. Where bulldozers plowed roads, land values doubled, tripled, rose tenfold. By the 1980s, the journalist Andrew Revkin reported, the Amazon had devolved into "a spreading grid of open space that, from the air, looked like ferns flattened between the pages of a book—main roads bristling with smaller feeder roads." Others said the forest resembled the skeleton of a monstrous fish, the highway its backbone, the auxiliary roads its ribs.

Like a fish, the Amazon was gutted. A natural phenomenon became unnaturally commonplace: "edge effects." The rainforest's floor is its dim basement, a dark, humid wilderness of columnar trunks and damp leaves. As roads gashed the forest, the corrosive forces of heat, light, and wind rushed in, and stricken trees dropped leaves to conserve water. The brittle branches and crispy leaf litter fueled groundfires, which opened more gaps, which killed more trees, which created more kindling, which ignited more fires—a fatal feedback loop that far transcended the dirt road that triggered it.

As edge effects eroded the rainforest, ecosystems fell into disarray. Among the most prominent victims were antbirds, hyper-specialized songbirds who flit behind voracious army ants and eat the panicked katydids, spiders, and scorpions the platoons flush. Antbirds are ant-following savants, capable of memorizing colony locations and communicating ants' whereabouts to other birds. One thing antbirds don't do, scientists discovered, is cross roads, even narrow dirt ones. Perhaps they feared the hawks that patrolled canopy gaps or distrusted the harsh novelty of the clearing, so unlike the interior's pleasant darkness. For deer it was traffic that imposed barriers; for antbirds forest edges formed the bars of their cage.

Even as antbirds declined, something counterintuitive was happening: the rainforest was becoming *more* diverse. At one research site nearly 140 new bird species turned up after clear-cuts. Yet it was a superficial diversity, augmented by opportunists, like rough-winged swallows and black vultures, that could have survived anywhere. By 1973 nonnative house sparrows had spread five hundred miles along one Brazilian highway, conquering the rainforest on humanity's coattails.

Other invaders were more pernicious. The Amazon's human colonists carried trillions of microscopic hitchhikers—dengue, herpes, Chagas, parasites. Mosquitoes bred in the stagnant craters gouged by highway equipment; by the early 1970s more than 10 percent of people along the Trans-Amazonian suffered from malaria. When maladies weren't entering forests via roads, they were emerging upon them. Near one highway, hantavirus sprang from soybean farms and cattle pastures as land conversion pressed people into ever-closer contact with rodents and their viral loads. This wasn't a unique event: where tropical environments change most dramatically, infectious diseases are likeliest to jump from wildlife into humans. (Nipah virus, to take one infamous example, emerged when deforestation displaced Malaysian fruit bats into orchards and farms.) Habitat loss not only forces wild animals into proximity with people; it stresses their bodies and thus causes them to shed viruses their immune systems once suppressed. In a 2021 article in the *Lancet*, the conservation biologist Gary Tabor and his colleagues described a concept they called "landscape immunity"—the notion that fewer zoonotic diseases spill from intact, diverse ecosystems than from fragmented, species-poor ones. And no force compromises immunity quite like a highway. "If you want to prevent pandemics, you have to think about roads," Tabor told me.*

In the Amazon, highway construction not only brought disease; it spelled genocide. Many Amazonian tribes experienced their first contacts with the outside world in the form of road workers. Sometimes these haphazard

---

* Once viruses infiltrate human bodies, infrastructure hastens their dispersal. The Chinese counties that suffered the worst SARS outbreaks, for instance, were those bisected by national highways. In *The Hot Zone*, Richard Preston noted that the Kinshasa Highway, the road that links the Democratic Republic of the Congo to Africa at large, was surfaced with crude gravel until the 1970s. When the highway was paved, traffic erupted and HIV with it. The AIDS pandemic emerged through "the transformation of a thread of dirt into a ribbon of tar."

introductions resulted in harmless misunderstandings—a squawking radio pierced by arrows, say. More often the outcome was tragic, as pathogens overwhelmed Indigenous immune systems. Within a year of the Trans-Amazonian's construction, nearly half the population of a tribe called the Parakanã had perished. In his memoir *The Falling Sky*, the Yanomami shaman Davi Kopenawa recalled learning of yet another highway proposal:

> Once again anger invaded my thought: "This white people's path is truly evil! The *xawarari* epidemic beings follow their machines and trucks on it. Will their hunger for human flesh really make all the rest of us die, one after another? Did they open this road to make the forest silent of our presence? To build their houses on the traces of our own after we disappear? Are these outsiders really evil beings to continue to mistreat us this way?"

As the world trickled in, Indigenous languages and customs were flattened by cultural bombardment. Anthropologists have found that linguistic diversity is lowest where South American road density is highest. Depression and suicide, too, haunt roads. "When the road is open, institutions arrive, the church arrives, merchants arrive, people that are not from the community arrive," the Peruvian anthropologist Rafael Mendoza told me. "Suddenly you realize that people don't like you because you're Indigenous. Women look at you like you're inferior because you don't speak Spanish well. Suddenly you realize that you are openly discriminated against."

Today, roadbuilding and land-clearing, the Amazon's scourges, still move in lockstep. Ninety-five percent of deforestation occurs near roads and rivers, and more than a third of the forest is considered "highly accessible." Illegal roads cut by gold prospectors have degraded hundreds of thousands of acres in Indigenous reservations. Philip Fearnside told me he especially dreaded the completion of BR-319, a road he'd branded the "highway to destruction." Although BR-319 was built in the 1970s to link the Amazon's industrial centers, traffic was light, and the road deteriorated into an orange morass, inadvertently sparing central Amazonia from deforestation. Bolsonaro, however, had declared his intent to pave and rebuild the highway, an announcement that immediately spurred illicit logging. The territories of sixty-three Indigenous tribes lay along the highway, yet the government hadn't consulted them, a violation of international convention. "Lots of things happen here that are not in accord with the law," Fearnside said. "And there are lots of laws on the books that are simply not enforced."

As Fearnside described the harm that upgraded Amazonian highways would inflict, I was struck, not for the first time, by road ecology's limitations. Wildlife crossings and fences might help anteaters and other creatures shuffle across new roads, but they were powerless to prevent the deforestation that would follow. In a tragic twist, wildlife passages could even become a form of greenwashing, a cynical tactic for laundering a road's environmental reputation. Some road ecologists have fretted that crossings, for all their benefits, may represent a form of "malicious restoration"—a fig leaf that conceals and rationalizes destruction. As one Brazilian ecologist put it, "Passages won't make any difference if we change the whole land use and destroy all the habitat."

Roads, any engineer would tell you, ostensibly exist to serve society. In truth, it's the other way around: once a road is built, our lives and landscapes bend to conform with it. Name an environmental problem, and it's exacerbated by the access that roads provide and the incentives they create. Roads are stalking horses for despoliation, the foot with which civilization holds ajar nature's door. Today edge habitats make up 70 percent of our planet's forests; they are no longer sporadic disruptions but the landscape itself. We inhabit a world as angular and broken as a corn maze, all edge and no heart. "The best thing you could do for the Amazon," the Brazilian scientist Eneas Salati once quipped, "is to bomb all the roads."

Rather than obliterating its highways, of course, Brazil, like most nations, is rapidly building new ones. As the infrastructure tsunami has washed ashore, the country's cadre of road ecologists has been forced to swim des-

perately to remain on its crest. From Mato Grosso do Sul, I flew to the city
of São Paulo to rendezvous with Fernanda Abra, one of the scientists trying to keep biodiversity afloat. "I don't miss one opportunity to talk about
roadkill," Abra told me over Skype before we met. "If they call me to talk
about roadkill at a baptism, I go."

The day after I arrived in São Paulo, Abra and I drove east in my rental
car, jolting forward in megalopolis traffic. Vendors weaved between lanes
to hawk popcorn and fruit, both exploiting and contributing to gridlock.
A capybara grazed a roadside canal. Capybaras, Abra said, were Brazil's
version of white-tailed deer, the species most often involved in dangerous
animal crashes. When I expressed surprise that a rodent, no matter its
unusual size, could imperil a driver's life, Abra explained that capybaras
were social creatures prone to blundering across highways in herds. "You
don't die when you roadkill one capybara—you die when you roadkill the
whole group," she said. In North America we consider such tragedies acts
of God, but Brazil, to its credit, lays the blame on road operators, who can
be sued for damages. This struck me as not only a legal innovation but
a philosophical one, a redirection of culpability from driver to engineer.
Installing fences and wildlife crossings was no less integral to public safety
than repaving lanes or keeping signals functional. Abra's best friend had
recently struck a fox and called her in tears. Abra advised her to report the
incident to the police. "She said, 'What? *I* killed the animal!'" Abra said.
"Her conscience is heavy; she doesn't want money. How many people think
like that? But this is a failure in their service!" Roadkill was not an accident
but a systemic crisis.

Soon the traffic thinned, and we found ourselves cruising through sunlit countryside. Toll booths interrupted our progress every few minutes,
the receipts accumulating in the center console like fallen leaves. The tolls
indicated that we'd pulled onto Tamoios Highway, one of São Paulo's many
private roads. In the mid-1990s the cash-strapped state had sold the operation of its decrepit highways to private concessionaires, under the condition
that they use toll revenues to upgrade them. The giveaway of a public asset
was not without critics, but there was little arguing that São Paulo's private

highways, smooth and broad, were far safer than Brazil's notorious public roads. Private highways also performed better from an ecological standpoint, since concessionaires, obligated to prevent roadkill, allocated some toll revenues to fences and wildlife crossings. We stopped at an underpass the Tamoios company had tasked Abra with monitoring and knelt to a muddle of tracks inscribed in the sandy floor. "This looks like a canid, because we have the nails," Abra said. "Ah, and this is puma. Here is *tamanduá*." Cars purred faintly overhead, travelers through a surface world far from the underpass's dim comforts.

Leaving the underpass to the anteaters, we continued to Tamoios headquarters, where Abra had arranged a meeting with the highway's operators. With a conspiratorial air, as though unveiling Oz, an official led us into a NASA-esque command center, in which a mosaic of screens relayed feeds from eighty-four roadside cameras. Technicians paged through maps, weather forecasts, and traffic reports, the only sound an electronic hum. "Everything that happens on the road—everything—we have a record," the official whispered. "Someone stopped because the tire is empty, or they have no gas, we have it written in the system." A fleet of tow trucks stood by to haul off breakdowns, and automated sensors ticketed speeders (I received a couple via email). "What we are doing here is organizing the environment," he added triumphantly. The company's proclivity for micromanagement extended to wildlife: it immediately dispatched crews to remove dead animals so that drivers wouldn't, say, strike the corpse of a six-hundred-pound tapir. Rather than landfilling these carcasses, Tamoios's workers stashed them in freezers for biologists' benefit. "It's super-sad, but when else would you have access to an ocelot—the brain, the stomach, the skin, the DNA?" Abra said.

We went into a back room, and Abra lifted a freezer's lid. Rigid bodies lay in an ignoble jumble: opossums, foxes, a freezer-burnt capybara. Abra snapped on plastic gloves and laid the carcasses on the concrete floor, ice chips flying. At last she reached the ocelot, his body curled into a hard comma. We gazed in wonder, hypnotized by the celestial black of his spots. I asked Abra how to say roadkill in Portuguese. *Atropelamento*, she said—

literally, "run over." I practiced it a few times, the syllables tripping over each other, a pretty word for an ugly event. Like roadkill, *atropelamento* had acquired nonliteral significance. "If we have the same goals and I do it faster, you can also say *atropelamento*," Abra said: one person crushed by another's ambition. It was a word that signified conflict and conquest, the victory of the strong over the weak, the swift over the slow. It seemed apt.

Brazilian highways are both the source of *atropelamento* and frontiers of power struggle in their own right. The country's infrastructure boom owes much to China, which, between 2010 and 2020, invested $66 billion in Brazil, primarily to secure access to crops and minerals. China has built Amazonian ports, snatched up concrete companies, and dumped money into transnational railroads. Some 80 percent of Brazil's soy exports end up in China, primarily to feed cows and chickens. The changes I'd seen in the *cerrado*—all those soyfields replacing scrubland—were driven largely by the Chinese consumer class's burgeoning, Americanized taste for meat.

Brazil is one of many countries being remade by Chinese investment. In 2013 China launched the Belt and Road Initiative (BRI), an incomprehensibly vast welter of highways, railroads, shipping lanes, power lines, and other linkages that would bind about seventy countries in Asia, Africa, Europe, and South America—the largest construction project in human history. (Although Brazil isn't an official BRI member, it's received nearly half of China's Latin American investments and is thus a *de facto* one.) The purpose of this New Silk Road is both to connect China with resources and to use the soft power of infrastructure to wrest less-developed trading partners away from the United States and Western Europe. Rather than a collection of discrete projects, the *New York Times* observed, the Belt and Road Initiative might be considered a "vaguely visible hand guiding all the interlocking developments in infrastructure, energy and trade where China plays any kind of role."

If the BRI is a gargantuan hand, roads are its fingers. The implicit function of the initiative is to assert Chinese influence in lightly developed countries; almost by definition, that means countries that still support large, intact habitats. More than 250 threatened species and 1,500 key biodiversity strongholds lie squarely in the BRI's path. In Kazakhstan and Mongolia the BRI's highways will disrupt the twice-yearly peregrinations of the saiga antelope; in Russia's far east they'll wend through pine forests that shelter the Amur tiger. Several Chinese-funded "development corridors" stand to literally bisect Kenya's national parks. "There's not much you can do," a Kenyan ecologist named Peter Kibobi told me. "You try to voice your concerns, but the government has already decided."*

And yet it's hard to condemn the Belt and Road outright. Many of its members are blighted by poor infrastructure; the dilapidated highways that ramble across sub-Saharan countries like Tanzania and Namibia, for example, prevent farmers from receiving fertilizer, depress crop yields, and worsen hunger. "The bottom line is that developing nations need to develop, both socially and economically," William Laurance, the road ecologist who coined the phrase "infrastructure tsunami," told me. "You can't even get in the game, you can't even enter the debate, if you brand yourself as anti-development." Instead of reflexively opposing every new highway, Laurance has proposed building the right roads in the right spatial patterns. Clustering roads together rather than letting them sprawl without limit can concentrate their impacts and spare land. Improving an existing road is vastly preferable to making the first cut in a roadless forest. A well-made highway can even serve as a "magnet," drawing farmers away from undamaged habitats and toward areas already under cultivation. Above all, Laurance has argued that the future of roads should be planned rather than spontaneous, that economic benefits and ecological costs be weighed and

---

* Not all countries crave the BRI's largesse. China's investments can be "debt traps": nations accept loans they can't repay, then default and cede control of their infrastructure to Beijing. In 2018 Malaysia's president, fearful of this "new version of colonialism," canceled projects worth almost $23 billion.

balanced, that roads become the products of sober accounting instead of happenstance. Rather than permitting roads to be civilization's masters, guiding our land-use patterns as they relentlessly self-replicate, it's time that we mastered our roads.

This, of course, will not be easy. Billions of yuan are an enticing carrot, and conservationists carry a flimsy stick. Most Belt and Road projects are funded by Chinese state banks that, unlike international lenders such as the World Bank, don't require recipients to protect biodiversity. Road ecologists are stuck wielding their own soft power: unable to prevent new projects altogether but still capable of nudging them in directions that, if not quite "right," are at least less wrong. "The way I see it, many of these roads are going to be built whether we like it or not," an Australian road ecologist named Rodney van der Ree told me. When we spoke, he'd recently helped persuade officials in Myanmar, another BRI member, to add underpasses to a planned highway that would divide the habitat of leopards, tigers, and elephants. "From a biodiversity standpoint, they shouldn't build the road through that area at all, but at least it's a better outcome than it was," he said. Other conservationists told me they've attempted to exploit the pride of vain government officials who hope to demonstrate their greatness through grandiose infrastructure. If you're determined to build a highway, why not build one that earns plaudits from the international community rather than condemnation?

The history of road ecology has, to date, been more reactionary than anticipatory. Not until America's interstates throttled deer migrations did we install underpasses; not until logging roads scarified our national forests did we notice the damage. The field has described the world better than it has shaped it: most road-ecology studies, van der Ree once glumly noted, "appear to have little influence on road planning and design." This isn't ecologists' fault; blame instead the agencies, planners, and engineers who have failed to solicit their input. If we expect developing countries flush with BRI funding to construct nature-sensitive highways out of the kindness of ministers' hearts, we'll be sorely disappointed. But nor are those nations beyond influence. The infrastructure tsunami, like the autonomous vehicle revolution, is an epochal transformation that we still have the power to shape.

"We have a government that doesn't care much about nature in general, but I'm an optimistic person," Fernanda Abra told me in São Paulo. "If you don't ask for something, they're not going to do it."

I spent my last few days in Brazil under Abra's wing, checking out underpasses and canopy bridges around São Paulo. Like all good road ecologists, Abra was hyperalert to the highway's wonders. She snipped whiskers from a roadkilled jaguarundi for genetic analysis one day, shepherded a disoriented tapir off the shoulder the next. One morning I asked her if she'd ever hit an animal. Never, she claimed, unless you counted the owl that smacked into her window, which she didn't because, technically, that bird hit her. Thus it was inevitable that a few hours later her tires thumped over the tail of a *caninana*, a black-and-yellow snake lounging on the sun-dappled road.

"Oh, fuck me, fuck me," Abra muttered, aghast. She pulled over and leapt out. With one hand she shooed the snake off the road; with the other she held up traffic like a crossing guard. The *caninana* slithered off.

On my last afternoon in São Paulo, Abra took me to see one of Brazil's most ingenious roads. We drove for hours through cornfields and sugarcane, the land crimping as we escaped the city's gravity. We slowed to pass horseback riders, their faces shadowed by straw hats. This, Abra said, was *onde o vento faz a curva*, an idiom that meant, roughly, *where the wind makes the curve*—the edge of the world.

At last we came to Carlos Botelho State Park, a ragged parcel of Atlantic forest that had avoided Portuguese axes centuries earlier. Like many of Latin America's protected areas—and many of the United States'—Carlos Botelho is gashed by a highway: SP-139, the road that connects São Paulo to points west. Rather than allowing SP-139 to rend Carlos Botelho entirely in two, however, the state had taken a progressive approach. As we neared the park, a sign warned that SP-139 was closed between 8:00 p.m. and 6:00 a.m., when animals were most active. Police cars, fire trucks, and ambulances could still drive through after dark, but otherwise the park belonged to ocelots and jaguarundis. There'd been some complaining when the closure was announced, but people had acclimated, and roadkill had plunged.

The nocturnal closure wasn't the park's only masterstroke. One of road ecology's great riddles is how to get drivers to slow down. Since Dayton Stoner's day, biologists have known that the simplest way to avoid collisions is to reduce vehicle speed. The problem is that motorists habitually ignore posted limits. Instead, we drive at a road's *design speed*, which is basically how fast the road itself wants you to go. Highways with gentle curves, wide lanes, and long sightlines have high design speeds; how often have you glanced at your speedometer on an open straightaway and been shocked to see the needle pushing ninety? Researchers who have experimentally lowered speed limits have found that drivers barely adjust and that animal collisions continue apace. Our highways and cars want us to go fast, and we're helpless to resist. "The virtue of slowness," the historian Gary Kroll lamented, "seems like an impossible dream."

In Brazil, however, the dream of slowness lived on. On entering Carlos Botelho Park, we found that SP-139 was curvaceous and undulating, its bed rising and falling in waves, like a gentle roller coaster. Signs warned of "Rodovia Sinuosa." We were prevented from speeding not only by law but by design. Practically every highway I'd ever driven on, I realized, was built to facilitate swift and seamless automobility. SP-139, by contrast, deliberately transgressed the conventional ideals of engineering, frustrating human users in service to wild animals.

We crept through the park, conscious of gathering dusk. Golden sun seeped through foliage. The roadbed's sine wave conveyed the impression that we were at sea. Monkey bridges dangled overhead like ziplines. Scrawny native fowl skittered ahead of us, heads pumping. Abra braked to inspect some tapir feces; soon we saw their globular scat everywhere. They were using the highway at night, Abra speculated. Without cars, what was a highway but an animal trail?

"In the world of road ecology, Brazil is a baby. We are a super-baby,"

Abra had said earlier. Yet SP-139 struck me as more radical than nearly any American road. Where had we closed a highway after dark or engineered one that prevented drivers from exceeding twenty-five miles per hour? Even our most dramatic interventions—our sweeping overpasses, our cavernous underpasses—tried to accommodate nature without sacrificing an iota of human mobility. SP-139 dared to inconvenience its passengers. Abra opened the windows, turned off the radio, and let the forest wash through the car, taking each bend slowly, following the earth's curve like the wind itself. On a planet preoccupied with speed, it felt good to go slow.

13

REPARATIONS

*Can we undo the damage highways have
wreaked on cities and human lives?*

Like anteaters and butterflies, we humans, too, live in the thrall of roads. Highways afflict natural habitats and urban ones in analogous ways, a parallel understood by their earliest critics. "[The engineer] does not hesitate to lay waste to woods, streams, parks, and human neighborhoods in order to carry his roads straight to their supposed destination," the historian Lewis Mumford observed in 1958. Most obviously, roads inflict what an ecologist would call direct mortality: every day some 3,600 people around the world perish in vehicle crashes, which means that in the thirty seconds or so it takes you to read this paragraph, at least one driver, passenger, cyclist, or pedestrian has lost her life. And then there are the subtler harms. The same engine noise that chases off birds ramps up our own cortisol production. The same pavement that sweeps motor oil into Puget Sound absorbs and discharges solar radiation, forming urban heat islands that make hot summers deadly. The same salt that's turning midwestern lakes into estuaries corroded pipes in Flint, Michigan, and released lead into its water supply. The same traffic that cleaves bobcat populations contributes to our social isolation; the same freeways that thwart Los Angeles's cougars make road-raging commuters more prone to committing domestic violence. Roads so transmute our lives that one scholar has described *Homo sapiens* as an "infrastructure species"—an organism whose essential trait is that it is shaped by its own construction.

Although nearly all humans are touched by roads, we're not touched equally. According to one Centers for Disease Control and Prevention study, Black people are 40 percent more likely to live near a highway than whites, Latinos are 60 percent more likely, and Asian Americans are 75 percent more likely. The noxious by-products of internal combustion—sulfur dioxide, nitrogen oxides, volatile organic compounds, and the vile gunk known as diesel particulate matter—are inordinately absorbed by Black and brown bodies. In Maryland, African American communities face higher exposure to airborne carcinogens; in the Bronx, a borough whose mostly Black and Latino populace is bracketed by three expressways, residents are three times more likely to die of asthma than the national average. So ghastly are air pollution's consequences, and so inequitable, that one journalist placed it on "the same ethical terrain as war, slavery, and genocide."

The asymmetry of air pollution isn't happenstance; instead, it is often the legacy of nakedly racist planners. The enormous mid-century freeways that cleave many American cities were designed, at least initially, with reasonable intent. Crooked, rough-hewn streets had become clotted with traffic; meanwhile, autocentric suburbs were popping up outside every metropolis. It seemed only logical to connect these two habitats, the cities and the suburbs, with new highways—safer, straighter, and wider than anything built before. Yet ulterior motives shaped urban freeways as well. By the 1930s discriminatory housing policies and official neglect had produced scores of segregated, dilapidated neighborhoods whose primarily Black residents couldn't receive loans to improve their properties or buy new ones. City planners fixated on "urban renewal," the demolition of communities that offended governmental sensibilities. If Black neighborhoods were nails, freeways were a convenient hammer. As early as 1938, Henry Wallace, then secretary of agriculture, claimed that urban highways could induce "the elimination of unsightly and unsanitary districts." The forces of Motordom seized on the idea; General Motors, for one, proposed using freeways to displace "undesirable slum areas." In 1956 the Federal Aid Highway Act, the law that authorized the interstates, placed a fearsome new cudgel into the hands of would-be renewers, who leapt at the chance to appease suburbanites and wipe out perceived blight in one stroke. The presence of Black people was enough to invite the bulldozers.

While highways shattered natural ecosystems inadvertently, their fragmentary effects were deliberately weaponized in cities. In St. Paul, I-94 devastated Rondo, a thriving middle-class neighborhood. In Miami, I-95 drove out thirty thousand residents of Overtown, a cultural hub known as the Harlem of the South. Interstate 10 came for New Orleans's Tremé, I-40 for North Nashville, I-244 for Tulsa's Greenwood. Freeways were the erasers with which cities rubbed out Black communities and the walls with which they partitioned them from white ones. Interstate 20's meandering route through Atlanta only makes sense when you consider that it established "the boundary between the white and Negro communities," as that city's mayor admitted. Interstate 85's course through Montgomery was plotted by Alabama's "foremost segregation leader," who plowed the freeway through the neighborhood of Oak Park to punish the civil rights leaders who called it home.

The national purge was breathtakingly swift and heartrendingly thorough. When John Williams, a Black journalist from Syracuse, New York, visited his childhood haunts in 1964, he found that a new interstate had obliterated the neighborhood. "You hear machines tearing down or building up; girders mount the low horizons," Williams wrote. "Yellow bulldozers groan back and forth making sure that every vestige of the unhappy Negro, who, toward the end, alone inhabited the Ward, is crushed out of sight, out of memory."

The freeway that expunged Williams's home is known as I-81, and, like all odd-numbered interstates, it runs south to north, from Tennessee to New York's border with Ontario. Along the way it passes through Syracuse or, more properly, passes over it atop a colossal, sun-blotting viaduct that many locals simply call the Bridge. It was this "boondoggle of concrete and steel" with which officials brushed away Syracuse's largest Black neighborhood in the 1960s and that today divides the city like a border wall. "Our parents didn't have the education to understand exactly what was happening," a Syracusan named Charlie Pierce-El told me when I visited the city. "In that era it destroyed us."

I met Pierce-El on a fall evening at Syracuse's South Side Community Center, a red folk-Victorian that housed the cramped offices of a neighborhood newspaper and some antediluvian desktop computers. He stood on the porch in dashiki and kufi cap and waved me inside to sit at a low plastic table. His thin white mustache looked to have been painted on with a fine brush. Pierce-El was born in 1946, one of twelve siblings. His parents, a mechanic and a dry cleaner, moved from Georgia, two of the six million African Americans who fled the South between 1915 and 1970. They'd gravitated to Syracuse, which once served as the Underground Railroad's "great central depot" and retained a reputation as a Black-friendly outpost. The Pierce-Els settled in the 15th Ward, the neighborhood that came to host 90 percent of the city's Black population. The 15th was the kind of place where you half-lived with your friends' parents and your parents' friends, where you could walk into a cousin's house down the block and smell garlic frying to ward off germs. The Pierce-Els kept a bountiful garden; an aunt supplied a chicken from her coop on Sundays. Jobs were abundant: the 15th Ward boasted cafés, canneries, breweries, a potato-chip factory, a nearby auto-parts factory. "We had bakeries just about on every other block," Pierce-El said. "We had about fourteen tailor shops." A young Sammy Davis Jr. lived behind a barbershop curtain; James Brown played the theaters. Pierce-El earned cash shining shoes at the restaurants, hotels, and bowling alleys owned by a local operator named Patsy Italiano. "I outraised my mom on a payday," Pierce-El said.

Pierce-El didn't know it, but the 15th Ward was doomed. In 1933, with America sagging beneath the Great Depression's weight, President Franklin Roosevelt had created a federal agency called the Home Owners' Loan Corporation, which would give government-backed mortgages to families facing foreclosure. The agency's purpose was munificent, but not all

Americans benefited from it. Federal examiners, worried that their loans wouldn't be repaid, graded lending risk in dozens of American cities using a simple color-code, a childish system with sinister repercussions. Affluent blocks, where homeowners could easily secure loans, were colored green. Blue and yellow areas were a couple of tiers down. Worst off were the red neighborhoods, where the government considered loans "hazardous" and whose residents had little hope of obtaining one. Nothing was a surer ticket for a red grade than a "population almost entirely Negro." When the loan corporation evaluated Syracuse, the 15th Ward was among the redlined neighborhoods.

As Charles Pierce-El came of age in the 1950s, the 15th Ward's community was as tightly knit as ever. But its physical context began to deteriorate. Barred from homeownership themselves, residents fell victim to absentee and predatory landlords. Roofs leaked; defective wiring sparked fires. "Literally thousands of roaches seethed over the walls . . . shrimp-like feelers moving back and forth," wrote one reporter with a shudder. The interstate highway system presented an opportunity to clear the deck. After Syracuse's mayor bridled against a proposal to route an elevated freeway through downtown, state engineers suggested a compromise: they would bump the freeway east, sparing the urban core and instead boring through the 15th Ward, the largely Black neighborhood targeted for renewal anyway. By 1961 the highway's path was settled: I-81 would rise, and the 15th would fall.

Syracusans arose in rebellion. Black activists marched on City Hall and pledged the "direct intervention of our physical selves" to halt construction. This was the era of the Freeway Revolts, national protests that briefly united Black Panthers, Haight-Ashbury hippies, and high-society garden clubs. In New York, Jane Jacobs vanquished the Lower Manhattan Expressway, the ten-lane monstrosity that Robert Moses had vowed to drill through Soho. San Francisco's revolutionaries managed to lop off several arms of the "cement octopus" with which state officials intended to wrap the city. Yet the movement, like the freeways, cleaved along racial lines. (There's a reason that the revolt triumphed in Beverly Hills and failed in Detroit.)

Syracuse's white residents were mostly indifferent to the 15th Ward's demolition. "It would be all right with me if they'd take it all," declared one local.

In the end the outcome was foreordained. As the viaduct proceeded in
phases throughout the 1960s, the 15th Ward was razed. (Although the state
compensated homeowners, it rarely paid market value.) Among the evicted
families were the Pierce-Els, who resettled in a suburb. Pierce-El and his
siblings were the only Black kids at their new school, and their classmates
didn't let them forget it. "I literally had to fight for months straight," Pierce-
El said. "That was the worst part of my life."

In the decades that followed, the viaduct became so deeply embedded
in Syracuse's fabric that people ceased to notice it. Asthma ran epidemic
at the Pioneer Homes, the public-housing project next to the freeway. "I
would use inhalers, I was on an albuterol machine, I was taking steroids. I
was in the hospital three to four times a month," a Pioneer resident named
Ryedell Davis told me. "I always thought it was something normal." As I-81
sickened the bodies in its ambit, it infected the city itself. Like roads everywhere, the interstate both connected and divided the landscape it crossed.
For suburbs like Cicero and Clay, it was a lifeline that linked white commuters to Syracuse's universities and hospitals. Meanwhile, the city's south
side, where Pierce-El and many other Black people ended up, withered. The
viaduct, a physical and psychological barrier, cut off Southsiders from the
"eds and meds" and funneled traffic away from the neighborhood, gradually strangling those businesses that had survived its initial incursion.

By the time I visited Pierce-El, a grim list of superlatives had attached
itself to Syracuse. It was among the country's most segregated cities, with
the worst concentration of Black and Latino poverty and sky-high rates of
asthma and lead poisoning. Interstate 81 wasn't the only culprit, but it was
the most conspicuous one, a fungus whose threads crept into every corner
of civic life. Community activists linked its pollution with low test scores
at nearby public schools and the demise of the 15th Ward's grocery stores
with obesity and diabetes. Pierce-El likened the interstate to the infamous
Tuskegee study, an appalling experiment in which researchers withheld
penicillin from hundreds of Black men with syphilis. The viaduct, like

Like so many urban freeways, Interstate 81 permanently transformed the city through which it ran. N. Scott Trimble

Tuskegee, was a cruel manipulation of Black lives, racism hiding behind a veneer of rationality.

"I didn't know it at the time," Pierce-El said, "but this whole thing was a test study."

Like most visitors to Syracuse, I arrived via I-81. I'd followed the highway north through rolling hardwood forest, autumn deepening with the rising latitude, the land shading from green to gold. Syracuse, when it arrived, felt sudden, with little southern sprawl to herald it, and I passed onto the viaduct before I truly realized it. Housing projects and parking lots unfolded below me, cars and people made minuscule by height. This was how the interstate's champions had intended travelers to experience cities: "Planing over the land at tree-top level—the roofs of houses below you—almost like flying," as one architect wrote. Cities weren't places to live but obstacles to be surmounted, the equivalent of flyover country.

I was keen to experience the viaduct from a perspective its designers hadn't much considered: the ground. One afternoon I solicited a tour from a

Syracuse city councilor named Joe Driscoll. He led me through traffic across Erie Boulevard, the former course of the Erie Canal, and into the viaduct's shadow. Although it was a bright fall day, the light was low and wintry beneath the concrete eclipse. The traffic noise was thunderous and multi-layered: it whined, thumped, hissed, roared. "Here it is," Driscoll shouted, spreading his arms like a symphony conductor. "The fuckin' debacle."

We turned south, plodding through urban flotsam: pigeon feathers, broken glass, Styrofoam clamshells, hubcaps. Rusty divots showed through the viaduct's peeling green paint; water damage seamed its pillars. Its underbelly was a dead zone of oily puddles and vacant pavement, a bleak milieu that could backdrop a zombie flick. We passed parking lots encircled by chain link, boarded-up gas stations, and a derelict garage. It occurred to me that the highway had created a landscape fit only for cars, a place that accommodated our machines but not us; that infrastructure duplicated itself like a malignant artificial intelligence.

"If you think about it, since we started walking, we haven't really seen anybody," Driscoll said. "We've seen people in *cars*. But we haven't seen any *people*."

We walked on—past towering hospitals and rundown churches, housing projects and campus buildings, poverty and institutional wealth uneasily juxtaposed. Driscoll, who's white, had a foot in Syracuse's many abutting worlds. He cherished its diversity, its stitched-togetherness, the way it mashed up and metabolized history. "If the city was a dog, it definitely wouldn't be a golden retriever. It would be, like, some kind of weird-looking pug," he said with affection. In 2015, after spending years making music abroad, Driscoll had moved home and thrown himself into city politics— organizing fundraisers, attending council meetings, and eventually winning office himself. Syracuse, he realized, wasn't the bastion of racial and socio-economic harmony he remembered from his youth. The city was segregated, poor, and balkanized. One report ranked it among the ten worst American cities for Black people. "There were all these historical wrongs that had been done," Driscoll said. And they all seemed to originate from one structure: I-81. "The more I researched it, the more outraged I became," he said.

Fortunately, the viaduct's tyrannical reign was drawing to a close. Steel bars peeked through its deck, and concrete sloughed from its pillars. Structural features that had been acceptable in the 1960s—tight S-curves, a lack of shoulders—now violated safety standards. The viaduct would have to be replaced, but with what? As the New York State Department of Transportation saw it, there were two options. The first was to build an even wider, taller, spiffier viaduct, a gargantuan skyway that would entrench the status quo for a century to come. The second was a radical urban overhaul, known as the Community Grid, that would free downtown Syracuse from the interstate's shackles. Under the Grid proposal, more than a mile of the existing viaduct would be torn down and supplanted by a street-level boulevard, graced with sidewalks and bike lanes and storefronts, that one advocate called the "the Champs-Elysees of Central New York." Cars bound for downtown would be dispersed through surface streets, while I-81's through-traffic would be shunted around the city rather than over it. If the viaduct had plunged a dagger into Syracuse's heart, the Community Grid would withdraw the blade.

To Driscoll the decision seemed obvious. His constituents, particularly his district's Black members, overwhelmingly supported the viaduct's removal. But the Grid's advocates faced staunch, well-heeled opposition from the suburbs, which had lined up behind a new viaduct. The interstate had spawned the 'burbs, after all; it seemed inconceivable that bedroom communities could survive without their infrastructural umbilicus. To many beleaguered Southsiders, it was a foregone conclusion that the skyway would prevail. Highways always won in America; surely one would again.

Thus it came as a shock when the state declared in 2019 that, after years of analysis, it favored the Grid. The decision heralded one of the largest urban makeovers in American history, a $1.9 billion reconstruction that would transform Syracuse's face. New York bigwigs like Senator Charles Schumer touted the plan, and U.S. Secretary of Transportation Pete Buttigieg called it a chance to "learn from our past and do it better." Almost overnight the Grid went from a political impossibility to a near certainty, and Driscoll still seemed incredulous. "This city has a horrible history of getting really simple problems profoundly wrong," he said. "When the state announced that the Grid was the preferred alternative, it was like—wait, *what?*"

Yet Driscoll was not entirely upbeat about Syracuse's future. Tearing down the viaduct was an essential precondition for its flourishing, but it wasn't sufficient by itself. Interstate 81 had so thoroughly reshaped Syracuse that its removal, too, would wreak upheaval. The viaduct's destruction would effectively birth up to eighteen acres of new land, unveiling real estate it had smothered decades earlier. Absent intervention, it was easy to imagine the former 15th Ward being colonized by fast-casual food chains, luxury condos, and urban development's other weedy pioneer species. "I'm realistic enough to know that it won't be shiny happy people holding hands and dancing in a circle after we knock down this barrier," Driscoll said.

And it wasn't just the land immediately beneath the viaduct at issue. Replacing a hideous, polluting freeway with a pedestrian-friendly boulevard, many locals feared, would shock property values around the city, accelerate gentrification, and precipitate what one columnist called "Negro Removal 2.0." You could yank a knife from a patient's heart and still watch him bleed to death.

The modern era of freeway removal began at 5:04 p.m., Pacific Daylight Time, on October 17, 1989, with the rupture of a fault six miles below California's Santa Cruz Mountains. The Loma Prieta earthquake, magnitude 6.9, touched off thousands of landslides around the Bay Area, shook buildings to rubble, and inflicted nearly $2 billion in damage to transportation systems. It reserved its worst devastation for the Cypress Viaduct, a double-decker freeway built on loose clay that carried I-880 over West Oakland. "The pavement started to move," recalled one awed trucker. "I had waves of asphalt come up over my windshield. . . . It was like a Disney ride." The viaduct's support columns burst, and its upper deck plunged onto its lower, crushing forty-two people beneath thousands of tons of concrete.

The freeway's collapse, however horrific, was also a *deus ex machina*. Like I-81 in Syracuse, the Cypress Viaduct had been an instrument of segregation. Its construction in the 1950s displaced hundreds of Black and Latino families, cut off thousands more from downtown, and coughed pollution across West Oakland. After the earthquake shook it to the ground, West Oaklanders mobilized to prevent California's transportation department from rebuilding it. To the community's satisfaction, the state agreed to reroute the freeway through an industrial corner of the city and replace the viaduct with a four-lane boulevard, called the Mandela Parkway, complete with bike lanes, walking paths, and acorn-shaped streetlights that paid homage to Oakland's namesake tree. "We now have sunshine, not shadows, grass not gas, trees not trucks, walking not ducking and butterflies not garbage flies," one Oaklander declared when the boulevard opened in 2005. Nelson Mandela himself gave it his blessing.

Other cities followed Oakland's lead. Milwaukee razed the Park East Freeway, Seattle the Alaskan Way Viaduct, and Rochester its Inner Loop, a depressed stretch of I-490 known as "the moat." By 2021 the Congress for the New Urbanism had tallied more than thirty prospective freeway teardowns in nearly as many cities. Like road ecology, however, urban freeway removal is a young science, and its early history is rife with surprises. In Oakland nitrogen oxide and soot pollution plummeted after the Cypress Viaduct was obliterated. But property values skyrocketed—who wouldn't want to live by a gorgeous boulevard adorned with acorn streetlights?—and the neighborhood's Black population fell by nearly 30 percent. It was a classic case of "green gentrification," whereby restoration projects push out the people they're intended to help. In Syracuse, Black residents were nervous that history would repeat itself. "They want to rebuild a city that's not for you," a community advocate named David Rufus told me. "They want one- and two-bedroom apartments when some of these families need three or four bedrooms. What happens to our churches? What happens to our businesses, those few that we have? They did this to us once before."

Rufus, whom I met on the porch of his late mother's house, had grown up in the viaduct's aftermath. "We saw a lot of things disappear," he told me as we swayed in rocking chairs. "Big M Market, Victory Market, Ebony Market, grocery stores, furniture stores. Victorian houses getting torn down. If you looked over the neighborhood it would look like somebody had dropped bombs on it." Rufus ended up working for the housing authority; his tenants often asked him to repaint their apartments, not realizing it was I-81's foul breath that chronically smudged their walls. When the state first floated the idea of tearing down the viaduct, however, Rufus felt disillusioned. He yearned to see it destroyed, yet he doubted that officials would include people like him in the complex process of reconfiguring his city. "There's so much to deal with that people become overwhelmed, and then the powers that be have the opportunity to swoop in and take you out," he said.

In 2020, Rufus had joined the New York Civil Liberties Union as a community liaison, soliciting Black Syracusans' input and representing their interests in meetings. He and his fellow advocates had influenced the viaduct's removal in a number of ways. They'd insisted on some design changes to the Community Grid: a busy roundabout near an elementary school, for example, would have to be relocated. But their most important asks weren't about physical infrastructure. In one report the Civil Liberties Union called for training and apprenticeship programs for Black construction workers and incentives that would encourage contractors to hire locals. The group also proposed that, instead of selling the acreage under the viaduct to the highest bidder, the city give it to a land trust, which would grant the community legal authority over the area and ward off gentrification.

Rufus and his colleagues were arguing for remediating the legacy of institutional racism—in a word, for reparations. The notion of repaying Black people for historical wrongs is an ancient one: in 1783 a freedwoman named Belinda Hall received fifteen pounds and twelve shillings, paid by an estate whose "immense wealth [had been] . . . augmented by her servitude." Over the following centuries a litany of horrific practices—redlining, educational disparities, predatory lending—exacerbated the wealth gap between Black and white Americans and strengthened the rationale for a compensatory system. Urban freeways represented another form of theft. When the state of New York forced out Black families to make way for I-81, it robbed them of the opportunity to grow their money through homeownership, producing what the Civil Liberties Union called a "generational loss of wealth accumulation." No one knew how much Black Syracusans had lost, but the number was undoubtedly immense. When I-94 punched through St. Paul in the 1960s, it cost the community of Rondo $157 million in home equity—enough missing wealth to purchase a college degree for every Black child in the county.*

"People want to run away from the word *reparations*," Rufus said. But reparations didn't necessarily mean writing a check for every resident. Perhaps it meant giving the descendants of 15th Ward evictees first shot at buying a home. Or maybe it meant providing seed money for Black-owned grocery stores, restaurants, and other businesses. "If you can create a budget that spends $1.9 billion to take a highway down, then you can reinvest in that community," Rufus said.

I asked Rufus to describe his vision for the post-viaduct era. He rhapsodized about mixed-use buildings—mom-and-pop commercial space on the bottom floors, apartments on the top—and a diversity of housing that accommodated everyone from young professionals to multigenerational families. "I see myself in twenty years being able to go down Salina Street to the grocery store, then walk down to State Street to get my hair cut at the barbershop," he said. "I see myself going over to Almond Street and visiting family and friends. If I want to go to church, I can walk to Tucker Missionary Baptist. We deserve those things that make this place livable, workable, lovable, and a place to worship."

As Rufus spoke, it occurred to me that what he was describing was connectivity. Humans, like virtually all animals, thrive by moving between discrete locales: we obtain food in different places than we seek shelter; we court our mates in different places than we attend to our hygiene. The great promise of automobility was that it would allow people to move smoothly

---

* Advocates in the Twin Cities have proposed a plan called Reconnect Rondo, a massive land bridge that would span I-94. Although other freeways have been "capped" by parks, the Rondo bridge would be covered not only with green space but with Black-owned businesses, homes, and cultural centers—a twenty-one-acre community woven from *terra nova*.

between these habitats, but in Syracuse, as in so many other cities, roads had made humans less free. The purpose of wildlife crossings is to create landscapes that are navigable by hoof or paw; the purpose of freeway obliteration is to allow people to walk through their communities. Road ecologists and urban advocates are engaged in the same epic project: creating a world that's amenable to feet.

The notion of reparations tends to attract knee-jerk outrage from all the predictable sources, and the idea of compensating Black people for freeways' harms is no different. When, for example, Buttigieg observed that "racism [was] physically built into some of our highways," conservative politicians feigned perplexity. How could a material as bland as concrete be guilty of a charge as explosive as racism? "To me," Florida governor Ron DeSantis scoffed, "a road's a road."

Ahistorical though DeSantis's comments may be, even the champions of freeway removal acknowledge that it's tough to comprehend the legacy of our roads. Just as road ecology pushes us to expand our view of how highways distort nature, we need to broaden our understanding of how they affect humans. Interstate 81 and similar freeways transformed some communities and eliminated others; these outcomes were deliberate, not random. The recognition that infrastructure has always been a force of social engineering compels us to rethink what a transportation project *is*, and to reimagine the scale of the remedies. As David Rufus put it, "Engineering is not just highways and roads."

And removing the viaduct wouldn't just liberate Syracuse from its oppressive shadow: it was also a chance to unburden the city of cars, period. For years global cities have been attempting to free their downtowns from vehicles; Oslo, for example, aims to clear cars from its core by jacking up tolls, building tram lines, and compelling businesses to deliver goods via electric bike. Nobody expected a small Rust Belt city to go full Scandinavia, but reorienting Syracuse around pedestrianism and mass transit was no less a social-justice issue than the viaduct demolition. As in most metropolises,

Black and Latino Syracusans were less likely to own a car and more likely to rely on buses than white residents. To rebuild the kind of urban connective tissue that Rufus envisioned, the city would have to reengineer a downtown that was, if not hostile to cars, not deferential toward them either. As one local blogger opined, the Grid's architects had to build streets that "*can't* carry the same amount of traffic that runs over the viaduct today."

In that sense the plans for the Community Grid concerned some of its supporters. The year before my visit, the state released a digital rendering of the surface streets that would replace the viaduct. The imagined boulevard was more attractive than the hideous freeway, but its manicured verges and medians looked sterile in their own way, like the landscaping outside a corporate park. Although the simulation showed pixelated families strolling down the sidewalk and pixelated cyclists rolling along the bike lanes, the boulevard's intimidating width conveyed the unmistakable impression of an automotive landscape. The rendering also failed to include Rufus's mixed-use development: no barbershops, no grocery stores, few apartments. "All these people walking on the sidewalk—where do they *live*?" one architect told me. "Are they all just tourists?" Renderings weren't reality, yet it was another reminder of the car's intransigence.

That the freeway was coming down at all, though, was extraordinary. On my last evening in Syracuse, I took I-81 to a suburb called Camillus, where the state was holding the last in a series of public meetings. From the Community Grid's earliest days, suburbanites had been its fiercest foes. "Damn if it helps the minority community, but it's gonna add two minutes to my drive," a Black legislator named Vernon Williams told me sarcastically. On my way to the meeting, I imagined fireworks: soccer moms and dads rising before a standing mic to shout their grievances, the enervated crowd cheering them on.

The reality was far more mellow. The Camillus municipal building could have been anywhere in America: dingy basketball court, cramped rows of wooden seats. The meeting was an open house, more science fair than courtroom drama. Card tables and posters explaining various aspects of the Grid stood around the room, each station manned by a technician.

There were displays about projected travel times and noise abatement and bike lanes and a jargony metric called "level of service," which apparently pertained to how much traffic a highway could handle without descending into total gridlock. I'd spent the week interviewing Syracusans about how the Grid's ramifications extended beyond concrete, but, as I browsed, I was nonetheless awestruck by its physical complexity. Every new interchange, exit ramp, and side street was a monumental undertaking requiring reams of blueprints and impact assessments. Like a tumor grown enmeshed with blood vessels, the viaduct could not simply be excised; as its presence had shaped Syracuse, the city would be rebuilt around its absence.

As I circulated, I eavesdropped on conversations between Camillans and state employees, hoping to overhear some quotable vitriol. But the suburbanites seemed resigned, and their questions were more technical than bombastic: How were commute times estimated? How were budgets calculated? The science-fair format, I realized, had been a canny choice. Locked in conversation with a khaki-clad engineer eager to explain the vagaries of travel-time modeling, it was hard to gripe that your concerns weren't being heard.

I sidled up to a monitor and watched images flicker across it. The pictures showed before-and-afters of Syracuse from various vantages: first a current photograph, the viaduct looming across the frame; the next, a rendering of the same spot, sunlit and leafy, after its demolition. The cycle was mesmeric and soothing; I could almost feel my alveoli expanding. I asked an engineer in a starched button-down how the attendees were reacting to the display. He cleared his throat. "People have different opinions, and I understand that," he said, a studiously neutral delegate of the state. Then he stepped closer, as though to reveal a secret. "My personal response is— which city would you rather live in?"

Epilogue

THE ANTHROPAUSE

Sometime in the middle of the twentieth century, human civilization, which for millennia had been lurching along at more or less the same listless pace, kicked into hyperdrive. Population and wealth spiked; water and fertilizer consumption exploded; car and truck production grew sixfold. International travel became prosaic, at least for the rich. As our footprint expanded, so did our influence. Oceans acidified, forests dwindled, and an invisible, odorless gas began to warm the earth. Scientists have described this boom as the Great Acceleration and posited it as the moment the Anthropocene began in earnest. The acceleration was both figurative—we caught more fish, dammed more rivers, planted more crops—and literal. Every year we traveled farther and faster, our journeys expedited by planes and ships and a cat's cradle of asphalt strings. We were on an unstoppable trajectory, the freest organisms that ever scurried over the earth.

And then, in March 2020, it all ground to a halt, a virus wedged in society's gears.

Among the countless ways the COVID-19 pandemic upended lives, the most dramatic was perhaps the loss of mobility. Every trip became a potentially disease-transmitting event; besides, with schools, offices, and restaurants closed, there wasn't anywhere to go. More evocative even than photos of shuttered storefronts or vacant stadiums, aerial images of empty California freeways captured the postapocalyptic lull. Stuck at home, we could newly empathize with an urban cougar or a roadside deer. For the first time, we, too, knew how it felt to have our movements curtailed by the exponential growth of a deviously prosperous organism.

With humanity sidelined, wildlife crept back into landscapes we'd abdicated. A pride of lions basked on the pavement in Kruger National Park; wild boars rooted in Barcelona's medians. A cougar strolled through San Mateo. On Twitter, #NatureIsHealing went viral. Some anecdotes were swiftly debunked: the dolphins supposedly frolicking in Venetian canals were actually filmed in Sardinia. But the broader phenomenon was real enough. A group of researchers, writing in *Nature*, dubbed it the Anthropause.

The Anthropause was the biggest inadvertent experiment in road ecology's history. In Costa Rica ocelot roadkill virtually ceased. In Maine twice as many frogs and salamanders survived spring migration. British hedgehog deaths fell by more than half. The most eye-catching data came from Fraser Shilling of the Road Ecology Center at the University of California, Davis, who analyzed carcass cleanup statistics from the few states that collected such figures. In California roadkill had fallen by 21 percent; in Idaho, 38 percent; in Maine, 44 percent. A year of reduced travel, Shilling estimated, would save twenty-seven thousand large animals in those states alone. Include smaller critters and extrapolate globally, and it wasn't a stretch to claim that the pandemic, strange as it sounded, would spare billions of lives. "This is the biggest conservation action that we've taken, possibly ever, certainly since the national parks were formed," Shilling told me that sedentary summer. "There's not a single other action that has saved that many animals."

The Anthropause manifested in other, more complex ways. San Francisco's white-crowned sparrows sang more softly than they had before the pandemic, yet their songs carried twice as far across the hushed city. They also sang *better*. With the low-frequency growl of traffic muted, male sparrows showed off their vocal range, crooning intricate melodies that veered into deeper registers—precisely the kind of expressive, full-bodied ballads that females find most attractive. COVID-19, ornithologists wrote, had "created a proverbial silent spring." Since Rachel Carson's day, those words had been synonymous with avicide. Now it was us who had fallen silent and the birds who rejoiced.

Among the pandemic's surprising effects was how ruthlessly it exposed the rot in America's status quo: its broken politics, its inadequate health care, its cultish individualism. It offered the same cruel clarity on our roads, laying bare the horrors of roadkill and traffic noise by abruptly turning them off. And it unmasked how cars had impinged on our own lives. When you have to steer six feet clear of other pedestrians for fear of contracting an infectious disease, you notice just how little room cities allocate for sidewalks and how much they surrender to cars. When indoor spaces become dangerous clusters of contagion, it feels perverse that municipalities devote more acreage to parking than to parks. When your air becomes breathable overnight, it throws into sharp relief its usual miasmatic condition.

As the pandemic revealed and exacerbated inequities, governments hastened to adapt. For more than a century, cities had contorted themselves to accommodate automobiles; now the arc of urban design bent toward livability. Oakland launched a program called Slow Streets, which excluded traffic from twenty-one road miles to give its residents room to bike, jog, and stroll. New York City suspended regulations on outdoor dining, allowing cafés to blossom atop parking spaces. Paris, Rome, and London expedited bike lanes and sidewalk-widening projects. A podcast called the War on Cars declared provisional victory.

It was, of course, hardly a triumph to celebrate: as one noise researcher put it, silence was the "sound of the city aching." But if the world was stricken with misery, it also seemed pregnant with potential. It was as though we were seeing roads for the first time, as though we could at last apprehend the bars of an invisible prison and glimpse the world beyond.

As the pandemic wore on, though, the respite waned. By June of 2020, traffic had rebounded to 90 percent of pre-COVID levels. Roadkill came roaring back, perhaps owing to what some researchers called "behavioral lags." In the absence of cars, this theory went, animals became accustomed to crossing roads and foraging along shoulders at their leisure. When traffic resumed, they failed to quickly adjust. Far from healing nature, the Anthropause may have fatally emboldened it, leaving jaywalking deer and bears ill-prepared to survive our return.

We, too, descended deeper into roads' thrall. Like N-95 masks, cars became personal protective equipment: virus-free bubbles into which drivers escaped the grubby dangers of buses and subways. The New York Stock Exchange banned its workers from riding public transportation; officials in San Francisco glumly predicted that plummeting train ridership would induce a "transit death spiral." As telework usurped the office, unmoored urbanites fled to exurbs near parks and hiking trails, and real estate boomed in America's hinterlands. New development spawned new roads. A biologist in New Mexico sent me a satellite photo of an unfinished subdivision in the Albuquerque outskirts: an eerie maze of prebuilt roads superimposed on the desert, silently waiting for the houses that would justify their existence. "It's an infinite desire," Jochen Jaeger, the road ecologist who promoted the term *landscape dissection*, told me. "Regardless of how many roads they build, they will always want more."

It was hard to dispute his point. In January 2021 the news site the *Intercept* published the results of its investigation into how Eastern Collier Property Owners, a mammoth Florida development group, was pursuing federal approval to convert thousands of acres of farms and forests into residential communities and other forms of development. According to conservationists, the plan would usher waves of traffic into Florida panther habitat and so kill more panthers than the fragile population could handle. Among the documents that the *Intercept* discovered in its inquiry was a letter from Eastern Collier's attorneys to the U.S. Fish and Wildlife Service, in which the lawyers averred that "future, offsite collisions between Florida panthers and vehicles operated by third parties" shouldn't be blamed on new houses and roads. The argument had a certain dangerous brilliance. If panther collisions were caused by pesky "third parties"—individual drivers like you and me—then roadkill must be *everybody's* fault. And if roadkill was every-body's fault, it was nobody's, not even that of the developer who had invited the traffic. It was something that just happened, a sad but inevitable out-come, a car accident.

In this, it recalled the way we've always assigned fault for crashes. As one oft-cited statistic has it, 94 percent of car crashes are caused by human error. Intuitively, that feels right; who hasn't cursed the idiot who changes lanes without checking his blind spot? But that framing deflects culpability from the car companies building ever more massive SUVs and the engi-neers designing unsafe streets. As we did fossil fuels, we came to rely on cars not by personal choice but by corporate design: as early as 1968, *Read-er's Digest* described the highway lobby as "a pressure packed alliance of all who promote highways for profit—from truckers and construction unions to billboard firms—[that is] riding roughshod over development of a sane transportation system in the United States." What could be more Ameri-can than blaming deep-rooted problems on individual failings rather than corporate power structures?

The most straightforward solution to the road's ills would be a collective rejection of automobility, à la the Jain monks who, deeming car travel "irre-deemably violent," navigate India exclusively by foot. I profoundly admire all the principled nondrivers who have concluded they want no part of the car's evils—people who do their shopping by e-bike, their commuting by bus, their cross-country travel by Amtrak—and the many advocates work-ing to wrest neighborhoods from Motordom's grip. In the course of writing this book, I've felt, at times, like a defeatist—as though, by extolling wild-life passages, I foreclose the possibility of a more radical, carless future.

But while I don't want the personal vehicle's dominance to become self-fulfilling prophecy, I also want to be realistic about the global future of transportation, which will almost surely feature more cars, not fewer. (Some projections hold that two billion motorized vehicles will prowl the planet's roads by 2030, twice as many as in 2010; the fact that many of them will be electric might bode well for the climate but doesn't offer much comfort for elk or ocelots.) Bikes and public transit alone can't save ecosystems from highways: transit primarily serves urban and suburban areas, whereas road-ecology problems are concentrated in rural places; it's hard to imagine the bus system that would spare Wyoming's mule deer from cars. We're largely stuck with the automobile, and we need to reconfigure our extant highways to manage its harms. By all means, slow down at night, brake for snakes, and shepherd salamanders across the pavement. Yet making roads lie lighter on the land isn't the job of individual drivers any more than swapping out light bulbs will solve climate change. Instead, it's a public works project, one of history's most colossal. Like reforestation, species reintroductions, and the planting of oyster reefs, it's part of what the historian Thomas Berry described as humanity's Great Work: "moving the human project from its devastating exploitation to a benign presence." A solar panel on every roof, a wetland on every floodplain, a wildlife crossing at every migration corridor. And our investment in crossings must be matched by our commitment to habitat preservation: an overpass whose surroundings get developed into strip malls and condos is just an expensive bridge to nowhere.

"A thing is right," Aldo Leopold famously wrote in his call for a land ethic, "when it tends to preserve the integrity, stability, and beauty of the biotic community." By that standard roads are the wrongest things imaginable, agents of chaos that shatter biotic integrity wherever they intrude. Perhaps we need an analogous *road* ethic, an aphorism that pithily encapsulates what makes a road right and what makes it wrong. *A road is right when its planners have done everything within their power to avoid disrupting the biotic and human communities through which it passes,* perhaps. Or *A road is right when it, like its masters, belongs to the land instead of conquering it.* Ultimately, it's hard to improve on the dictum that the Confederated Salish and Kootenai Tribes applied to Highway 93, one of North America's finest road-ecology projects: "The road is a visitor."

By the pandemic's second autumn, road ecology's Great Work was belatedly under way. Repairing our crummy infrastructure, that perpetual pipedream, now seemed like the only way to pull ourselves from malaise. In November 2021, Congress passed a $1.2 trillion infrastructure bill, the largest since Eisenhower authorized the interstates, that allotted billions for the usual concrete potpourri, like bridge repairs and highway repaving. Tucked in among all the pork, however, was a more radical provision: a program allocating $350 million for wildlife crossings, fences, and other road-ecology interventions, easily the largest investment in animal-friendly infrastructure in American history.

It was easy to feel cynical about this development. As the road ecologist Patricia Cramer pointed out to me, $350 million was mere "decimal dust" in the context of transportation budgets. The act also committed billions to highway expansions, exacerbating the landscape dissection that road ecology was meant to solve. Nonetheless, the wildlife-crossing provision was a welcome sign of changing priorities, or at least additional priorities. Animals would no longer have to compete against potholes and rusty bridges for loose change. "We can't treat every mile of highway" with the funding, said Rob Ament, a road ecologist at the Center for Large Landscape Conservation, one of the groups who'd crafted and pushed for the provision. "But we can take care of a lot of the areas that are seriously affecting wildlife populations."

And the infrastructure act wouldn't just support more animal crossings: it also augured new *types* of road-ecology projects. The bill vowed to prioritize "innovative technologies" and "advanced design techniques," an intriguing stipulation that meant, well, nobody knew what. Perhaps it would fund more sophisticated animal detection systems, arrays of roadside sensors that notice the ingress of animals and illuminate flashing signs.* Ament hoped it would include wildlife overpasses made of fiber-reinforced polymers, composites that are already lighter and stronger than steel and concrete and should soon be cheaper. Such wildlife overpasses could, in

---

* Unlike traditional, static deer signs, animal detection systems only light up when there's actually an animal to detect and thus prevent driver habituation. If you see the sign flash, you know to slam the brakes. These systems typically reduce roadkill by about 50 percent.

theory, be assembled from modular units, which would allow engineers to snap them together on-site without disrupting traffic. They could be made from plant-based materials, like flax, or recycled plastics. And, most exciting of all, Ament suggested that future bridges could be mobile, capable of being disassembled and rebuilt as climate change displaces wildlife. As animals' ranges shift, crossings may someday move with them, lumbering across the land like polymerized elephants.

Around the time of the act's passage, I drove on I-90 across Washington to see the Great Work in progress. The Inland Northwest scrolled past, space and time annihilated by speed: scablands, wheatfields, pine forest. The land tilted as I drove west, sagebrush steppe crumpling into foothills and then mountains. The clouds descended to dust the spruce with snow. I'd reached Snoqualmie Pass.

Snoqualmie Pass is Washington's granitic brow, an ecological crossroads where disparate habitats—glaciers, sagebrush, forest—converge and meld. Elk blunder through the woods, mountain goats clack across stone, wolverines shuffle on stealthy journeys. The outcroppings are alive with jumping mice and pikas; the streams thrum with bull trout and giant salamanders. Native peoples—the Snoqualmie, the Muckleshoot, the Yakama—long crossed the pass on foot, hunting deer on the east side and harvesting salmon on the west, their soles beating paths into moss and needles. Wagon roads came through in the 1860s, hitching the interior West to a muddy tidal backwater called Seattle. The first car crept over in 1905, and a two-lane highway followed. Even after the feds expanded the road to four lanes and rechristened it I-90, it remained tortuous, badly lit, and unevenly paved. Avalanches and rockslides engulfed motorists. One winter I nearly became a Snoqualmie fatality myself: white-knuckling down the pass during a blizzard, I'd spun out and smashed into a guardrail. I was okay, my car's alignment less so.

Snoqualmie's danger wasn't lost on Washington's transportation department, which, in the late 1990s, began to plan a long-overdue reconstruction of I-90. The state, however reluctantly, was obliged to consult with the U.S. Forest Service, which managed the surrounding lands. At the first awkward meeting between the agencies, a Forest Service biologist named Patty Garvey-Darda declared her preference for the rebuild: a humongous, miles-long bridge that would allow animals to wander beneath I-90 anywhere they wanted. The cost-conscious engineers stared in incredulous silence. Negotiations remained tense over the following years; before one meeting Garvey-Darda threw up in the bathroom. Still, she pushed forward, cheerful and tireless. She enlisted colleagues to scour the snow for animal tracks to identify crossing spots, paid the requisite visit to Banff, won over the engineers. Conservationists campaigned for a modest gas tax hike to pay for crossings, among the only environmental campaigns ever to request *more* funding for construction. Soon the former foes were singing kumbaya, and engineers had designed more than twenty wildlife crossings in a fifteen-mile stretch of highway.

Befitting Snoqualmie Pass's biodiversity, the crossings ran the gamut. There were spacious underpasses for bears and cougars, upgraded culverts for trout and salamanders, tunnels for toads and mice. It would be Washington's own Great Work: two decades of accumulated road-ecology knowledge applied to a single highway. The flagship structure was a Banffian double-arch bridge that towered thirty-five feet above the highway's deck. When a coyote became the first critter to use the overpass in December 2018, thermal cameras captured him trotting up one side and down the other, head high and feet quick, an apparition scuttling through the night.

I'd come to Snoqualmie Pass to cross the overpass myself, with Garvey-Darda my guide. We rendezvoused at a pullout with a couple of her colleagues and carpooled to the bridge. A sign reading "Wildlife Corridor: Closure Area" hung from a locked gate in the roadside fence, warning away hikers, hunters, and COVID vaccine protesters, who had planned to hang anti-vax banners from the overpass before the Forest Service dissuaded them. Garvey-Darda unlocked the gate, and we slipped into the forest. Several years ago, she explained, the land that surrounded the highway had been a paved rest area. When the transportation department built the overpass, it also peeled away that concretized scab; now wetlands puddled in its place. The crossings were embedded within a larger restoration effort that sought to shrink the highway's footprint, a manufactured Anthropause—the wild world returning to a space abandoned by humans, nature claiming our infrastructure for its own ends.

We climbed to the bridge's apex. It was my first time atop an overpass since 2013, when I'd strolled up the U.S. 93 crossing in Montana, and the state of road ecology's art had advanced in the intervening years. Concrete walls blocked the sight and sound of traffic, and scattered rockpiles and logs afforded cover for rodents and amphibians. Nursery-grown seedlings—snowberry, wild rose, maple—poked from aromatic mulch. Silver tags denoted spots where the soil had been inoculated with endemic fungi, and Douglas-fir snags beckoned to woodpeckers. Elk scat lay everywhere. Whole herds often bedded on the overpass, cows nursing calves and bulls jousting as trucks hurtled beneath them. "That, to me, is the coolest thing that I didn't anticipate," Garvey-Darda said. "I thought they'd start on one side and make a run for it. But no, they're hanging out here, doing their behaviors. It's part of their home range now."

Although elk were the bridge's most enthusiastic occupants, smaller creatures used it, too. The overpass was kitted out with pitfall traps, low aluminum strips that funneled jumping mice, shrews, toads, and other tiny crossers into buried buckets, where they could be counted and studied. Most of the critters who'd tumbled into the buckets were habitat generalists, intrepid explorers who could persist in almost any environment. But as shrubs and trees flourished atop the overpass, rarer forest specialists, like flying squirrels and snowshoe hares, might turn up as well. Some small animals—the red-backed vole, say—would spend their entire lives darting among the rockpiles. Like Los Angeles's Liberty Canyon overpass, the Snoqualmie crossing was less a bridge than an environment, a span both alive and lived upon.

Atop the crossing, I tried to conjure its future. The canopy thickening, firs scraping the sky. The black bear padding through in fall, bound for a den in the volcanic debris of a distant mountain. The bobcat gliding over powder, tufted ears perked to a hare's munching. The fisher, tail bushy and eyes avid, scrabbling up a hemlock; the Douglas squirrel chittering his staccato alarm. The western toad, the Cascades frog, the alligator lizard, making their slow and secret progress. The wolverine stalking south toward the rumor of a female, all curved claws and bottled ferocity, the forest quailing with his passage. The shrew skittering between logs, noticed only by thermal cameras and pygmy owls. Lives not on the road but over and beyond it, a road that animals would never meet, a road the land would never notice.

ACKNOWLEDGMENTS

The hardest part of writing about road ecology is, I think, capturing the intricacy of its human ecosystem. Every wildlife crossing, from the grandest overpass to the humblest culvert, is the product of collaboration. There are the conservationists and concerned locals who advocate for crossings, the scientists whose data inform them, the engineers who design them, the contractors who build them, and the politicians who fund them, to name only a few players. Although the field's breadth and interdisciplinarity are strengths, they also make matters hard on journalists, or at least they did on this one. Writing this book unavoidably entailed abridgement, and to those people and entities I left out, I beg forgiveness.

There are also so *many* road-ecology projects, each innovative and influential in its own right, and more being built every month. I didn't even mention Alligator Alley, or Pigeon River Gorge, or the Tobin Land Bridge, or dozens of other story-worthy ventures. I could have written another version of this book that focused on an entirely different set of structures, figures, and fulcra in road-ecology history. *Crossings* is anything but the final word on roads and nature, and I fervently hope that it prompts its readers to further explore the subject.

In the course of reporting this book, I spoke with approximately 250 sources—some of whom are named in its pages, many of whom are not. Each of these experts, named or not, contributed to my understanding of road ecology's history and practice. Aerin Jacob, Amanda Hardy, Clayton Lamb, Greg Nickerson, Kari Gunson, Norris Dodd, and Travis DeVault were notably generous in sharing their expertise. Alexander Lawson and Braeden Van Deynze gave helpful notes. Renee Callahan, executive director of ARC Solutions, a group that's done much to advance the science of wildlife crossings, connected me with resources and contacts as effectively as any overpass. Adam Switalski gave me an eye-opening tour of road decommissioning sites in Montana; Sarah Zwissler and Garshaw Amidi-Abraham facilitated my visit to Tillamook County; Jessica Moreno took me to the scenic Oracle Road crossings in Tucson; Daniel Scognamillo drove me around ocelot habitats in South Texas; Dennis and Renata Freedman introduced me to their pademelons; and Keith Chung escorted me on one of his patented penguin walks in Tasmania. Yes, penguins can be roadkill—or at least they could, until Keith installed fences between their nesting grounds and a coastal highway.

Over the last decade, my reporting has received assistance from several institutions. Sections of this book first appeared in *Atlantic*, *Believer*, *bioGraphic*, and *High Country News*; thanks to the editorial staff at those and other publications for boosting my work and sharpening my words. The Solutions Journalism Network funded a memorable two-month trip throughout the Yellowstone to Yukon region in 2013, when I first drove Highway 93 with Marcel Huijser. (Pat Basting, then with the Montana Department of Transportation, was also on that tour, and helped pique my interest in road ecology.) A 2019 fellowship from the Alicia Patterson Foundation supported travel to Brazil and Australia, among other locations, and gave me the time and space to explore this book's themes in magazine articles. And a 2020 Creative Nonfiction Grant from the Whiting Foundation allowed me to complete my reporting when the Anthropause threatened to disrupt it.

Wendy Strothman, my agent, saw the potential in this book and found it a worthy home. Rob Rich, a writer and naturalist whose ear for language is matched only by his eye for animal sign, provided incisive suggestions. Many good people at W. W. Norton & Company helped shepherd *Crossings* to completion; thanks especially to Huneeya Siddiqui and Matt Weiland, the latter a human GPS device whose counsel guided my writing back onto the highway whenever it threatened to veer into a cul-de-sac.

Finally, my wife, Elise Rose, not only gave me invaluable notes on this manuscript; she accompanied me on many of the adventures described in it. For her years of support, and the support of my family and friends, I'm ever and always grateful.

NOTES

INTRODUCTION: THE WING OF THE SWALLOW

1   **"Once the environment is ruined"**: Charles Brown, interview with the author, January 11, 2019.

2   **car-struck swallows had longer wings**: Charles R. Brown and Mary Bomberger Brown, "Where Has All the Road Kill Gone?" *Current Biology* 23, no. 6 (2013): 233–234.

3   **"Everything in life is somewhere else"**: E. B. White quoted in the seminal textbook on road ecology: Richard T. T. Forman et al., *Road Ecology: Science and Solutions* (Washington, D.C.: Island Press, 2003), 49.

4   the **"architecture of our restlessness"**: Rebecca Solnit, *Savage Dreams, Twentieth Anniversary Edition* (Berkeley: University of California Press, [1994] 2014), 365.

4   **"two lanes [that] take us anywhere"**: "Thunder Road," by Bruce Springsteen, released 1975, track 1 on *Born to Run*, Columbia PC 33795.

4   **"the leading direct human cause of vertebrate mortality"**: Richard T. T. Forman and Lauren E. Alexander, "Roads and Their Major Ecological Effects," *Annual Review of Ecology and Systematics* 29 (1998): 212.

4   **More birds die on American roads every week**: The *Deepwater Horizon* spill killed around a million birds; meanwhile, bird roadkill has been estimated at 80 million birds per year, or 220,000 birds per day. Wallace P. Erickson, Gregory D. Johnson, and David P. Young Jr., "A Summary and Comparison of Bird Mortality from Anthropogenic Causes with an Emphasis on Collisions," in *Bird Conservation Implementation and Integration in the Americas: Proceedings of the Third International Partners in Flight Conference,*

ed. C. John Ralph and Terrell D. Rich, vol. 2, Gen. Tech. Rep. PSW-GTR-191, March 20–24, 2002, Asilomar, CA (Albany, CA: U.S. Department of Agriculture, Forest Service, Pacific Southwest Research Station, 2005), 1029–1042.

4   **the toll had quadrupled:** Jacob E. Hill, Travis L. DeVault, and Jerrold L. Belanta, "Research Note: A 50-Year Increase in Vehicle Mortality of North American Mammals," *Landscape and Urban Planning* 197 (2020): 103746.

4   **its surface began to shed sediment:** G. Evelyn Hutchinson et al., "Ianula: An Account of the History and Development of the Lago di Monterosi, Latium, Italy," *Transactions of the American Philosophical Society* 60, no. 4 (1970): 1–178.

4   **spread through Gabon:** Peter D. Walsh et al., "Logging Speeds Little Red Fire Ant Invasion of Africa," *Biotropica* 36, no. 4 (2004): 637–641.

4   **the "road-effect zone":** Richard T. T. Forman, "Estimate of the Area Affected Ecologically by the Road System in the United States," *Conservation Biology* 14, no. 1 (2000): 31–35.

4   **you'll still see fewer birds:** Ana Benítez-López, Rob Alkemade, and P. A. Verweij, "The Impacts of Roads and Other Infrastructure on Mammal and Bird Populations: A Meta-Analysis," *Biological Conservation* 143 (2010): 1307–1316.

4   **permitted the ingress of wolves:** Hillary Rosner, "Pulling Canada's Caribou Back from the Brink," *Atlantic*, December 17, 2018.

5   **soften desert soils for pocket gophers:** Laurence M. Huey, "Mammalian Invasion via the Highway," *Journal of Mammalogy* 22, no. 4 (1941): 383–385.

5   **"I noticed the long slice":** Richard Forman, interview with the author, July 31, 2019.

6   **"life change[s] for plants and animals":** Forman et al., *Road Ecology*, 7.

6   **a translation of *straßenökologie*:** Heinz Ellenberg et al., "Straßen-Ökologie: Auswirkungen von Autobahnen und Straßen auf Ökosysteme deutscher Landschaften" (Road ecology: Effects of motorways and roads on ecosystems in German landscapes), *Ecology and Road: Pamphlet Series of the German Road League* 3 (1981): 19–22.

7   **"The road is a visitor":** Quoted in Mark Matthews, "Montana Tribes Drive the Road to Sovereignty," *High Country News*, August 13, 2001.

8   **national defragmentation plan:** Edgar A. van der Grift, "Defragmentation in the Netherlands: A Success Story?" *GAIA* 14, no. 2 (2005): 144–147.

9   **tens of thousands of successful traversals:** Marcel P. Huijser et al., "Effectiveness of Short Sections of Wildlife Fencing and Crossing Structures along Highways in Reducing Wildlife–Vehicle Collisions and Providing Safe Crossing Opportunities for Large Mammals," *Biological Conservation* 197 (2016): 61–68.

10   **"black, smooth and straight":** Richard Adams quoted in Forman et al., *Road Ecology*, 113.

10   **"We treat the attrition":** Barry Lopez, *Apologia* (Athens: University of Georgia Press, 1998).

11   **"infrastructure tsunami":** William F. Laurance, "Conservation and the Global Infrastructure Tsunami: Disclose, Debate, Delay!" *Trends in Ecology & Evolution* 33, no. 8 (2018): 568–571.

11   **three-quarters of the infrastructure:** Hans-Peter Egler and Raul Frazao, "Sustainable Infrastructure and Finance: How to Contribute to a Sustainable Future," United Nations Environment Programme, 2016, https://wedocs.unep.org/20.500.11822/7756.

12   **"Infrastructure isn't sexy":** John Oliver, "Infrastructure," *Last Week Tonight with John Oliver*, HBO, March 2, 2015.

12   **"other nations, caught with ourselves":** Henry Beston, *The Outermost House, Seventy-Fifth Anniversary Edition* (New York: Henry Holt, 2003), 25.

1: AND NOW THE DEVIL-WAGON!

15   **"reaped a rich entomological harvest":** Charles C. Nutting, "Barbados-Antigua Expedition," *University of Iowa Studies in Natural History* 8 (1920): 165, 118.

16   **"a considerable number":** Dayton Stoner, "The Toll of the Automobile," *Science* 61, no. 1568 (1925): 56–57.

16   **"ruled rather strictly":** L. L. Snyder, "Dayton Stoner—1883 to 1944," *Journal of Mammalogy* 26, no. 2 (1945): 111–113.

16   **"The mole's tunnelled chambers":** Thomas Hardy, "The Field of Waterloo," quoted in Claire Tomalin, *Thomas Hardy* (London: Penguin, 2007), Google Books.

16   **entered the lexicon in 1943:** Robert A. McCabe, *Hungarian Partridge* (Perdix perdix Linn.) *Studies in Wisconsin* (Madison: University of Wisconsin Press, 1943), 56.

16   **"carrion or road-kills":** William Safire, "History Is Toast," *New York Times Magazine*, April 1997.

16   **"one of the important checks":** Stoner, "Toll of the Automobile," 57, 56.

17   **"sagacious selection of the most sure":** Archer Butler Hulbert, *Historic Highways of America, Volume 1: Paths of the Mound-Building Indians and Great Game Animals* (Cleveland: Arthur H. Clark, 1902), 137.

17   **Indigenous roads paralleled the Columbia:** Roxanne Dunbar-Ortiz, *An Indigenous Peoples' History of the United States* (Boston: Beacon, 2014), 29–30.

17   **"in the most stony and rockie places":** Quoted in Sidney Smith Rider, *The Lands of Rhode Island: As They Were Known to Caunounicus and Miantunnomu When Roger Williams Came in 1636—An Indian Map of the Principal Locations Known to the Nahigansets, and Elaborate Historical Notes* (Providence: Sidney Smith Rider, 1904), 23.

17   **"saddened by a flitting presentiment":** Nathaniel Hawthorne, *Tales in Two Volumes: Vol. 1* (London: Bell and Daldy, 1866), 39.

18   **"Let us then bind the Republic":** John Calhoun quoted in U.S. Federal Highway Administration, *America's Highways: 1776–1796* (Washington, D.C.: U.S. Department of Transportation, 1977), 19.

18   **strong-armed the Potowatomi:** Federal Highway Administration, *America's Highways*, 23.

18   **"had reverted to a trace":** Federal Highway Administration, *America's Highways*, 15.

18   **"Dark Age of the Rural Road":** Federal Highway Administration, *America's Highways*, 51.

18 **Engineers in coastal Texas:** Mary Jo O'Rear, *Barrier to the Bays* (College Station: Texas A&M University Press, 2022), 77.

18 **"straw day":** M. O. Eldridge, "Road Improvement in the Pacific Northwest," *Good Roads Magazine*, January 1904, 4.

18 **"the very highest, fullest and completest":** Quoted in Livia Gershon, "The Moral Threat of Bicycles in the 1890s," *JSTOR Daily*, February 22, 2016.

19 **"eastern bicycle fellers":** Harvey Ingham, "Practical Road Reform in Iowa," *Good Roads Magazine*, July 1893.

19 **"object lesson roads":** Earl Swift, *The Big Roads* (New York: Houghton Mifflin Harcourt, 2011), 16.

19 **"more has been done":** Eldridge, "Road Improvement in the Pacific Northwest," 3.

19 **"of a kind wholly undreamed of":** Quoted in Christopher Wells, *Car Country: An Environmental History* (Seattle: University of Washington Press, 2012), 42.

19 **only 8 percent of America's 2.2 million:** Swift, *Big Roads*, 24.

19 **"The American who buys an automobile":** Albert Pope quoted in Wells, *Car Country*, 40.

19 **his siblings called him "sir":** Swift, *Big Roads*, 55.

20 **"make the gravel fly":** Thomas MacDonald, "Our Iowa Roads," *Webster City Freeman*, August 29, 1916.

20 **forty thousand highway miles:** Swift, *Big Roads*, 82.

20 **"organic entities governed by wind":** Wells, *Car Country*, 33.

21 **manicure their nails:** Peter Norton, *Fighting Traffic* (Cambridge, MA: MIT Press, 2008), 70.

21 **"road hogs," "speed maniacs":** Bill Loomis, "1900–1930: The Years of Driving Dangerously," *Detroit News*, April 26, 2015.

21 **"the foot touches the accelerator":** "The Man in the Street," *Times Herald* (Port Huron, MI), August 20, 1923.

21 **"Milwaukee mushrooms":** Norton, *Fighting Traffic*, 61.

21 **vehicles killed 23,600 people:** C. F. Hardwood quoted in *Long Island Railroad Information Bulletin* 3, no. 5 (1924): 24.

21 **"mingled her thick dark blood":** F. Scott Fitzgerald, *The Great Gatsby* (New York: Charles Scribner's Sons, 1925), 165.

21 **lugubrious anti-car demonstrations:** Norton, *Fighting Traffic*, describes urban safety parades, 38–45.

22 **"turn turtle":** Loomis, "Years of Driving Dangerously."

22 **"distance rule":** Travis DeVault et al., "Speed Kills: Ineffective Avian Escape Responses to Oncoming Vehicles," *Proceedings of the Royal Society B* 282, no. 1801 (2015).

22 **"The Haunted Auto":** Bryant Baker, "The Haunted Auto," *Puck* 67, no. 1729 (1910).

23 **"The surfaced roads":** Stoner, "Toll of the Automobile," 57.

23 **a practice . . . called "dead-listing":** Gary Kroll, "An Environmental History of Roadkill," *Environmental History* 20 (2015): 8. Kroll's paper is the best academic account of the history of roadkill studies and features many of this chapter's prominent figures, among them the Stoners, James Raymond Simmons, Jerry Chiappetta, and Edward Bellis.

23   **coots, red-winged blackbirds:** Ernest D. Clabaugh, "Bird Casualties Due to Automo-
      biles," *Condor* 30, no. 2 (1928): 157.

23   **"colossal tragedies":** William H. Davis, "The Automobile as a Destroyer of Wildlife,"
      Science (1934), quoted in Kroll, "Environmental History of Roadkill," 8.

24   **"estimated that the carcasses":** Kenneth Gordon, "Rabbits Killed on an Idaho High-
      way," *Journal of Mammalogy* 13, no. 2 (1932): 169.

24   **"Verily, the menaces of civilisation":** Charles L. Whittle, "And Now the Devil-
      Wagon!" *Bulletin of the Northeastern Bird-Banding Association* 2, no. 3 (1926): 59.

24   **"soft, hot tar":** James Raymond Simmons, *Feathers and Fur on the Turnpike* (Boston:
      Christopher Publishing House, 1938): 14–16.

24   **"Would some of my cooperators":** Simmons, *Feathers and Fur*, 21–22.

24   **"a vanished sky":** Simmons, *Feathers and Fur*, 57, 89.

25   **"inexperienced young appear[ing] on the road":** Simmons, *Feathers and Fur*, 36, 30.

25   **Male puff adders, for instance:** G. J. Alexander and B. Maritz, data cited in *Handbook
      of Road Ecology*, ed. Rodney van der Ree, Daniel J. Smith, and Clara Grilo (Oxford: John
      Wiley & Sons, 2015), 442.

25   **"kill rapidly":** Simmons, *Feathers and Fur*, 13.

25   **"preponderance of creeping and crawling things":** Dayton Stoner, "Automobiles and
      Animal Mortality," *Science* 69, no. 1800 (1929): 670–671.

25   **0.029 skunks per mile:** Dayton Stoner, "Highway Mortality among Mammals," *Science*
      81, no. 2104 (1935): 401–402, cited in Kroll, "Environmental History of Roadkill."

25   **"water contents of ants":** "William A. Dreyer," *Bulletin of the Ecological Society of America*
      37, no. 4 (1956): 93–94.

25   **"exceptional cases of destruction":** William A. Dreyer, "The Question of Wildlife
      Destruction by the Automobile," *Science* 82, no. 2132 (1935): 439–440. The debate between
      Dreyer and Stoner is described in Kroll, "Environmental History of Roadkill."

26   **"soft and yielding":** Dayton Stoner, "Wildlife Casualties on the Highways," *Wilson Bul-
      letin* 48, no. 4 (1936): 279.

26   **"characteristically thorough and detailed":** Snyder, "Dayton Stoner," 111.

26   **dubbed itself Motordom:** "Year of Great Promise Lies before Organized Motordom,"
      Brooklyn Daily Eagle, January 6, 1929.

26   **"whang-doodle of a charlatan":** Norton, *Fighting Traffic*, 97.

26   **"a stereotypical stupidity":** Mike Michaels, "Roadkill: Between Humans, Non-Human
      Animals, and Technologies," *Society and Animals* 12, no. 4 (2004): 285.

26   **"All animals, if given a reasonable time":** Stoner, "Toll of the Automobile," 57.

27   **"A safety campaign on behalf of [wildlife]":** Simmons, *Feathers and Fur*, 9.

27   **"Something soft rolled against my foot":** Jerry Chiappetta, "Our Deadly Highway
      Game," *Field and Stream* 67, no. 1 (1962): 25.

27   **"You never can relax":** Chiappetta, "Our Deadly Highway Game," 126.

28   **"Taxes. Development. Deer":** Peter Swiderski, interview with the author, February 14,
      2020.

28   **a deer meets a car every eight minutes:** Sara Cline, "A Deer Is Struck Every 8 Minutes in NY—and It's Worse in the Fall," *Albany Times Union*, November 6, 2018.

28   **more than a million crashes:** Michael R. Conover et al., "Review of Human Injuries, Illnesses, and Economic Losses Caused by Wildlife in the United States," *Wildlife Society Bulletin* 23, no. 3 (1995): 407–414.

28   **A more recent estimate roughly doubled:** Michael R. Conover, "Numbers of Human Fatalities, Injuries, and Illnesses in the United States Due to Wildlife," *Human-Wildlife Interactions* 13, no. 2 (2019): 264–76.

29   **costing Virginia more than $500 million:** Bridget M. Donaldson, "Improving Animal-Vehicle Collision Data for the Strategic Application of Mitigation," FHWA/VTRC report no. 18-R16, Virginia Transportation Research Council, U.S. Department of Transportation, December 2017.

29   **"We had no idea how bad":** Bridget Donaldson, interview with the author, May 5, 2020.

29   **a staggering 6,723 dead animals:** H. Elliott McClure, "An Analysis of Animal Victims on Nebraska's Highways," *Journal of Wildlife Management* 15, no. 4 (1951): 410–420.

29   **"[kept] constant protecting watch":** James Mooney, "Myths of the Cherokee" (Washington, D.C.: U.S. Government Printing Office, 1902).

29   **exporting a half-million deerskins:** Shepard Krech, *The Ecological Indian* (New York: W. W. Norton, 1999), 160.

30   **"From the islands of cover":** James B. Trefethen, "The Return of the White-Tailed Deer," *American Heritage* 21, no. 2 (1970).

30   **"so close to Babylon":** Peter Whoriskey, "Life as Art," *Chicago Tribune*, May 9, 1997.

30   **drive-up funeral parlors:** Kenneth T. Jackson, *Crabgrass Frontier: The Suburbanization of the United States* (New York: Oxford University Press, 1985), 263.

31   **a "minimum count":** Fred A. Thompson, "Deer on Highways—1966 Supplement" (Santa Fe: New Mexico Department of Game and Fish, 1966), described in Daniel L. Leedy, "Highway-Wildlife Relationships, Volume 1: A State of the Art Report," Urban Wildlife Research Center, U.S. Federal Highway Administration, 1975, p. 28.

31   **quadrupled in a decade:** Laurence R. Jahn, "Highway Mortality as an Index of Deer-Population Change," *Journal of Wildlife Management* 23, no. 2 (1959): 187–197.

31   **even Joseph McCarthy hit a deer:** Associated Press, "McCarthys Miss Hurt in Deer, Car Crash," republished in the *Sacramento Bee*, October 23, 1956.

31   **the body count skyrocketed:** Joseph P. Vaughan, "Influence of Environment on the Activity and Behavior of White-Tailed Deer (*Odocoileus virginianus*) along an Interstate Highway in an Agricultural Area of Pennsylvania," PhD dissertation, Pennsylvania State University, 1970.

31   **"close brush with death":** "Close Call for 28 as Plane Kills Deer," *Standard Speaker* (Hazleton, PA), June 15, 1964.

31   **"going over Niagara Falls in a steel barrel":** J. C. Furnas, "And Sudden Death," *Reader's Digest*, August 1935.

31 **"death, injury and the most inestimable sorrow"**: Ralph Nader, *Unsafe at Any Speed* (New York: Grossman, 1965).

32 **"Atomic Deer"**: Associated Press, "Atomic Deer," republished in *Herald-Palladium* (Benton Harbor, MI), December 31, 1956.

32 **"There are so many signs"**: Chiappetta, "Our Deadly Highway Game," 126.

32 **warning signs in Colorado**: Thomas M. Pojar et al., "Effectiveness of a Lighted, Animated Deer Crossing Sign," *Journal of Wildlife Management* 39, no. 1 (1975): 87–91.

32 **"litter on sticks"**: Quoted in Josh Crane, "The Secret Life of Moose . . . Crossing Signs" (podcast), Vermont Public Radio, June 24, 2021, https://www.vermontpublic.org/programs/2021-06-24/the-secret-life-of-moose-crossing-signs.

32 **"These signs serve as targets"**: Quoted in Paul Hanna, "The Impact of Interstate Highway 84 on the Sublett-Black Pine Migratory Deer Population" (Coeur d'Alene: Idaho Department of Fish and Game, October 1982), 43.

33 **outfitted a dangerous road**: "Reflectors Save Deer," *Douglas County Herald*, August 12, 1971.

33 **"optical warning fence"**: Jim Arpy, "Strieter Product Saves Bucks," *Quad City Times* (Davenport, IA), March 24, 1986.

33 **three hundred thousand Swarovski reflectors**: Jim Arpy, "An Optical Fence Keeps Deer Jumping," *Quad City Times* (Davenport, IA), March 9, 1980.

33 **An analysis commissioned by the Strieter Corporation**: Robert Grenier, "A Study of the Effectiveness of Strieter-Lite Wild Animal Highway Warning Reflector Systems," Strieter Corporation, June 28, 2002.

33 **"road planners should not take these claims"**: Trina Rytwinski et al., "How Effective Is Road Mitigation at Reducing Road-Kill? A Meta-Analysis," *PloS ONE* 11, no. 11 (2016).

33 **their fear wore off**: Marianne Ujvári, Hans J. Baagøe, and Aksel B. Madsen, "Effectiveness of Wildlife Warning Reflectors in Reducing Deer-Vehicle Collisions: A Behavioral Study," *Journal of Wildlife Management* 62, no. 3 (1998): 1094–1099.

34 **"a mighty network of highways"**: Dwight. D. Eisenhower, "Address at the University of Kentucky Coliseum in Lexington," October 1, 1956, quoted in the American Presidency Project, https://www.presidency.ucsb.edu.

34 **"move enough dirt and rock"**: Quoted in Jane Holtz Kay, *Asphalt Nation* (Berkeley: University of California Press, 1997), 260.

34 **"float like magic carpets"**: Dan Albert, *Are We There Yet?* (New York: W. W. Norton, 2019), 124.

35 **forty-four deer died**: Frank W. Peek and Edward D. Bellis, "Deer Movements and Behavior along an Interstate Highway," *Highway Research News* 34 (1969): 39.

35 **"The rights-of-way may be visualized"**: Edward D. Bellis and H. B. Graves, "Collision of Vehicles with Deer Studied on Pennsylvania Interstate Section," *Highway Research News* 43 (1971): 13.

35 **a well-made fence along I-94**: John Ludwig and Timothy Bremicker, "Evaluation of

2.4-m Fences and One-Way Gates for Reducing Deer-Vehicle Collisions in Minnesota," *Transportation Research Record* 913 (1983): 19–21.

35 **"probably of little value":** Edward D. Bellis and H. B. Graves, "Highway Fences as Deterrents to Vehicle-Deer Collisions," *Transportation Research Record* 674 (1978): 56.

35 **deer mannequins along the highway:** H. B. Graves and Edward D. Bellis, "The Effectiveness of Deer Flagging Models as Deterrents to Deer Entering Highway Rights-of-Way," report to Federal Highway Administration, Institute for Research on Land and Water Resources, Pennsylvania State University, University Park, 1978.

36 **"drastically increased":** Bellis and Graves, "Highway Fences as Deterrents," 57.

36 **"The traffic itself forms a moving fence":** Bellis and Graves, "Highway Fences as Deterrents," 57.

2: THE MOVING FENCE

39 **"We have rarely found the mule deer":** Meriwether Lewis, "May 10, 1805," Journals of the Lewis and Clark Expedition (website), https://lewisandclarkjournals.unl.edu/item/lc.jrn.1805-05-10#lc.jrn.1805-05-10.01.

39 **"great harsh sweep of wind":** Annie Proulx, *Close Range: Wyoming Stories* (New York: Scribner, 1999), 97.

39 **more than a million Wyoming ungulates:** For a comprehensive description of the state's ungulate movements, see Matthew J. Kauffman et al., *Wild Migrations: Atlas of Wyoming's Ungulates* (Corvallis: Oregon State University Press, 2018).

41 **Monique the Space Elk:** Ben Goldfarb, "Monique the Space Elk and the Wild History of Wildlife Tracking," *High Country News*, April 21, 2020.

42 **They travel three hundred miles round-trip:** Hall Sawyer et al., "The Red Desert to Hoback Mule Deer Migration Assessment," Wyoming Migration Initiative, University of Wyoming, Laramie, 2014.

42 **"We had no idea":** Hall Sawyer, interview with the author, March 18, 2020.

42 **"old-time jollification to include bonfires":** Joseph M. Carey quoted in the excellent John Richard Waggener, *Snow Chi Minh Trail: The History of Interstate 80 between Laramie and Walcott Junction* (Wheatland: Wyoming State Historical Society, 2018), 26.

43 **"the worst stretch of interstate in the nation":** Quoted in Waggener, *Snow Chi Minh Trail*, 131, 127.

43 **a thousand muleys died:** A. Lorin Ward, "Mule Deer Behavior in Relation to Fencing and Underpasses on Interstate 80 in Wyoming," *Transportation Research Record* 859 (1982): 8.

43 **"either unable or unwilling":** Hanna, "Impact of Interstate Highway 84," 31.

43 **"Berlin Wall for Wildlife":** Bill Andree quoted in Laura Peterson, "Building a Better Crosswalk—for Moose, Bear and Elk," *New York Times*, January 10, 2011.

43 **anxiously pawing at the snow:** For a fuller account of this incident and of I-80's impacts generally, see Gregory Nickerson, "Repairing a Fragmented Landscape," *West-*

ern Confluence, University of Wyoming, September 27, 2021, https://westernconfluence
.org/repairing-a-fragmented-landscape/.

44   **They pooled against the interstate:** Hall Sawyer, "Seasonal Distribution Patterns
     and Migration Routes of Mule Deer in the Red Desert and Jack Morrow Hills Planning
     Area," Western Ecosystems Technology, for Bureau of Land Management, March 10,
     2014, p. 19. For starvation rate, see Sawyer et al., "Red Desert to Hoback Mule Deer
     Migration Assessment," 16.

47   **hanging out at "stopover sites":** Hall Sawyer and Matthew Kauffman, "Stopover Ecol-
     ogy of a Migratory Ungulate," *Journal of Animal Ecology* 80 (2011): 1078–1087.

47   **"surfing the green wave":** Ellen O. Aikens et al., "The Greenscape Shapes Surfing of
     Resource Waves in a Large Migratory Herbivore," *Ecology Letters* 20 (2017): 741–750.

47   **hustle to catch up:** Hall Sawyer et al., "A Framework for Understanding Semi-Permeable
     Barrier Effects on Migratory Ungulates," *Journal of Applied Ecology* 50, no. 1 (2012): 68–78.

47   **losing their *vagility*:** Marlee Tucker et al., "Moving in the Anthropocene: Global Reduc-
     tions in Terrestrial Mammalian Movements," *Science* 359, no. 6374 (2018): 466–469.

48   **split animals into four categories:** Sandra L. Jacobson et al., "A Behavior-Based Frame-
     work for Assessing Barrier Effects to Wildlife from Vehicle Traffic Volume." *Ecosphere* 7,
     no. 4 (2016): e01345.

49   **intervals as short as five seconds:** Digvijay Pawar and Gopal Patil, "Pedestrian Tem-
     poral and Spatial Gap Acceptance at Mid-Block Street Crossing in Developing World,"
     Journal of Safety Research 52 (2015): 39–46.

49   **every thirty seconds or less:** Corinna Riginos et al., "Traffic Thresholds in Deer Road-
     Crossing Behavior," Northern Rockies Conservation Cooperative, for Wyoming Depart-
     ment of Transportation, pub. no. WY-1807F, May 1, 2018.

50   **"As a rule of thumb":** Andreas Seiler and Manisha Bhardwaj, "Wildlife and Traffic: An
     Inevitable but Not Unsolvable Problem?" in *Problematic Wildlife II: New Conservation and
     Management Challenges in Human-Wildlife Interactions*, ed. F. M. Angelici, 171–190 (Springer,
     2020), 174.

50   **"Sometimes when you see them":** Corinna Riginos, interview with the author,
     November 15, 2019.

50   **"Curtailed movement of some herds":** Bill Hepworth quoted in Nickerson, "Repair-
     ing a Fragmented Landscape."

51   **"particularly dangerous deer crossings":** Chiappetta, "Our Deadly Highway Game,"
     125.

51   **They measured ten feet wide:** Ward, "Mule Deer Behavior," 11.

51   **"When they got really close":** Hank Henry, interview with the author, April 14, 2020.

52   **shunned the culverts:** Kelly M. Gordon and Stanley H. Anderson, "Mule Deer Use of
     Underpasses in Western and Southeastern Wyoming," *Proceedings of the 2003 Interna-
     tional Conference on the Ecology of Transportation*, Road Ecology Center, University of Cali-
     fornia, Davis, 2003, p. 316.

52 **more than a hundred deer perished:** Gordon and Anderson, "Mule Deer Use of Underpasses," 309.

53 **"That just collected all the dead deer":** John Eddins, interview with the author, March 9, 2020.

53 **Wide, well-lit, appealing ones:** Mule deer prefer an openness ratio above 0.6. See Anthony P. Clevenger and Nigel Waltho, "Factors Influencing the Effectiveness of Wildlife Underpasses in Banff National Park, Alberta, Canada," *Conservation Biology* 14, no. 1 (2000): 47–56.

53 **almost ten victims monthly:** Hall Sawyer, Chad Lebeau, and Thomas Hart, "Mitigating Roadway Impacts to Migratory Mule Deer: A Case Study with Underpasses and Continuous Fencing," *Wildlife Society Bulletin* 36, no. 3 (2012): 492–498.

54 **Within three years nearly fifty thousand muleys:** Sawyer, Lebeau, and Hart, "Mitigating Roadway Impacts to Migratory Mule Deer," 2012.

54 **set society back $6,600:** Marcel P. Huijser et al., "Cost–Benefit Analyses of Mitigation Measures Aimed at Reducing Collisions with Large Ungulates in the United States and Canada: A Decision Support Tool," *Ecology and Society* 14, no. 2 (2009): 15.

54 **Adjusting Huijser's figures for inflation:** "Reducing Wildlife Vehicle Collisions by Building Crossings: General Information, Cost Effectiveness, and Case Studies from the U.S.," prepared by the Center for Large Landscape Conservation for the Pew Charitable Trusts, 2020, https://largelandscapes.org/wp-content/uploads/2021/01/Reducing-Wildlife-Vehicle-Collisions-by-Building-Crossings.pdf.

55 **at least three miles:** Marcel P. Huijser et al., "Effectiveness of Short Sections of Wildlife Fencing and Crossing Structures along Highways in Reducing Wildlife–Vehicle Collisions and Providing Safe Crossing Opportunities for Large Mammals," *Biological Conservation* 197 (2016): 61–68.

55 **"We could finally show the engineers":** Patricia Cramer, interview with the author, November 9, 2021.

56 **"I have photos of deer":** Donaldson, interview with the author.

56 **roadkill declined by more than 90 percent:** Bridget Donaldson and Kaitlyn E. M. Elliott, "Enhancing Existing Isolated Underpasses with Fencing Reduces Wildlife Crashes and Connects Habitat," *Human–Wildlife Interactions* 15, no. 1 (2021):148–161.

57 **Wildlife crossings, with rare exceptions:** One of those exceptions occurred in Island Park, Idaho, where residents rejected a proposed overpass on Highway 20. Some locals objected, reasonably enough, that roadside fencing would impair the community's viewshed; others muttered darkly that the overpass was the vanguard of a government conspiracy to seize private property. See Ben Goldfarb, "When Wildlife Safety Turns into Fierce Political Debate," *High Country News*, January 1, 2020.

57 **"A license plate fits perfectly":** Joshua Coursey, interview with the author, March 9, 2020.

57 **Wyoming's mule deer population was crashing:** Mule Deer Working Group, "2021 Range-Wide Status of Black-Tailed and Mule Deer," Western Association of Fish and

Wildlife Agencies, 2021, https://wafwa.org/wpdm-package/2021-range-wide-status
-of-black-tailed-and-mule-deer/.

59   **relocated into unfamiliar landscapes:** Brett R. Jesmer et al., "Is Ungulate Migration
Culturally Transmitted? Evidence of Social Learning from Translocated Animals," *Science* 361, no. 6406 (2018): 1023–1025.

59   **the six-thousand-year-old bones of pronghorn:** Mark E. Miller and Paul H. Sanders,
"The Tappers Point Site (48SU1006): Early Archaic Adaptations and Pronghorn Procurement in the Upper Green River Basin, Wyoming," *Plains Anthropologist* 45, no. 174 (2000):
39–52.

60   **nearly sixty thousand mule deer:** Hall Sawyer and Patrick Rodgers, "Pronghorn and
Mule Deer Use of Underpasses and Overpasses along US Highway 191, Wyoming,"
Western Ecosystems Technology, for Wyoming Department of Transportation, pub. no.
FHWA-WY-06/01F, September 1, 2015. For nondirectional crossings, see p. 15.

3: HOTEL CALIFORNIA

63   **he'd killed his longtime mate:** For biographical details on individual mountain lions,
see "Puma Profiles," Santa Monica Mountains National Recreation Area, National Park
Service, https://www.nps.gov/samo/learn/nature/puma-profiles.htm.

64   **"a long-established pattern":** David Brodsly, *L.A. Freeway, An Appreciative Essay* (Berkeley: University of California Press, 1981), 7–9.

64   **"a single comprehensible place":** Reyner Banham, *Los Angeles: The Architecture of Four
Ecologies* (Berkeley: University of California Press, 1971), 195.

64   **"sprawling complex of suburban housing":** Dave Siddon, "Ventura Freeway Vital
Link for West Valley," *Valley Times*, April 2, 1960.

65   *Landschaftszerschneidung:* Jochen Jaeger, *Landschaftszerschneidung: Eine transdisziplinäre
Studie gemäss dem Konzept der Umweltgefährdung* (Stuttgart: Eugen Ulmer, 2002).

65   **saw their breeding curtailed:** Kathleen Semple Delaney, Seth P. D. Riley, and Robert
N. Fisher, "A Rapid, Strong, and Convergent Genetic Response to Urban Habitat Fragmentation in Four Divergent and Widespread Vertebrates," *PloS ONE* 5, no. 9 (2010):
e12767.

66   **"exhilarating sport":** Thomas Curwen, "A Week in the Life of P-22, the Big Cat Who
Shares Griffith Park with Millions of People," *Los Angeles Times*, February 8, 2017.

66   **"These animals are constantly moving":** Jeff Sikich, interview with the author,
November 4, 2021.

67   **"lion TMZ":** Dana Goodyear, "Lions of Los Angeles," *New Yorker*, February 5, 2017.

67   **"The female bobcats that we've tracked":** Seth Riley, interview with the author, January 7, 2022.

68   **"Islands . . . are where species go to die":** David Quammen, *The Song of the Dodo* (New
York: Scribner, 1997), 258.

68   **a bombshell study in *Nature***: William Newmark, "A Land-Bridge Island Perspective on Mammalian Extinctions in Western North American Parks," *Nature* 325 (1987): 430–432.

69   **"[listen] politely"**: Peter Ling, "Sex and the Automobile in the Jazz Age," *History Today*, November 1989.

69   **so disunited by highways**: Michael F. Proctor et al., "Genetic Analysis Reveals Demographic Fragmentation of Grizzly Bears Yielding Vulnerably Small Populations," *Proceedings of the Royal Society B* 272, no. 1579 (2005): 2409–2416.

69   **encircled by a highway exit loop**: I. Keller, W. Nentwig, and C. R. Largiader, "Recent Habitat Fragmentation Due to Roads Can Lead to Significant Genetic Differentiation in an Abundant Flightless Ground Beetle," *Molecular Ecology* 13, no. 10 (2004): 2983–2994.

69   **seldom reproduced on the other side**: Seth P. D. Riley et al., "Effects of Urbanization and Habitat Fragmentation on Bobcats and Coyotes in Southern California," *Conservation Biology* 17, no. 2 (2003): 566–576.

70   **"extinction vortex"**: John F. Benson et al., "Interactions between Demography, Genetics, and Landscape Connectivity Increase Extinction Probability for a Small Population of Large Carnivores in a Major Metropolitan Area," *Proceedings of the Royal Society B* 283, no. 1837 (2016).

70   **atrial septal defects**: Warren E. Johnson et al., "Genetic Restoration of the Florida Panther," *Science* 329, no. 5999 (2010): 1641–1645.

70   **the panther's numbers ticked upward**: This isn't to say that Florida panther populations are recovered. In 2018, twenty-six cats died on the state's roads, a terrible casualty rate for a creature whose total population hovers around two hundred animals.

70   **nearly all specimens were abnormal**: Audra A. Huffmeyer et al., "First Reproductive Signs of Inbreeding Depression in Southern California Male Mountain Lions (*Puma concolor*)," *Theriogenology* 177, no. 1 (2022): 157–164.

71   *passages à faune sauvage*: Forman et al., *Road Ecology*, 17.

72   **"You would go to engineers"**: Trisha White, interview with the author, November 6, 2020.

73   **"In the mid-nineties, there was really"**: Terry McGuire, interview with the author, July 15, 2020.

73   **"It was the cheapest fieldwork"**: Tony Clevenger, interview with the author, November 19, 2021.

74   **overpasses were expensive failures**: Another myth about the Banff crossings was the "prey-trap hypothesis": the idea that wolves and cougars habitually lie in wait at passages to ambush elk and deer. For a thorough debunking, see Adam Ford and Anthony Clevenger, "Validity of the Prey-Trap Hypothesis for Carnivore-Ungulate Interactions at Wildlife-Crossing Structures," *Conservation Biology* 24, no. 6 (2014): 1679–1685.

74   **"The first time I met Tony"**: Adam Ford, interview with the author, January 3, 2020.

74   **Bear movements began to climb**: Adam T. Ford, Kathy Rettie, and Anthony P. Clevenger, "Fostering Ecosystem Function through an International Public–Private Partnership: A Case Study of Wildlife Mitigation Measures along the Trans-Canada Highway

in Banff National Park, Alberta, Canada," *International Journal of Biodiversity Science & Management* 5, no. 1 (2005): 181–189.

74 **more than eighty thousand wild commuters:** Anthony Clevenger, "Highways through Habitats," *Transportation Research News* 249 (2007): 14–17.

74 **"There is no evidence that wildlife overpasses":** Luca Corlatti, Klaus Hacklander, and Fredy Frey-Roos, "Ability of Wildlife Overpasses to Provide Connectivity and Prevent Genetic Isolation," *Conservation Biology* 23, no. 3 (2009): 548–556.

75 **Eight males and seven females:** Michael A. Sawaya, Steven T. Kalinowski, and Anthony P. Clevenger, "Genetic Connectivity for Two Bear Species at Wildlife Crossing Structures in Banff National Park," *Proceedings of the Royal Society B* 281, no. 1780 (2014).

75 **the most "family-friendly" structures:** Adam T. Ford, Mirjam Barrueto, and Anthony P. Clevenger, "Road Mitigation Is a Demographic Filter for Grizzly Bears," *Wildlife Society Bulletin* 41, no. 4 (2017): 712–719.

76 **"Tony is basically the Brad Pitt":** Robert Rock, interview with the author, May 7, 2021.

77 **"It was like seeing Bigfoot":** Miguel Ordeñana, interview with the author, October 23, 2021.

78 **"For those of us in L.A.":** Christopher Weber, "California to Build Largest Wildlife Crossing in the World," Associated Press, August 20, 2019.

82 **"architecturally solid terrain":** Clark Stevens quoted in Marianna Guernieri, "How to Build a Highway for Animals," *Domus*, September 12, 2019, https://www.domusweb.it/en/architecture/2019/09/12/how-to-build-a-highway-for-animals.html.

83 **"The default has been":** Rock, interview with the author.

83 **"faunal furniture":** See, for example, Miriam Goosem, "Wildlife Surveillance Assessment Compton Road Upgrade 2005," prepared for the Brisbane City Council by the Cooperative Research Centre for Tropical Rainforest Ecology and Management, 2005.

83 **one-third of a wildlife bridge's ideal width:** Liam Brennan, Emily Chow, and Clayton Lamb, "Wildlife Overpass Structure Size, Distribution, Effectiveness, and Adherence to Expert Design Recommendations," *PeerJ* 10 (2022): e14371.

83 **"a piece of spaghetti":** Patricia Cramer, interview with the author, November 9, 2021.

85 **"anthropogenic resistance":** Arash Ghoddousi et al., "Anthropogenic Resistance: Accounting for Human Behavior in Wildlife Connectivity Planning," *One Earth* 4, no. 1 (2021): 39–48.

85 **jaguars, ocelots, or Iberian lynx:** Ana Ceia-Hasse et al., "Global Exposure of Carnivores to Roads," *Global Ecology and Biogeography* 26, no. 5 (2017): 592–600.

86 **wolves have used seven green bridges:** Mike Plaschke et al., "Green Bridges in a Recolonizing Landscape: Wolves (*Canis lupus*) in Brandenburg, Germany," *Conservation Science* 3, no. 3 (2021).

86 **"renaturalized, reenchanted city":** Jennifer Wolch, *Animal Geographies* (New York: Verso, 1998), 124.

86 **prevent 700,000 deer-vehicle collisions:** Sophie L. Gilbert et al., "Socioeconomic Benefits of Large Carnivore Recolonization through Reduced Wildlife-Vehicle Collisions,"

Conservation Letters 10 (2017): 431–439; and Jennifer L. Raynor, Corbett A. Grainger, and Dominic P. Parker, "Wolves Make Roadways Safer, Generating Large Economic Returns to Predator Conservation," *Proceedings of the National Academy of Sciences USA* 118, no. 22 (2021).

87   **"signs of distress":** "California Department of Fish and Wildlife and National Park Service Team Up to Evaluate P-22," California Department of Fish and Wildlife News Room, December 8, 2022.

4: IN COLD BLOOD

92   **"looped and useless as an old bicycle tire":** Mary Oliver, "The Black Snake," in *Twelve Moons* (New York: Little, Brown, 1979), 9.

92   **at least twenty-one critters:** Marcel P. Huijser et al., "Wildlife-Vehicle Collision Reduction Study: Report to Congress," Western Transportation Institute, Federal Highway Administration, report no. FHWA-HRT-08-034, 2008.

92   **dwindled by an average of 60 percent:** World Wildlife Fund, "Living Planet Report, 2018: Aiming Higher" (Gland, Switzerland: WWF, 2018).

92   **a third of our vertebrates:** Gerardo Ceballos, Paul R. Ehrlich, and Rodolfo Dirzo, "Biological Annihilation via the Ongoing Sixth Mass Extinction Signaled by Vertebrate Population Losses and Declines," *Proceedings of the National Academy of Sciences USA* 114, no. 30 (2017).

93   **"the silence of the frogs":** David M. Carroll, *Swampwalker's Journal* (New York: Houghton Mifflin, 1999), 49.

94   **"massive squishing":** Forman et al., *Road Ecology*, 19.

94   **nearly 28,000 leopard frogs:** E. Paul Ashley and Jeffrey T. Robinson, "Road Mortality of Amphibians, Reptiles and Other Wildlife on the Long Point Causeway, Lake Erie, Ontario," *Canadian Field-Naturalist* 6, no. 6 (1996): 403–412.

94   **10,000 red-sided garter snakes:** David Seburn and Carolyn Seburn, *Conservation Priorities for the Amphibians and Reptiles of Canada* (Toronto: World Wildlife Fund Canada and Canadian Amphibian and Reptile Conservation Network, September 2000).

94   **2,500 toads:** Trevor J. C. Beebee, "Effects of Road Mortality and Mitigation Measures on Amphibian Populations," *Conservation Biology* 27, no. 4 (2013): 657–668.

94   *95 percent* **were reptiles and amphibians:** David J. Glista, Travis L. DeVault, and J. Andrew DeWoody, "Vertebrate Road Mortality Predominantly Impacts Amphibians," *Herpetological Conservation and Biology* 3, no. 1 (2008): 77–87.

94   **"I love to see that Nature":** Henry David Thoreau, *Walden* (Boston: Ticknor and Fields), 340.

94   **"When you deal with a group":** Lenore Fahrig, interview with the author, March 3, 2020.

95   **the poorest remnant amphibian communities:** Lenore Fahrig et al., "Effect of Road Traffic on Amphibian Density," *Biological Conservation* 73, no. 3 (1995): 177–182.

95   **a population of black rat snakes:** Jeffrey R. Rowa, Gabriel Blouin-Demersa, and Pat-

rick J. Weatherhead, "Demographic Effects of Road Mortality in Black Ratsnakes (*Elaphe obsoleta*)," *Biological Conservation* 137, no. 1 (2007): 117–124.

95 **up to three-quarters of the region's populations:** James P. Gibbs and W. Gregory Shriver, "Can Road Mortality Limit Populations of Pool-Breeding Amphibians?" *Wetlands Ecology and Management* 13 (2005): 281–289.

95 **vehicle-killed elk are healthier:** Kari E. Gunson, Bryan Chruszcz, and Anthony Clevenger, "Large Animal-Vehicle Collisions in the Central Canadian Rocky Mountains: Patterns and Characteristics," in *Proceedings of the International Conference on Ecology and Transportation*, Lake Placid, NY, 2003.

95 **small salamander egg masses:** Nancy E. Karraker and James P. Gibbs, "Contrasting Road Effect Signals in Reproduction of Long- versus Short-Lived Amphibians," *Hydrobiologia* 664, no. 1 (2011): 213–218.

95 **"synergistic threats":** James E. Paterson et al., "Individual and Synergistic Effects of Habitat Loss and Roads on Reptile Occupancy," *Global Ecology & Conservation* 31 (2021): e01865.

96 **called such losses *defaunation*:** Rodolfo Dirzo et al., "Defaunation in the Anthropocene," *Science* 345, no. 6195 (2014): 401–406.

96 **others know it as *biological annihilation*:** Ceballos, Ehrlich, and Dirzo, "Biological Annihilation."

96 **The biologist E. O. Wilson:** E. O. Wilson, "Beware the Age of Loneliness," *Economist*, November 18, 2013.

96 **"bred from the putrefaction":** Patricia Dale-Green, "*Bufo bufo*: A Study in the Symbolism of the Common Toad," *British Homeopathic Journal* 49, no. 1 (1960): 64.

96 **"large live toad . . . in a vessel of water":** Voltaire, *A Philosophical Dictionary* (London: W. Dugdale, 1843), 160.

97 **"phase of intense sexiness":** George Orwell, "Some Thoughts on the Common Toad," Orwell Foundation, originally published in 1946, https://www.orwellfoundation.com/the-orwell-foundation/orwell/essays-and-other-works/some-thoughts-on-the-common-toad/.

97 **"right out in the middle":** W. H. Hudson, *The Book of a Naturalist*, excerpted in *Nature's Fading Chorus*, ed. Gordon L. Miller (Washington, D.C.: Island Press, 2000).

97 **"Toad the terror":** Kenneth Graeme, *The Wind in the Willows* (London: Methuen, 1908).

97 **90 percent of the toads:** J. J. van Gelder, "A Quantitative Approach to the Mortality Resulting from Traffic in a Population of *Bufo bufo* L.," *Oecologia* 13, no. 1 (1973): 93–95.

97 **"several people were killed":** Glynn Mapes, "The Toad People," in *Herd on the Street: Animal Stories from the* Wall Street Journal, 102–103 (New York: Free Press, 2003).

97 **"There were actually quite a lot":** Tom Langton, interview with the author, January 28, 2020.

98 **conveying a quarter-million toads:** Mapes, "Toad People."

98 **"The toad is an inoffensive":** Graham Heathcote, "Britain Opens Tunnel to Save Toads," *Associated Press*, March 14, 1987.

98   **"an historic event"**: "Tunnel of Love Fails Amherst Salamanders," *North Adams Transcript*, March 26, 1988.

98   **color-coded paper dots**: Scott D. Jackson and Thomas F. Tyning, "Effectiveness of Drift Fences and Tunnels for Moving Spotted Salamanders *Ambystoma maculatum* under Roads," in *Amphibians and Roads: Proceedings of the Toad Tunnel Conference*, ed. Thomas E. S. Langton, 93–99 (Shefford, UK: ACO Polymer Products, 1989).

99   **"Spring is in the air"**: Trudy Tynan, "Salamanders Do Love Dance," Associated Press, republished in *Daily News Leader* (Staunton, VA), April 14, 1993.

99   **As more herp tunnels sprouted**: For tunnel design principles, see Thomas E. S. Langton, "A History of Small Animal Road Ecology," in *Roads and Ecological Infrastructure*, ed. Kimberly M. Andrews, Priya Nanjappa, and Seth P. D. Riley, 7–20 (Baltimore: Johns Hopkins University Press, 2015).

100   **"Probably on a subtle psychic level"**: Quoted in Stephen Colbert, "Tunnel Vision," *Daily Show with Jon Stewart*, Comedy Central, April 27, 1999.

100   **"national laughing stock"**: Mike Fitch, *Growing Pains: Thirty Years in the History of Davis* (Davis, CA: City of Davis, 1998).

100   **a $318,000 turtle fence**: The congressman in question was Peter Hoekstra. "The Original 'Turtle Fence' Speech!" YouTube (originally on C-SPAN), November 6, 2009, https://www.youtube.com/watch?v=7krh9Bpy1-A.

100   **salamander sex jokes on Fox News**: Chris Slesar, "Movin' Lizards," *Indigo Magazine*, Orianne Society, 2020.

100   **"toad welfare scam"**: John Kelso, "Amphibian Amenities Show County's Fondness for Toads," *Austin-American Statesman*, February 16, 1995.

100   **around $3,000 in "passive-use value"**: John Duffield and Chris Neher, "Incorporating Deer and Turtle Total Value in Collision Mitigation Benefit-Cost Calculations," prepared by Bioeconomics, Inc., for Western Transportation Institute et al., 2021, p. xi.

101   **"They always seemed so vulnerable"**: Matthew Aresco, interview with the author, March 8, 2021.

101   **Few of the refugees**: Any turtle who ventured onto the highway had a *98 percent* chance of being crushed. Matthew J. Aresco, "Highway Mortality of Turtles and Other Herpetofauna at Lake Jackson, Florida, USA, and the Efficacy of a Temporary Fence/Culvert System to Reduce Roadkills," in *Proceedings of the International Conference on Ecology and Transportation*, Lake Placid, NY, 2003.

102   **"When they get into a rut"**: Henry David Thoreau, *The Writings of Henry David Thoreau in Twenty Volumes: Volume 16*, ed. Bradford Torrey (New York: Houghton Mifflin, 1906), 481.

102   **"His front wheel struck the edge"**: John Steinbeck, *The Grapes of Wrath* (New York: Viking, 1939), 22.

102   **swerved to flatten rubber turtles**: E. Paul Ashley, Amanda Kosloski, and Scott A. Petrie, "Incidence of Intentional Vehicle-Reptile Collisions," *Human Dimensions of Wildlife* 12 (2007): 137–143.

102   **"They've lived here longer than"**: Chris Smith, interview with the author, July 14, 2021.

103 **fell to 177 in less than two decades:** Morgan L. Piczak, Chantel E. Markle, and Patricia Chow-Fraser, "Decades of Road Mortality Cause Severe Decline in a Common Snapping Turtle (*Chelydra serpentina*) Population from an Urbanized Wetland," *Chelonian Conservation and Biology* 18, no. 2 (2019): 231–240.

103 **"potentially cause irreversible declines":** Matthew J. Aresco, "Mitigation Measures to Reduce Highway Mortality of Turtles and Other Herpetofauna at a North Florida Lake," *Journal of Wildlife Management* 69, no. 2 (2005): 557.

104 **roadkilled female turtles outnumbered:** Paul S. Crump, Stirling J. Robertson, and Rachel E. Rommel-Crump, "High Incidence of Road-Killed Freshwater Turtles at a Lake in East Texas, USA," *Herpetological Conservation and Biology* 11, no. 1 (2016): 181–187.

104 **"perception of persistence":** Jeffrey E. Lovich et al., "Where Have All the Turtles Gone, and Why Does It Matter?" *BioScience* 68, no. 10 (2018): 771–781.

104 **The first EcoPassage:** C. Kenneth Dodd Jr., William J. Barichivich, and Lora L. Smith, "Effectiveness of a Barrier Wall and Culverts in Reducing Wildlife Mortality on a Heavily Traveled Highway in Florida," *Biological Conservation* 118, no. 5 (2004): 619–631.

105 **In that era of bailouts:** See U.S. Department of the Treasury, Troubled Assets Relief Program (TARP), https://home.treasury.gov/data/troubled-assets-relief-program.

105 **"saves a lot of our four legged friends":** Tom Coburn, "100 Stimulus Projects: A Second Opinion," U.S. Senate, 111th Congress, 2009.

106 **"species loneliness":** Robin Wall Kimmerer, *Braiding Sweetgrass* (Minneapolis: Milkweed, 2015), 358.

108 **moved more than a million toads:** Anna Bonardi et al., "Usefulness of Volunteer Data to Measure the Large Scale Decline of 'Common' Toad Populations," *Biological Conservation* 144, no. 9 (2011): 2328–2334.

109 **In one much-publicized case:** Fabrice G. W. A. Ottburg and Edgar van der Grift, "Effectiveness of Road Mitigation for Common Toads (*Bufo bufo*) in the Netherlands," *Frontiers in Ecology and Evolution* 7 (February 12, 2019).

109 **The city of Marquette, Michigan:** Hani Barghouthi, "Salamanders Had to Cross the Road, So U.P. City Closed It," *Detroit News*, April 4, 2022.

109 **"Snake Road":** Kevin S. Held, "Famed 'Snake Road' Closes in Illinois for Reptile, Amphibian Crossing," *FOX 2 NOW*, March 14, 2022.

109 **"If there is to be an answer":** Elizabeth Kolbert, *Under a White Sky* (New York: Crown, 2022), 8.

5: ROADS UNMADE

113 **"a beckoning, a strangeness":** William Least Heat-Moon, *Blue Highways: A Journey into America* (New York: Little, Brown, 1983), Kindle.

115 **"Improving access was prerequisite":** Bud Moore, *The Lochsa Story* (Missoula, MT: Mountain Press, 1996), 300.

116 **"the junk of American industrialization":** Moore, *Lochsa Story*, 314–322.

116   **"The road has its own reasons"**: Cormac McCarthy, *The Crossing* (New York: Vintage, [1994] 1995), 230.

117   **"tank traps"**: Todd Wilkinson, "The Forest Service Sets Off into Uncharted Territory," *High Country News*, November 8, 1999.

117   **"travel management"**: See the Forest Service's Travel Management Rule, 2005, which required each national forest unit to "identify the minimum road system (MRS) needed for safe and efficient travel." U.S. Department of Agriculture, Forest Service, https://www.fs.usda.gov/science-technology/travel-management.

117   **"expressions of ideology"**: Jedediah Rogers, *Roads in the Wilderness* (Salt Lake City: University of Utah Press, 2013), 6.

117   **"the greatest good for the greatest number"**: Robert Westover, "Forest Service Celebrates 150th Birthday of Founder," U.S. Department of Agriculture, Forest Service, August 11, 2015, https://www.fs.usda.gov/features/forest-service-celebrates-150th-birthday-founder.

118   **"Oh, this is bully!"**: For Roosevelt's quote and early Forest Service history, see Timothy Egan, *The Big Burn* (New York: Houghton Mifflin Harcourt, 2009), 70.

118   **"all of it . . . practically unknown"**: Elers Koch, *Forty Years a Forester*, ed. Char Miller (Lincoln: University of Nebraska Press, 2019), 64.

118   **"Two undesirable prostitutes"**: Koch, *Forty Years a Forester*, 79.

118   **"When one has spent many arduous hours"**: Koch, *Forty Years a Forester*, 74.

118   **"At least six or eight men"**: Koch, *Forty Years a Forester*, 94.

119   **"along the face of precipitous slopes"**: Arthur E. Loder, "The Location and Building of Roads in the National Forests," *Public Roads* 1, no. 4 (1918): 12.

119   **the corps erected 48,000 bridges**: Gerald W. Williams, "The USDA Forest Service: The First Century," USDA Forest Service Office of Communication, Washington, D.C., 2005.

120   **"roads parallel, roads crisscross"**: Rosalie Edge, "Roads and More Roads in the National Parks and National Forests" (1936), in *A Road Runs through It*, ed. Thomas Reed Petersen (Boulder, CO: Johnson Books, 2006), 13.

120   **"the print of the automobile tire"**: Koch, *Forty Years a Forester*, 189.

120   **countered Koch's laissez-faire philosophy**: Andrew J. Larson, "Introduction to the Article by Elers Koch: *The Passing of the Lolo Trail*," *Fire Ecology* 12 (2016): 1–6.

120   **"Roads in the western forests"**: Richard E. McArdle, "Report of the Chief of the Forest Service: America's Stake in World Forestry" (Washington, D.C.: U.S. Department of Agriculture, Forest Service, 1952), 22.

120   **the service's road network doubled**: David Havlick, *No Place Distant* (Washington, D.C.: Island Press, 2002), note on p. 224.

121   **"essentially trad[ing] national forest trees"**: Havlick, *No Place Distant*, 67.

121   **"Some loggers admit"**: Robert E. Miller, "Lincoln Group Opposes Forest Service Objectives," *Independent-Record* (Helena, MT), June 10, 1962.

121   **"I would that I could turn the clock back"**: Koch, *Forty Years a Forester*, 191.

121   **A new generation of conservationists**: For the definitive account of the wilderness

movement's antipathy for roads and cars, see Paul Sutter, *Driven Wild* (Seattle: University of Washington Press, 2002).

121   **"propaganda spread by the Automobile Association of America"**: Bob Marshall quoted in Sutter, *Driven Wild*, 233.

121   **"development of the Coney Island type"**: Benton MacKaye quoted in Sutter, *Driven Wild*, 183.

121   **"There is No God but Gasoline"**: Aldo Leopold quoted in Sutter, *Driven Wild*, 78.

121   **"To build a road"**: Aldo Leopold, *A Sand County Almanac*, special commemorative edition (New York: Oxford University Press, [1949] 1987), 101.

121   **"badly cut up"**: Aldo Leopold quoted in Curt Meine, *Aldo Leopold: His Life and Work* (Madison: University of Wisconsin Press, 1988), 185.

121   **"kept devoid of roads"**: Aldo Leopold quoted in Marybeth Lorbiecki, *A Fierce Green Fire: Aldo Leopold's Life and Legacy* (New York: Oxford University Press, [1996] 2016), 80.

122   **"the impending motorization"**: Aldo Leopold quoted in Sutter, *Driven Wild*, 47.

122   **"untrammeled by man"**: The Wilderness Act, Public Law 88-577 (16 U.S.C. 1131-1136), 88th Congress, 2d session, September 3, 1964.

122   **"Recreational development"**: Leopold, *Sand County Almanac*, 176.

122   **black bears hid deep**: Allan J. Brody and Michael R. Pelton, "Effects of Roads on Black Bear Movements in Western North Carolina," *Wildlife Society Bulletin* 17, no. 1 (1989): 5–10.

123   **shot or trapped nearly half the wolves**: L. David Mech, "Wolf Population Survival in an Area of High Road Density," *American Midland Naturalist* 121, no. 2 (1989): 387–389.

123   **Even a measly half-kilometer**: Michael F. Proctor et al., "Effects of Roads and Motorized Human Access on Grizzly Bear Populations in British Columbia and Alberta, Canada," *Ursus* 2019, no. 30e2 (2019): 16–39.

123   **fragmented many forests more egregiously**: Rebecca A. Reed, Julia Johnson-Barnard, and William L. Baker, "Contribution of Roads to Forest Fragmentation in the Rocky Mountains," *Conservation Biology* 10, no. 4 (1996): 1098–1106.

123   **"rather see cut-over lands"**: Aldo Leopold quoted in Sutter, *Driven Wild*, 206.

123   **more than a hundred thousand locations**: Mary M. Rowland et al., "Effects of Roads on Elk: Implications for Management in Forested Ecosystems," in *The Starkey Project: A Synthesis of Long-Term Studies of Elk and Mule Deer*, ed. Michael J. Wisdom, 42–52 (Lawrence, KS: Alliance Communications Group, 2005). Reprinted from *Transactions of the 69th North American Wildlife and Natural Resources Conference*, 2004.

123   **"No one else in the world"**: Mary Rowland, interview with the author, January 15, 2021.

124   **"low volume" roads**: Alisa W. Coffin et al., "The Ecology of Rural Roads: Effects, Management and Research," *Issues in Ecology*, report no. 23 (Washington, D.C.: Ecological Society of America, 2021).

124   **"daggers thrust into the heart of nature"**: Michael Soulé quoted in Havlick, *No Place Distant*, 36.

124   **the agency employed 24 fisheries biologists**: John Fedkiw, *Managing Multiple Use on*

National Forests, 1905–1995 (Washington, D.C.: U.S. Forest Service, 1998), chap. 3 (see table 1).

124 **"There was no limit"**: Mike Dombeck, interview with the author, March 18, 2020.

125 **"a rudderless ship"**: Quoted in Tom Turner, *Roadless Rules: The Struggle for the Last Wild Forests* (Washington, D.C.: Island Press, 2009), 29.

125 **"hand grenade rolled under my door"**: Quoted in Turner, *Roadless Rules*, 31.

125 **he requested $22 million**: T. H. Watkins, "The End of the Road," in Petersen, ed., *A Road Runs through It*, 170.

126 **More than nine hundred landslides**: Douglas E. McClelland et al., "Assessment of the 1995 and 1996 Floods and Landslides on the Clearwater National Forest," A Report to the Regional Forester, Northern Region, U.S. Forest Service, December 1997.

127 **"as ugly as a war zone"**: Watkins, "End of the Road," 171.

128 **rich in nutrients and organic matter**: Rebecca A. Lloyd, Kathleen Ann Lohse, and Ty P. A. Ferré, "Influence of Road Reclamation Techniques on Ecosystem Recovery," *Frontiers in Ecology and the Environment* 11, no. 2 (2013): 75–81.

128 **feasted on new thimbleberry**: T. Adam Switalski and Cara R. Nelson, "Efficacy of Road Removal for Restoring Wildlife Habitat: Black Bear in the Northern Rocky Mountains, USA," *Biological Conservation* 144 (2011): 2666–2673.

128 **where a fire will erupt**: Alexandra D. Syphard and Jon E. Keeley, "Location, Timing and Extent of Wildfire Vary by Cause of Ignition," *International Journal of Wildland Fire* 24, no. 1 (2015): 37–47.

129 **"The timber guys believe"**: Adam Rissien, interview with the author, August 7, 2020.

130 **the legacy of a nineteenth-century statute**: For a solid overview of R.S. 2477, see Jonathan Thompson, "R.S. 2477 and the Utah Road-Fetish," *Land Desk*, February 1, 2021, https://www.landdesk.org/p/rs-2477-and-the-utah-road-fetish.

130 **"wasteland to 'productive' Eden"**: Rogers, *Roads in the Wilderness*, 34, 17.

131 **"If they can control the roads"**: Jenna Whitlock, interview with the author, June 3, 2021.

131 **the protection of 30 percent**: Eric Dinerstein et al., "A Global Deal for Nature: Guiding Principles, Milestones, and Targets," *Science Advances* 5, no. 4 (2019).

131 **"a relatively easy and cost-efficient step"**: Matthew S. Dietz et al., "The Importance of U.S. National Forest Roadless Areas for Vulnerable Wildlife Species," *Global Ecology and Conservation* 32 (2021): e01943.

131 **"the destruction of our nation"**: Quoted in Aaron Weiss, "The 30x30 Disinformation Brigade," Center for Western Priorities, 2022, https://stop30x30disinformation.org/wp-content/uploads/2022/04/30x30_Disinfo_Brigade_2.2.pdf.

131 **"road closures, road decommissioning"**: See, for example, "Resolution Opposing the Federal Government's '30x30' Land Preservation Goal," County of Garfield, State of Colorado, 2021, https://garfield-county.granicus.com/MetaViewer.php?view_id=3&clip_id=1721&meta_id=204239.

131 **"How the hell can you have an economy"**: Quoted in Juliet Eilperin, "Trump to Strip

Protections from Tongass National Forest, One of the Biggest Intact Temperate Rainforests," *Washington Post*, October 28, 2020.

132 **dead-end "cherry-stem" roads:** Evan Girvetz and Fraser Shilling, "Decision Support for Road System Analysis and Modification on the Tahoe National Forest," *Environmental Management* 32, no. 2 (2003): 218–233.

132 **"well-to-do city folks":** For the most famous rebuke of wilderness, see William Cronon, "The Trouble with Wilderness, or Getting Back to the Wrong Nature," in *Uncommon Ground: Rethinking the Human Place in Nature*, ed. William Cronon, 69–90 (New York: W. W. Norton, 1995).

132 **"most important biotic areas":** Colby Loucks et al., "USDA Forest Service Roadless Areas: Potential Biodiversity Conservation Reserves," *Conservation Ecology* 7, no. 2 (2003): 5.

132 **a "siege strategy":** Stephen Blake et al., "Roadless Wilderness Area Determines Forest Elephant Movements in the Congo Basin," *PloS ONE* 3, no. 10 (2008): e3546.

132 **planetary buffers against extinction:** Moreno Di Marco et al., "Wilderness Areas Halve the Extinction Risk of Terrestrial Biodiversity," *Nature* 573 (2019): 582–585.

132 **Decommissioning just *1 percent*:** Carlos Carroll et al., "Defining Recovery Goals and Strategies for Endangered Species: The Wolf as a Case Study," *BioScience* 56, no. 1 (2006): 25–37.

133 **"future fossils":** David Farrier, *Footprints: In Search of Future Fossils* (New York: Farrar, Straus and Giroux, 2020), 31.

6: THE BLAB OF THE PAVE

134 **lies within a kilometer:** Kurt H. Riitters and James D. Wickham, "How Far to the Nearest Road?" *Frontiers in Ecology and the Environment* 1, no. 3 (2003): 125–129.

134 **"By the time I was ten":** Helen Macdonald, *Vesper Flights* (New York: Grove, 2020), 23.

135 **"sufficient to wake the dead":** Quoted in Garret Keizer, *The Unwanted Sound of Everything We Want: A Book about Noise* (New York: PublicAffairs, 2010), 88.

135 **"blab of the pave":** Walt Whitman, *Song of Myself*, Section 8, University of Iowa International Writing Program, https://iwp.uiowa.edu/whitmanweb/en/writings/song-of -myself/section-8.

135 **truncates lives by more than *three years*:** "Health Impact of Transport Noise in the Densely Populated Zone of Ile-de-France Region," Bruitparif, February 2019.

135 **"Once you notice noise":** Rachel Buxton, interview with the author, November 30, 2020.

135 **a forty-decibel wedge:** Omid Ghadirian et al., "Identifying Noise Disturbance by Roads on Wildlife: A Case Study in Central Iran," *SN Applied Sciences* 1, no. 8 (2019): 808.

135 **pushes prairie dogs back:** Rachel T. Buxton et al., "Varying Behavioral Responses of Wildlife to Motorcycle Traffic," *Global Ecology and Conservation* 21 (2020): e00844.

136 **"windshield wildernesses":** David Louter, *Windshield Wilderness: Cars, Roads, and Nature in Washington's National Parks* (Seattle: University of Washington Press, 2006).

136 **Nearly half the total area:** Rachel T. Buxton et al., "Anthropogenic Noise in US National Parks—Sources and Spatial Extent," *Frontiers in Ecology and the Environment* 17, no. 10 (2019): 559–564.

136 **"demon on wheels":** For this Burroughs quote and an entertaining account of the Vagabonds' travels, see Jeff Guinn, *The Vagabonds: The Story of Henry Ford and Thomas Edison's Ten-Year Road Trip* (New York: Simon and Schuster, 2019).

136 **"Out of that automobile":** Henry Ford quoted in Shannon Wianecki, "When America's Titans of Industry and Innovation Went Road-Tripping Together," *Smithsonian Magazine,* January 26, 2016.

136 **"We cheerfully endure wet, cold, smoke":** John Burroughs, "A Strenuous Holiday," in *Under the Maples* (New York: Houghton Mifflin, 1913), 122.

138 **"laryngitis complicated with tuberculosis":** Marguerite Sands Shaffer quoted in Richard F. Weingroff, "The National Old Trails Road Part 2: See America First in 1915," U.S. Department of Transportation, Federal Highway Administration, 2013.

138 **"the blessing of hours of pleasure":** Quoted in Greg Botelho, "The Car That Changed the World," CNN, August 10, 2004.

138 **"Camping out used to be done":** Nina Wilcox Putnam, "Auto Camping Is the Life!" *Sioux City Journal,* August 3, 1924.

138 **"finding lodgment in Eden":** James Bryce quoted in Louter, *Windshield Wilderness,* 23.

139 **"mingle their gas-breath":** John Muir quoted in Albert, *Are We There Yet?* 140.

139 **"Here we get a free shower!":** Scott Einberger, "The Triumph of Manic Mather," *Psychology Today,* March 27, 2018.

139 **"Scenery is a hollow enjoyment":** "Vocation plus Avocation Equals Preservation," *National Park Service 75th Anniversary,* U.S. Department of the Interior, 1991, http://npshistory.com/publications/nps-75-1.pdf.

139 **"Dear Steve, if you don't like":** Franklin K. Lane quoted in Kate Siber, "The Visionaries," *National Parks Magazine,* Fall 2011.

139 **"roads problem" . . . would "enable the motorists":** Stephen K. Mather, *Ideals and Policy of the National Park Service: Handbook of Yosemite National Park,* U.S. National Park Service, 1921.

139 **filigreed by 1,300 road miles:** For Mather's developmentalist tendencies, see Sutter, *Driven Wild,* 38; and Havlick, *No Place Distant,* 23.

140 **"rocking back and forth":** Horace Albright, *Creating the National Park Service* (Norman: University of Oklahoma Press, 1999). Available online through the Crater Lake Institute, https://www.craterlakeinstitute.com/index-of-general-cultural-history-books/people-and-organizations/creating-the-national-park-service-the-missing-years/.

140 **"begets contentment":** Mather quoted in *Mapping the Future of America's National Parks,* ed. Mark Henry and Leslie Armstrong (Washington, D.C.: ESRI, in cooperation with the National Park Service, 2004), 3.

140 **"would mean the extinction of the moose":** Mather quoted in Havlick, *No Place Distant,* 34.

140 **"It is like the route to the fabled Olympus":** "Going-to-the-Sun Highway in Glacier Park Dedicated Today: Scenic Automobile Route over Divide to Be Thrown Open," *Great Falls Tribune*, July 15, 1933.

140 **"They're built to be scenic":** Kurt Fristrup, interview with the author, November 19, 2020.

141 **Bats seize upon the crunch:** For an overview of noise's impacts, see Jesse R. Barber, Kevin R. Crooks, and Kurt M. Fristrup, "The Costs of Chronic Noise Exposure for Terrestrial Organisms," *Trends in Ecology & Evolution* 25 (2010): 180–189.

141 **halves the "listening area":** Buxton et al., "Anthropogenic Noise in US National Parks."

141 **in Portugal's oak woodlands:** Joana Farrusco Araújo, "Roads as a Driver of Changes in the Bird Community and Disruptors of Ecosystem Services Provision," PhD dissertation, Mestrado em Biologia da Conservação, University of Lisbon, 2020.

142 **right whales' floating feces:** Rosalind M. Rolland et al., "Evidence That Ship Noise Increases Stress in Right Whales," *Proceedings of the Royal Society B* 279, no. 1737 (2012): 2363–2368.

142 **"Lombard effect" in Japanese quail:** Henrik Brumm and Sue Anne Zollinger, "The Evolution of the Lombard Effect: 100 Years of Psychoacoustic Research," *Behaviour* 148 (2011): 1173–1198.

142 **male tree frogs croak:** Kirsten M. Parris, Meah Velik-Lord, and Joanne M. A. North, "Frogs Call at a Higher Pitch in Traffic Noise," *Ecology and Society* 14, no. 1 (2009): 25.

142 **"relatively rapid development of dialects":** Kirsten M. Parris and Angela Schneider, "Impacts of Traffic Noise and Traffic Volume on Birds of Roadside Habitats," *Ecology and Society* 14, no. 1 (2008): 29.

142 **"primary cause for avian community changes":** Richard T. T. Forman and Robert D. Deblinger, "The Ecological Road-Effect Zone of a Massachusetts (U.S.A.) Suburban Highway," *Conservation Biology* 14, no. 1 (2000): 36–46.

142 **"People had been guessing":** Jesse Barber, interview with the author, June 8, 2020.

143 **avoided the Phantom Road altogether:** Heidi E. Ware et al., "A Phantom Road Experiment Reveals Traffic Noise Is an Invisible Source of Habitat Degradation," *Proceedings of the National Academy of Sciences USA* 112, no. 39 (2015): 12105–12109; and Christopher J. W. McClure et al., "An Experimental Investigation into the Effects of Traffic Noise on Distributions of Birds: Avoiding the Phantom Road," *Proceedings of the Royal Society B* 280, no. 1773 (2013).

143 **"the chipmunk next door":** Heidi Ware Carlisle, interview with the author, November 6, 2019.

143 **"foraging-vigilance tradeoff":** Ware et al., "Phantom Road Experiment," 2015.

144 **"sealed in their metallic shells":** Edward Abbey, *Desert Solitaire* (New York: Simon and Schuster, 1968), 233, 52.

144 **collided over Grand Canyon National Park:** This crash and the Natural Sounds Program's origins are described in Kim Tingley, "Whisper of the Wild," *New York Times Magazine*, March 18, 2012.

145   **"a significant adverse effect":** National Parks Overflights Act of 1997, sponsored by
      Sen. John McCain, 105th Congress, 1st session, https://www.congress.gov/bill/105th
      -congress/senate-bill/268/text.

145   **"There are no whales":** Fristrup, interview with the author.

145   **"We have a lot of measurements":** Emma Brown, interview with the author, Decem-
      ber 4, 2020.

145   **noise contaminated more than a third:** Buxton et al., "Anthropogenic Noise in US
      National Parks."

146   **visitors heard more birds:** Mitchell J. Levenhagen et al., "Does Experimentally Quiet-
      ing Traffic Noise Benefit People and Birds?" *Ecology and Society* 26, no. 2 (2021): 32.

147   **"giant vacuum-cleaner-like device":** Shirley Wang, "Quest for Quiet Pavement Is No
      Easy Road," *Wall Street Journal*, May 30, 2012.

148   **"avoid long straight lines":** Quoted in Erik K. Johnson, "The 'High Line' Road,"
      National Park Service, U.S. Department of the Interior, Cultural Resource Report 2019-
      DENA-014, June 2019.

148   **"You sit down at a table":** Quoted in Jane Bryant, *Snapshots from the Past: A Roadside
      History of Denali National Park and Preserve*, National Park Service, U.S. Department of the
      Interior, 2011, p. 40.

148   **"purity of [the park's] wilderness atmosphere":** Quoted in Frank Norris, "Drawing a
      Line in the Tundra: Conservationists and the Mount McKinley Park Road," in *Rethinking
      Protected Areas in a Changing World*, ed. Samantha Weber and David Harmon, 167–173
      (Hancock, MI: George Wright Society, 2008).

149   **"a Washington Monument with no elevator":** Gary A. Crabb, "McKinley Village Pres-
      ident Writes on Park Road Closing," *Fairbanks Daily News-Miner*, April 8, 1972.

149   **Denali's Vehicle Management Plan:** For some light beach reading, check out
      "Denali Park Road Final Vehicle Management Plan and Environmental Impact State-
      ment," National Park Service, U.S. Department of Interior, July 2, 2012, https://www
      .federalregister.gov/documents/2012/07/02/2012-16070/final-environmental-impact
      -statement-on-the-denali-park-road-vehicle-management-plan-denali-national.

150   **"oriented toward the road":** John Dal-Molle and Joseph Van Horn, "Observations
      of Vehicle Traffic Interfering with Migration of Dall's Sheep, *Ovis dalli dalli*, in Denali
      National Park, Alaska," *Canadian Field Naturalist* 105 (1991): 409–411.

150   **falls short of maintaining the sheep gap:** William C. Clark, "Results of Denali Park
      Road Vehicle Management Plan (VMP) Monitoring: Results from the 2019 Field Season,"
      U.S. National Park Service, last updated 2020, https://www.nps.gov/articles/denali-crp
      -park-road.htm#:~:text=Results%20from%20the%202019%20Field,Stop%20and%20
      Eielson%20Visitor%20Center.

151   **"There's a lot of solitude":** Davyd Betchkal, interview with the author, January 15,
      2021.

152   **"localized spruce decline":** Sarah E. Stehn and Carl Roland, "Effects of Dust Palliative

Use on Roadside Soils, Vegetation, and Water Resources (2003–2016)," National Park Service, U.S. Department of the Interior, January 2018.

152 **"near-wilderness":** Robert Manning, William Valliere, and Jeffrey Hallo, "Busing through the Wilderness: Managing the 'Near-Wilderness' Experience at Denali," *Alaska Park Science* 13, no. 1 (2014): 59–65.

153 **"for those too old and too sickly":** Abbey, *Desert Solitaire*, 54.

153 **"the fossiliferous racket":** David G. Haskell, "The Voices of Birds and the Language of Belonging," *Emergence Magazine*, May 26, 2019.

153 **crash of waves calms heart-surgery patients:** Mohammad Javad Amiri, Tabandeh Sadeghi, and Tayebeh Negahban Bonabi, "The Effect of Natural Sounds on the Anxiety of Patients Undergoing Coronary Artery Bypass Graft Surgery," *Perioperative Medicine* 6, no. 17 (2017).

153 **boosts the cognition of test-takers:** Stephen C. Van Hedger et al., "Of Cricket Chirps and Car Horns: The Effect of Nature Sounds on Cognitive Performance," *Psychonomic Bulletin & Review* 26 (2019): 522–530.

7: LIFE ON THE VERGE

155 **"aerial plankton":** Lopez, *Apologia*.

155 **the windshield phenomenon:** Both the windshield phenomenon and the collapse of insects are described in Brooke Jarvis, "The Insect Apocalypse Is Here," *New York Times Magazine*, November 27, 2018.

155 **insects are going extinct *eight times faster*:** Francisco Sánchez-Bayo and Kris A. G. Wyckhuys, "Worldwide Decline of the Entomofauna: A Review of Its Drivers," *Biological Conservation* 232 (2019): 8–27.

156 **cars kill billions:** James H. Baxter-Gilbert et al., "Road Mortality Potentially Responsible for Billions of Pollinating Insect Deaths Annually," *Journal of Insect Conservation* 19, no. 5 (2015): 1029–1035.

156 **"I drive a Land Rover":** Martin Sorg quoted in Gretchen Vogel, "Where Have All the Insects Gone?" *Science*, May 10, 2017, https://www.science.org/content/article/where-have-all-insects-gone.

156 **among the "last prairie relics":** Margaret A. Kohring, "Saving Michigan's Railroad Strip Prairies," *Ohio Biological Survey, Biological Notes* 15 (1981): 150–151.

159 **"a disturber of the peace":** John Brinckerhoff Jackson, *A Sense of Place, a Sense of Time* (New Haven, CT: Yale University Press, 1994), 190.

159 **Pike and walleye migrate:** Kaitlin Stack Whitney, "How Grizzlies, Monarchs and Even Fish Can Benefit from U.S. Highways," *Ensia*, May 19, 2017, https://ensia.com/features/highways/.

159 **"keep their pelage from becoming matted":** Rachel E. Brock and Douglas A. Kelt, "Influence of Roads on the Endangered Stephens' Kangaroo Rat (*Dipodomys stephensi*): Are Dirt and Gravel Roads Different?" *Biological Conservation* 118 (2004): 633–640.

159   exploiting the "human shield"; "learned that the nearer the railroad": Joel Berger,
      "Fear, Human Shields and the Redistribution of Prey and Predators in Protected Areas,"
      Biological Letters 3, no. 6 (2007): 620–623.

159   covers at least seventeen million acres: Doreen Cubie, "Habitat Highways," *National
      Wildlife* (April–May 2016), March 30, 2016.

159   "All along the cleared roadside": This and subsequent Waugh quotes come from
      Frank A. Waugh, "Ecology of the Roadside," *Landscape Architecture Magazine* 21, no. 2
      (1931): 81–92.

160   "front yard of the nation": J. M. Bennett, *Roadsides: The Front Yard of the Nation* (Boston:
      Stratford, 1936). The influence of Bennett's front-yard paradigm was described in Bonnie
      Harper-Lore, "The Roadside View of a National Weed Strategy," in *Proceedings of the
      California Exotic Pest Plant Council*, Concord, CA, 1997.

160   "Nature . . . cannot produce the desired results alone": Bennet, *Roadsides*, 172, 83, 171, 82.

160   made "driving hazardous": John W. Zukel and C. O. Eddy, "Present Use of Herbicides
      on Highway Areas," *Weeds* 6, no. 1 (1958): 61–63.

160   "mowing, burning, blading, dragging": O. K. Normann, "Weed Control and Eradica-
      tion on Roadsides," *Public Roads* 17, no. 12 (1937): 281–282.

160   "Keep cow, plow, and mower": Leopold, *Sand County Almanac*, 45.

160   "chemical salesman and the eager contractors": Rachel Carson, *Silent Spring*, Forti-
      eth Anniversary Edition (New York: Houghton Mifflin, [1962] 2002), 68–69.

161   downy mallard chicks: Robert B. Oetting, "Right-of-Way Resources of the Prairie
      Provinces," *Blue Jay* 29, no. 4 (1971): 179–183.

161   "The manicured lawn look is out": Charles Bullard, "Counties Look to Prairie
      Grasses, Flowers to Control Roadside Weeds," *Des Moines Register*, June 11, 1989.

161   "long, ribbon-like habitat": Hendrik J. W. Vermeulen, "Corridor Function of a Road
      Verge for Dispersal of Stenotopic Heathland Ground Beetles Carabidae," *Biological Con-
      servation* 69, no. 3 (1994): 339–349.

161   the meadow vole expanded its range: Lowell L. Getz, Frederick R. Cole, and David L.
      Gates, "Interstate Roadsides as Dispersal Routes for *Microtus pennsylvanicus*," *Journal of
      Mammalogy* 59, no. 1 (1978): 208–212.

162   "attracted to the street by waste": Homer Dill, "Is the Automobile Exterminating the
      Woodpecker?" *Science* 63, no. 1620 (1926): 69.

162   "filled with weed seed": Timothy Gollob and Warren M. Pulich, "Lapland Longspur
      Casualties in Texas," *Bulletin of the Texas Ornithological Society* 11, no. 2 (1978): 45.

162   "with several birds being hit": Robert C. Dowler and Gustav A. Swanson, "Highway
      Mortality of Cedar Waxwings Associated with Highway Plantings," *Wilson Bulletin* 94,
      no. 4 (1982): 602.

162   red-winged blackbirds and goldfinches: Gerald L. Roach and Ralph D. Kirkpatrick,
      "Wildlife Use of Roadside Woody Plantings in Indiana," *Transportation Research Record*
      1016 (1985): 11–15.

163 **"so heavy on the branches"**: Sue Halpern, *Four Wings and a Prayer* (London: Weidenfeld & Nicolson, 2001), 7.

164 **"Seventy percent of the butterflies"**: Chip Taylor, interview with the author, March 1, 2021.

164 **thirty million acres of milkweed**: Chip Taylor, "Monarch Population Status," *Monarch Watch Blog*, January 29, 2014, https://monarchwatch.org/blog/2014/01/29/monarch-population-status-20/.

164 **84 percent decline in the monarch population**: Brice X. Semmens et al., "Quasi-Extinction Risk and Population Targets for the Eastern, Migratory Population of Monarch Butterflies (*Danaus plexippus*)," *Scientific Reports* 6, no. 23265 (2016).

165 **450 million stems endured**: John Pleasants, "Milkweed Restoration in the Midwest for Monarch Butterfly Recovery: Estimates of Milkweeds Lost, Milkweeds Remaining and Milkweeds That Must Be Added to Increase the Monarch Population," *Insect Conservation and Diversity* 10, no. 1 (2017): 42–53.

165 **"long and lonesome I-35"**: Larry McMurtry, *Roads: Driving America's Great Highways* (New York: Touchstone, 2000), 23.

165 **foster thousands of acres**: Pollinator Health Task Force, "National Strategy to Promote the Health of Honey Bees and Other Pollinators," The White House, Washington, D.C., 2015.

165 **"protect, plant and manage pollinator habitats"**: "Memorandum of Understanding: Agreement for the Support of a Monarch Highway," MnDOT Contract # 1003329, Minnesota Department of Transportation, May 2016.

166 **"symbolic highway"**: Laura Lukens, interview with the author, April 14, 2020.

168 **"a high likelihood" of providing enough habitat**: "Questions and Answers: 12-Month Finding on a Petition to List the Monarch Butterfly," U.S. Fish and Wildlife Service, December 2020.

170 **"Mrs. Housewife to do a bit of storming"**: "Use of Salt on Roads Discussed," *Newport Daily Express*, January 12, 1948.

170 **"disastrous effect on nylon stockings"**: "State Studying Use of Chemicals and Salt to Combat Ice on Roads," *Rutland Daily Herald*, January 1, 1949.

170 **"bare-pavement concept"**: *Highway Deicing: Comparing Salt and Calcium Magnesium Acetate* (Washington, D.C.: Transportation Research Board, 1991), 18.

170 **cut accidents by nearly 90 percent**: David Kuemmel and Rashad Hanbali, "Accident Analysis of Ice Control Operations," *Transportation Research Center: Accident Analysis of Ice Control Operations*, 1992, https://epublications.marquette.edu/transportation_trc-ice/2/.

170 **"Some say snow fighters"**: Quoted in Brian Clark Howard, "The Surprising History of Road Salt," *National Geographic*, February 14, 2014.

170 **"a large road which the buffaloes have"**: Croghan quoted in Hulbert, *Historic Highways of America*, 116.

171 **Do Not Let Moose Lick Your Car**: Yan Zhuang, "Why Were Canadians Warned Not to Let Moose Lick Their Cars?" *New York Times*, November 23, 2020.

171 **"sweet sensation of joy"**: Elizabeth Bishop, "The Moose," 1980, Poetry Foundation (website), https://www.poetryfoundation.org/poems/48288/the-moose-56d22967e5820.

171 **officials drained the pools**: Paul D. Grosman et al., "Reducing Moose–Vehicle Collisions through Salt Pool Removal and Displacement: An Agent-Based Modeling Approach," *Ecology and Society* 14, no. 2 (2009): 17.

171 **rototilling human and dog hair**: Roy V. Rea et al., "The Effectiveness of Decommissioning Roadside Mineral Licks on Reducing Moose (*Alces alces*) Activity Near Highways: Implications for Moose–Vehicle Collisions," *Canadian Journal of Zoology* 99, no. 12 (2021).

171 **"long-term salinization"**: Hilary A. Dugan et al., "Salting Our Freshwater Lakes," *Proceedings of the National Academy of Sciences USA* 114, no. 17 (2017): 4453–4458.

172 **"increasing nutrients beyond some level"**: Emilie C. Snell-Rood et al., "Anthropogenic Changes in Sodium Affect Neural and Muscle Development in Butterflies," *Proceedings of the National Academy of Sciences USA* 111, no. 28 (2014): 10221–10226.

173 **"have the disconcerting habit"**: May Berenbaum, "Road Worrier," *American Entomologist* 61, no. 1 (2015): 5–8.

174 **"cricket menace"**: "Mormon Crickets Menace Traffic for 300 Miles of DeWitts' Trip," *Lime Springs Herald*, August 10, 1939.

174 **"Sometimes you're the windshield"**: "The Bug," by Mark Knopfler, originally performed by Dire Straits, released 1991, track number 5 on *On Every Street*, Vertigo 510 160-1.

174 **Duane and Katherine McKenna**: Duane McKenna and Katherine McKenna, "Mortality of Lepidoptera along Roadways in Central Illinois," *Journal of the Lepidopterists' Society* 55, no. 2 (2001): 63–68.

174 **killed around two million butterflies**: Tuula Kantola et al., "Spatial Risk Assessment of Eastern Monarch Butterfly Road Mortality during Autumn Migration within the Southern Corridor," *Biological Conservation* 231 (2019): 150–160.

174 **Even the "wind vortices"**: Blanca Xiomara Mora Alvarez, Rogelio Carrera-Treviño, and Keith A. Hobson, "Mortality of Monarch Butterflies (*Danaus plexippus*) at Two Highway Crossing 'Hotspots' during Autumn Migration in Northeast Mexico," *Frontiers in Ecology and Evolution* 7 (2019).

174 **"It's staggering, absolutely staggering"**: Andy Davis, interview with the author, June 29, 2021.

174 **as many as twenty-five million monarchs**: Andy Davis, "The Scariest Paper about Monarchs That I've Ever Read (and Most People Haven't)," *Monarch Science*, September 24, 2015.

175 **The highway's din clearly bothered**: Andrew K. Davis et al., "Effects of Simulated Highway Noise on Heart Rates of Larval Monarch Butterflies, *Danaus plexippus*: Implications for Roadside Habitat Suitability," *Biology Letters* 14, no. 5 (2018).

175 **"gigantic experiment"; "Once again, Dr Davis"**: This exchange occurred on Dplex-L, a popular listserv for monarch discussion, on May 9, 2021.

175   **a national Pollinator Plan:** "All-Ireland Pollinator Plan 2021–2025," National Biodiversity Data Centre, no. 25, March 2021, https://pollinators.ie/.

175   **"road reserves":** Greg Nicolson, "Wildflowers of the N7 Road Reserve: A Walk from Vioosldrift to Capetown," *Veld & Flora* 95, no. 2 (2009).

175   **"responsibility species":** Jan-Olof Helldin, Jörgen Wissman, and Tommy Lennartsson, "Abundance of Red-Listed Species in Infrastructure Habitats: 'Responsibility Species' as a Priority-Setting Tool for Transportation Agencies' Conservation Action," *Nature Conservation* 11 (2015): 143–158.

175   **"largest unofficial nature reserve":** Edward Chell, "Soft Estate: Exhibition Proposal," University for the Creative Arts, Canterbury College, U.K. Research and Innovation (website), https://gtr.ukri.org/projects?ref=AH%2FJ005797%2F1.

176   **monarchs are less productive:** See, for example, Grace M. Pitman, D. T. Tyler Flockhart, and D. Ryan Norris, "Patterns and Causes of Oviposition in Monarch Butterflies: Implications for Milkweed Restoration," *Biological Conservation* 217 (2018): 54–65.

176   **"The reality is that":** Iris Caldwell, interview with the author, April 3, 2020.

176   **mowing roadsides a couple of weeks before:** Samantha M. Knight et al., "Strategic Mowing of Roadside Milkweeds Increases Monarch Butterfly Oviposition," *Global Ecology and Conservation* 19 (2019).

176   **make roadsides *more* attractive:** Leslie Ries, Diane M. Debinski, and Michelle L. Wieland, "Conservation Value of Roadside Prairie Restoration to Butterfly Communities," *Conservation Biology* 15, no. 2 (2001): 401–411.

8: THE NECROBIOME

179   **"we could have lost that shuttle":** Mike Leinbach, interview with the author, July 6, 2021.

179   **"like hungry diners":** John Johnson Jr., "Ready for Launch—If Vultures Keep Away," *Los Angeles Times*, July 1, 2006.

179   **"roadkill posse":** Todd Halvorson, "NASA's Roadkill Plan Accomplishes Mission," *Florida Today*, May 5, 2006.

181   **the *necrobiome*:** Mark Eric Benbow et al., "Seasonal Necrophagous Insect Community Assembly during Vertebrate Carrion Decomposition," *Journal of Medical Entomology* 50, no. 2 (2013): 440–450.

181   **"skeletonized" carcasses:** Rebecca L. Antworth, David A. Pike, and Ernest E. Stevens, "Hit and Run: Effects of Scavenging on Estimates of Roadkilled Vertebrates," *Southeastern Naturalist* 4, no. 4 (2009): 647–656.

181   **"A blackbird was seen to take":** Fred Slater, "An Assessment of Wildlife Road Casualties: The Potential Discrepancy between Numbers Counted and Numbers Killed," *Web Ecology* 3 (2002): 33–42.

181   **"disgusting bird":** Charles Darwin, *A Naturalist's Voyage* (London: John Murray, 1889), 344.

181 **"wonderful quickness and power"**: John James Aubudon, "Black Vulture, or Carrion Crow," in *John J. Audubon's Birds of America*, Audubon.org, https://www.audubon.org/birds-of-america/black-vulture-or-carrion-crow.

182 **"the gentle recyclers of the animal kingdom"**: Katie Fallon, *Vulture: The Private Life of an Unloved Bird* (Lebanon, NH: ForeEdge, 2017), 2.

183 **"new, manufactured thermal corridor"**: Keith Bildstein quoted in T. Edward Nickens, "Vultures Take Over Suburbia," *Audubon Magazine*, November–December 2008.

183 **the region's most-whacked bird**: D. Vidal-Vallés, A. Rodríguez, and E. Pérez-Collazos, "Bird Roadkill Occurrences in Aragon, Spain," *Animal Biodiversity and Conservation* 41, no. 2 (2018): 379–388.

183 **Vultures . . . are poor judges of speed**: Travis L. DeVault et al., "Effects of Vehicle Speed on Flight Initiation by Turkey Vultures: Implications for Bird-Vehicle Collisions," *PLoS ONE* 9, no. 2 (2014).

183 **city slickers learn to cross safely**: Michael J. Evans et al., "Spatial Genetic Patterns Indicate Mechanism and Consequences of Large Carnivore Cohabitation within Development," *Ecology and Evolution* 8, no. 10 (2018): 4815–4829.

183 **"If bears can learn to be careful"**: Mark Ditmer, interview with the author, October 10, 2019.

183 **the animals' pulses rose**: Mark A. Ditmer, "Behavioral and Physiological Responses of American Black Bears to Landscape Features within an Agricultural Region," *EcoSphere* 6, no. 3 (2015): 1–21.

184 **by watching comrades get smacked**: Ronald L. Mumme et al., "Life and Death in the Fast Lane: Demographic Consequences of Road Mortality in the Florida Scrub-Jay," *Conservation Biology* 14, no. 2 (2000): 501–512.

184 **"sensible crow family"**: Roger Tabor, "Earthworms, Crows, Vibrations and Motorways," *New Scientist*, May 23, 1974.

184 **originated near a driving school**: Yoshiaki Nihei and Hiroyoshi Higuchi, "When and Where Did Crows Learn to Use Automobiles as Nutcrackers?" *Tohoku Psychologica Folia* 60 (2001): 93–97.

184 **"arguably the most important carcass consumers"**: Bernd Heinrich, *Life Everlasting: The Animal Way of Death* (Boston: Houghton Mifflin Harcourt, 2012), 70–71.

184 **"Colleagues would rib us"**: Roy Lopez, interview with the author, December 12, 2019.

185 **"lunged upward snapping"**: Teryl G. Grubb, Roy G. Lopez, and Martha M. Ellis, "Winter Scavenging of Ungulate Carrion by Bald Eagles, Common Ravens, and Coyotes in Northern Arizona," *Journal of Raptor Research* 52, no. 4 (2018): 479.

186 **Our "aseptic" approach**: Antoni Margalida and Marcos Moleón, "Toward Carcass-Free Ecosystems?" *Frontiers in Ecology and the Environment* 14, no. 4 (2016): 183–184.

186 **"They're just not very agile"**: Steve Slater, interview with the author, September 7, 2021.

187 **dragging deer just forty feet**: Steven J. Slater, Dustin M. Maloney, and Jessica M. Taylor, "Golden Eagle Use of Winter Roadkill and Response to Vehicles in the Western United States," *Journal of Wildlife Management* 86, no. 6 (2022): e22246.

188 **pled guilty to illegal hunting:** Lyssa Beyer, "Driver Accused of Hitting Deer on Purpose Faces Charges," *Patch*, May 8, 2013.

188 **eighty thousand cows or eight million chickens:** Donald W. Bruckner, "Strict Vegetarianism Is Immoral," in *The Moral Complexities of Eating Meat*, ed. Ben Bramble and Bob Fischer, 30–47 (Oxford: Oxford University Press, 2015).

188 **"Extremely well done":** John McPhee, "Travels in Georgia," *New Yorker*, April 28, 1973.

188 **"Doesn't a person who eats":** Dave Barry, "Just Wait until Oprah Hears about This," syndicated in the *Dispatch*, April 3, 1998.

189 **"I don't know how many times":** Laurie Speakman, interview with the author, September 21, 2021.

190 **"Before salvage was legal":** Jason Day, interview with the author, March 5, 2021.

192 **a noble, timeless pursuit:** For scavenging's role in human evolution, see Marta Zaraska, *Meathooked* (New York: Basic Books, 2016).

9: THE LOST FRONTIER

193 **migratory fish populations have collapsed:** Stefan Lovgren, "Many Freshwater Fish Species Have Declined by 76 Percent in Less Than 50 Years," *National Geographic*, July 27, 2020.

194 **the hollowed-out trunks of gum trees:** "Federal Aid Projects Approved," *Public Roads* 1, no. 2 (1918): 30.

194 **"ran silver":** For the history of anadromous fish, see John Waldman, *Running Silver* (Guilford, CT: Globe Pequot Press, 2013).

194 **"that it is almost impossible to ride through":** Beverley quoted in *Annual Report of the Fish Commissioners of the State of Virginia* (Richmond: Virginia Fish Commission, 1875), 8.

194 **more than ten thousand crossings mutilate:** Cecilia Gontijo Leal, "Amazonian Dirt Roads Are Choking Brazil's Tropical Streams," *Conversation*, March 1, 2018.

194 **impassable culverts are among the reasons:** John Koehn and Pam Clunie, "National Recovery Plan for the Murray Cod *Maccullochella peelii peelii*," National Murray Cod Recovery Team, Victorian Government Department of Sustainability and Environment, Melbourne, October 2010.

195 **spilled onto a wagon track:** J. Edward Norcross, "Random Jottings," *Vancouver Sun*, August 25, 1933.

196 **a "lost frontier":** "The Salmon Crisis," ed. Don Gooding and Les Hatch (Seattle: Washington Department of Fisheries, 1949), 6.

196 **"anywhere a salmon can get to":** Timothy Egan, *The Good Rain* (New York: Vintage Departures, [1990] 1991).

196 **"Really . . . we should have the whole salmon":** Charlene Krise, interview with the author, January 18, 2021.

197 **By day's end the tribes:** For the Treaty of Medicine Creek, see Squaxin Island Tribe (website), "Who We Are," https://squaxinisland.org/government/who-we-are/.

198 **police raided a fishing camp:** For the fish-ins, see Gabriel Chrisman, "The Fish-In

Protests at Franks Landing," Seattle Civil Rights and Labor History Project, University of Washington, 2008, https://depts.washington.edu/civilr/fish-ins.htm.

198 **Native people had the right to fish:** For a history of this ruling, known as the Boldt Decision, see Charles Wilkinson, *Blood Struggle: The Rise of Modern Indian Nations* (New York: W. W. Norton, 2005).

199 **Scientists would eventually pin decades:** Zhenyu Tian et al., "A Ubiquitous Tire Rubber-Derived Chemical Induces Acute Mortality in Coho Salmon," *Science* 371, no. 6526 (2021): 185–189.

199 **produce an additional two hundred thousand adult fish:** Richard Du Bey, Andrew S. Fuller, and Emily Miner, "Tribal Treaty Rights and Natural Resource Protection: The Next Chapter. United States v. Washington—the Culverts Case," *American Indian Law Journal* 7, no. 2 (2019).

199 **"the most recurrent and correctable obstacles":** This passage appeared in a 1997 report jointly prepared by the state's Department of Fish and Wildlife and Department of Transportation. For a discussion of the report, see, for example, Billy Frank Jr., "State's Duty: Fix the Culverts," Northwest Indian Fisheries Commission, December 10, 2000, https://nwifc.org/states-duty-fix-the-culverts/.

200 **"We need to start fixing them right now":** Billy Frank Jr. quoted in Trova Heffernan, *Where the Salmon Run: The Life and Legacy of Billy Frank, Jr.* (Olympia: Washington State Heritage Center, 2012), 259.

201 **"Whatever happens to that salmon":** Testimony of Charlene Krise, *United States of America, et al., v. State of Washington, et al.,* Case No. C70-9213, October 13, 2009.

201 **Samuel Alito grappled with the semantic difference:** *State of Washington,* Petitioner, *v. United States, et al.,* Respondents, No. 17-269, April 18, 2018, https://www.supremecourt.gov/oral_arguments/argument_transcripts/2017/17-269_o75q.pdf.

203 **north of $3 billion:** "2020 State of Salmon in Watersheds: Executive Summary," Governor's Salmon Recovery Office, Washington State Recreation and Conservation Office, Olympia, December 2020.

203 **"The fish don't care":** Braeden van Deynze, interview with the author, April 8, 2021.

207 **"They were literally still hauling":** Dave Harris, Tillamook Estuaries Partnership, interview with the author, August 31, 2020.

207 **just waiting for a few cheap tweaks:** For a comprehensive overview of how to turn culverts and bridges into wildlife crossings, see Julia Kintsch and Patricia Cramer, "Permeability of Existing Structures for Terrestrial Wildlife: A Passage Assessment System," Report to the Washington State Department of Transportation, Research Report No. WA-RD 777.1, July 2011.

208 **"We had no idea whether":** Kerry Foresman, interview with the author, September 22, 2014.

208 **his cameras captured thousands of crossings:** Kerry Foresman, "How Does the Small Mammal Cross the Road?" *Montana Outdoors,* July–August 2006.

210 **"climate shield":** Daniel J. Isaak et al., "The Cold-Water Climate Shield: Delineating

Refugia for Preserving Salmonid Fishes through the 21st Century," *Global Change Biology* 21 (2015): 2540–2553.

210 **"The more culvert work we do":** Harris, interview with the author.

210 **Even the Sierra Club:** Rachel Frazin, "Green Group Proposes Nearly $6T Infrastructure and Clean Energy Stimulus Plan," *The Hill*, May 26, 2020.

10: THE GRACIOUSNESS AT THE HEART OF CREATION

215 **nearly three car-struck ungulates stagger off:** Tracy S. Lee et al., "Developing a Correction Factor to Apply to Animal–Vehicle Collision Data for Improved Road Mitigation Measures," *Wildlife Research* 48, no. 6 (2021): 501–510.

215 **"curled on the ground":** Leath Tonino, "The Doe's Song," *Orion Magazine*, April 13, 2017.

216 **obviated "the wild":** Emma Marris, *Wild Souls* (New York: Bloomsbury, 2021), 11.

216 **birth control for pigeons:** Dylan Matthews, "The Wild Frontier of Animal Welfare," *Vox*, April 21, 2021.

216 **"design[ing] our buildings, roads, and neighborhoods":** Sue Donaldson and Will Kymlicka, *Zoopolis: A Political Theory of Animal Rights* (Oxford: Oxford University Press, 2011), 5–6.

217 **"wouldn't survive a proper cost-benefit analysis":** Quoted in Emily Sohn, "Every Living Thing," *Aeon Magazine*, August 31, 2015.

217 **"mutual isolation":** Daniel Lunney, "Wildlife Roadkill: Illuminating and Overcoming a Blind Spot in Public Perception," *Pacific Conservation Biology* 19 (2013): 236.

217 **a half-million marsupial babies:** Bruce Englefield, Melissa Starling, and Paul McGreevy, "A Review of Roadkill Rescue: Who Cares for the Mental, Physical and Financial Welfare of Australian Wildlife Carers?" *Wildlife Research* 42, no. 2 (2018): 108.

217 **"bloody carpet"; Roadkill Capital of the World; "something [Tasmanians] grow used to":** Donald Knowler, *Riding the Devil's Highway* (Australia: self-published, 2014), Kindle.

218 **"People just don't care":** Glenn Albrecht, interview with the author, June 18, 2019.

222 **"are a national asset that requires strategic nurturing":** For this quote and the statistics in this paragraph, see Englefield, Starling, and McGreevy, "Review of Roadkill Rescue," 109–113.

222 **"species perfectly distinct from":** Quoted in Quammen, *Song of the Dodo*, 281.

222 **"dreadful havoc among the flocks":** Quoted in Brooke Jarvis, "The Obsessive Search for the Tasmanian Tiger, *New Yorker*, June 25, 2018.

223 **"antipodean England":** Sharon Morgan, *Land Settlement in Early Tasmania: Creating an Antipodean England* (Cambridge: Cambridge University Press, 1992).

223 **one animal every 2.7 kilometers:** Alistair J. Hobday and Melinda Minstrell, "Distribution and Abundance of Roadkill on Tasmanian Highways: Human Management Options," *Wildlife Research* 35, no. 7 (2008): 715.

227 **the "lead gift":** Robinson Jeffers, "Hurt Hawks," 1928, Poetry Foundation (website), https://www.poetryfoundation.org/poems/51675/hurt-hawks.

227 **"Animal caregivers are among the most susceptible"**: Charles R. Figley and Robert G. Roop, *Compassion Fatigue in the Animal-Care Community* (Washington, D.C.: Humane Society Press, 2006), 35.

227 **combat depression or hopelessness**: Benjamin Marton, Teresa Kilbane, and Holly Nelson-Becker, "Exploring the Loss and Disenfranchised Grief of Animal Care Workers," *Death Studies* 44 (2020): 31–41.

227 **employees of American dog pounds**: Hope M. Tiesman et al., "Suicide in U.S. Workplaces, 2003–2010: A Comparison with Non-Workplace Suicides," *American Journal of Preventative Medicine* 48, no. 6 (2015): 674–682.

228 **identifying tags, microchips, or collars**: Bruce Englefield et al., "The Demography and Practice of Australians Caring for Native Wildlife and the Psychological, Physical and Financial Effects of Rescue, Rehabilitation and Release of Wildlife on the Welfare of Carers," *Animals* 9, no. 1127 (2019).

228 **"The paradox here"**: Englefield, Starling, and McGreevy, "Review of Roadkill Rescue," 112.

229 **permits to kill 136 wombats**: Englefield, Starling, and McGreevy, "Review of Roadkill Rescue," 112.

229 **"strategy to silence"**: Joe Hinchliffe, "Wild at Heart: Rescuers Oppose Caring Ban Proposal," *Sydney Morning Herald*, May 6, 2018.

229 **"love tunnels"**: Ian M. Mansergh and David J. Scotts, "Habitat Continuity and Social Organization of the Mountain Pygmy Possum Restored by Tunnel," *Journal of Wildlife Management* 53, no. 3 (1989): 701–707.

229 **"We are a lucky country"**: Kylie Soanes, interview with the author, April 17, 2018.

230 **one widely publicized study**: Samantha Fox et al., "Roadkill Mitigation: Trialing Virtual Fence Devices on the West Coast of Tasmania," *Australian Mammalogy* 41, no. 2 (2018): 205–211.

230 **skeptics soon poked holes**: Graeme Coulson and Helena Bender, "Roadkill Mitigation Is Paved with Good Intentions: A Critique of Fox et al. (2019)," *Australian Mammalogy* 42, no. 1 (2019): 122–130.

230 **virtual fences didn't actually reduce roadkill**: Bruce Englefield et al., "A Trial of a Solar-Powered, Cooperative Sensor/Actuator, Opto-Acoustical, Virtual Road-Fence to Mitigate Roadkill in Tasmania, Australia," *Animals* 9, no. 10 (2019): 752.

230 **"Mitigation isn't impossible"**: Josie Stokes, interview with the author, April 23, 2020.

230 **releasing injured turtles and bats**: James E. Paterson, Sue Carstairs, and Christina M. Davy, "Population-Level Effects of Wildlife Rehabilitation and Release Vary with Life-History Strategy," *Journal for Nature Conservation* 61 (2021): 125983.

231 **more than a billion**: Brigit Katz, "More Than One Billion Animals Have Been Killed in Australia's Wildfires, Scientist Estimates," *Smithsonian Magazine*, January 8, 2020.

231 **the *New York Times* ran an apocalyptic photo**: Matthew Abbott as told to Mark Shimabukuro, "In One Photo, Capturing the Devastation of Australia's Fires," *New York Times*, January 9, 2020.

231 **"Because whenever people come together"**: Fred Rogers quoted in Tom Junod, "What Would Mr. Rogers Do?" *Atlantic*, December 2019.

232 **"extending social justice and reparations"**: Heather Pospisil, "Perspectives on Wildlife from the Practice of Wildlife Rehabilitation," master's thesis, California Institute of Integral Studies, San Francisco, 2014, p. 78.

11: SENTINEL ROADS

233 **"He runs a finger down the smooth arc"**: Lopez, *Apologia*.

234 **"the most useful single wildlife observation"**: Amy L. W. Schwartz, Fraser M. Shilling, and Sarah E. Perkins, "The Value of Monitoring Wildlife Roadkill," *European Journal of Wildlife Research* 66, no. 18 (2020).

234 **the Road Deaths Enquiry**: N. L. Hodson and D. W. Snow, "The Road Deaths Enquiry, 1960–61," *Bird Study* 12, no. 2 (1965): 90–99.

234 **roped in road-trippers to count**: "Final Survey of Animal Highway Deaths Planned; Volunteers Still Needed," *News of the Humane Society of the United States* 11, no. 3 (1966): 12.

235 **separates the word "citizen"**: Mary Ellen Hannibal, *Citizen Scientist* (New York: The Experiment, 2016).

236 **"large culvert with side walks"**: E. Jane Ratcliffe, *Through the Badger Gate* (London: G. Bell & Sons, 1974), 104.

236 **"ghost hedgehogs"**: Patrick Barkham, "'Ghost Hedgehogs' on Dorset Roads Highlight Animals' Plight," *Guardian*, September 8, 2020.

237 **"hot moments"**: Frederic Beaudry, Phillip G. Demaynadier, and Malcolm J. Hunter, "Identifying Hot Moments in Road-Mortality Risk for Freshwater Turtles," *Journal of Wildlife Management* 74, no. 1 (2010): 152–159.

237 **Kangaroo kills spike during the full moon**: Jai M. Green-Barber and Julie M. Old, "What Influences Road Mortality Rates of Eastern Grey Kangaroos in a Semi-Rural Area?" *BMC Zoology* 4, no. 11 (2019).

237 **surged after Daylight Saving Time**: W. A. N. U. Abeyrathna and Tom Langen, "Effect of Daylight Saving Time Clock Shifts on White-Tailed Deer-Vehicle Collision Rates," *Journal of Environmental Management* 292 (2021): 112774.

237 **spare the lives of thirty-three drivers**: Calum X. Cunningham et al., "Permanent Daylight Saving Time Would Reduce Deer-Vehicle Collisions," *Current Biology* 32, no. 22 (2022): P4982–4988.

237 **Polecats perish in a "bimodal distribution"**: Schwartz, Shilling, and Perkins, "Value of Monitoring Wildlife Roadkill."

237 **Pheasant roadkill also climbs in winter**: Joah R. Madden and Sarah E. Perkins, "Why Did the Pheasant Cross the Road? Long-Term Road Mortality Patterns in Relation to Management Changes," *Royal Society Open Science* 4, no. 10 (2017).

238 **compared roads to motion-activated trail cameras**: James Baxter-Gilbert, "Turning

the Threat into a Solution: Using Roadways to Survey Cryptic Species and to Identify Locations for Conservation," *Australian Journal of Zoology* 66, no. 1 (2017).

238   **a computational error once changed:** Scott Sprague et al., "Carcass Reporting Arizona Streets and Highways (CRASH): A Digital Toolsuite for Capturing Wildlife Carcass Data," International Conference on Ecology and Transportation (virtual), September 22–29, 2021.

238   **a roadkill hotspot at Tim Hortons:** Stephen Legaree, interview with the author, October 5, 2021.

239   **volunteers had accurately identified 97 percent:** David P. Waetjen and Fraser M. Shilling, "Large Extent Volunteer Roadkill and Wildlife Observation Systems as Sources of Reliable Data," *Frontiers in Ecology and Evolution* 5, no. 89 (2017).

239   **"in a reliable and robust manner":** Stéphanie Périquet et al., "Testing the Value of Citizen Science for Roadkill Studies: A Case Study from South Africa," *Frontiers in Ecology and Evolution* 6, no. 15 (2018).

240   **alarming concentrations of long-banned pesticides:** Samantha K. Carpenter et al., "River Otters as Biomonitors for Organochlorine Pesticides, PCBs, and PBDEs in Illinois," *Ecotoxicology and Environmental Safety* 100 (2014): 99–104.

240   **the frightening tenacity of toxic flame retardants:** Angela Pountney et al., "High Liver Content of Polybrominated Diphenyl Ether (PBDE) in Otters (*Lutra lutra*) from England and Wales," *Chemosphere* 118 (2015): 81–86.

240   **tougher laws have reduced lead pollution:** Elizabeth A. Chadwick et al., "Lead Levels in Eurasian Otters Decline with Time and Reveal Interactions between Sources, Prevailing Weather, and Stream Chemistry," *Environmental Science and Technology* 45, no. 5 (2011): 1911–1916.

240   **invasive pythons had plundered Everglades National Park:** Michael E. Dorcas et al., "Severe Mammal Declines Coincide with Proliferation of Invasive Burmese Pythons in Everglades National Park," *Proceedings of the National Academy of Sciences USA* 109, no. 7 (2012): 2418–2422.

240   **in Washington a dead wolf:** Tom Sowa, "Gray Wolves Are State Residents," *Spokesman-Review*, July 18, 2008.

240   **The changing seasonality of green whip snake roadkill:** Massimo Capula et al., "Long-Term, Climate Change-Related Shifts in Monthly Patterns of Roadkilled Mediterranean Snakes (*Hierophis viridifavus*)," *Herpetological Journal* 24, no. 6 (2014): 97–102.

241   **leprosy in nine-banded armadillos:** Heather L. Montgomery, *Something Rotten* (New York: Bloomsbury Children's Book, [2018] 2019), 95.

241   **"Be sure to cook your 'dillo thoroughly":** Chad Garrison, "Leprosy: Another Reason to Avoid Those Armadillos Invading from the South," *Riverfront Times*, May 2, 2011.

241   **rediscovered the pygmy blue-tongue lizard:** G. Armstrong and J. Reid, "The Rediscovery of the Adelaide Pygmy Bluetongue *Tiliqua adelaidensis* (Peters, 1863)," *Herpetofauna* 22 (1992): 3–6.

241   **the crew spotted a nightjar carcass:** R. J. Safford et al., "A New Species of Nightjar from Ethiopia," *Ibis* 137, no. 3 (1995): 301.

241 **Around a quarter of British otters:** Willow Smallbone et al., "East-West Divide: Temperature and Land Cover Drive Spatial Variation of *Toxoplasma gondii* Infection in Eurasian Otters (*Lutra lutra*) from England and Wales," *Parasitology* 144, no. 11 (2017): 1433–1440.

241 **Car-killed pademelons are nearly three times likelier:** Tracey Hollings et al., "Wildlife Disease Ecology in Changing Landscapes: Mesopredator Release and Toxoplasmosis," *International Journal for Parasitology: Parasites and Wildlife* 2 (2013): 110–118.

242 **they registered more than two hundred:** Fernanda Zimmermann Teixeira et al., "Vertebrate Road Mortality Estimates: Effects of Sampling Methods and Carcass Removal," *Biological Conservation* 157 (2013): 317–323.

242 **"monitor pond-breeding amphibians":** Fernando Ascensão, Cristina Branquinho, and Eloy Revilla, "Cars as a Tool for Monitoring and Protecting Biodiversity," *Nature Electronics* 3 (2020): 295.

242 **"The End of Roadkill":** Malia Wollan, "The End of Roadkill," *New York Times Magazine*, November 8, 2017.

242 **"put a lot of effort in seeing how animals moved":** Quoted in Eric Adams, "Volvo's Cars Now Spot Moose and Hit the Brakes for You," *Wired Magazine*, January 27, 2017.

243 **"When it's in the air":** Quoted in Naaman Zhou, "Volvo Admits Its Self-Driving Cars Are Confused by Kangaroos," *Guardian*, June 30, 2017.

243 **"down to the size of a small dog":** Ponz Pandikuthira, email communication with the author, November 12, 2020.

243 **an Italian physicist described a rule:** Cesare Marchetti, "Anthropological Invariants in Travel Behavior," *Technological Forecasting and Social Change* 47 (1994): 75–88.

244 **"A fractal web of farms, factories, and towns":** Anthony Townsend, *Ghost Road: Beyond the Driverless Car* (New York: W. W. Norton, 2020), 206.

244 **"but only if conservation biologists play a role":** Amanda C. Niehaus and Robbie S. Wilson, "Integrating Conservation Biology into the Development of Automated Vehicle Technology to Reduce Animal–Vehicle Collisions," *Conservation Letters* 11, no. 3 (2018).

245 **the system captured nearly 80 percent:** Diana Sousa Guedes et al., "An Improved Mobile Mapping System to Detect Road-Killed Amphibians and Small Birds," *ISPRS International Journal of Geo-Information* 8, no. 12 (2019).

248 **the citizen study both confirmed:** Kim Rondeau and Tracy Lee, "Collision Count: Improving Human and Wildlife Safety on Highway 3," Miistakis Institute, Calgary, Alberta, April 30, 2018.

248 **"It's a really powerful shift":** Tracy Lee, interview with the author, May 8, 2020.

248 **"a powerful act of reciprocity":** Kimmerer, *Braiding Sweetgrass*, 252.

12: THE TSUNAMI

250 **"We think we're such leaders":** Rob Ament, interview with the author, November 12, 2019.

251   "**the most explosive era of infrastructure expansion**": William F. Laurance et al., "Reducing the Global Environmental Impacts of Rapid Infrastructure Expansion," *Current Biology* 25, no. 7 (2015): R259–R262.

251   **the word for development is *thara'gee***: Luke Heslop, "Runways to the Sky," *Allegra Lab*, March 2020.

251   **so many cheetahs have been hit by cars**: John Lee Anderson, "A Kenyan Ecologist's Crusade to Save Her Country's Wildlife," *New Yorker*, January 25, 2021.

251   **threaten to splinter tiger habitat**: Neil Carter et al., "Road Development in Asia: Assessing the Range-Wide Risks to Tigers," *Science Advances* 6, no. 18 (2020).

251   **"No Planet for Apes"**: Fernando Ascensão, Marcello D'Amico, and Rafael Barrientos, "No Planet for Apes? Assessing Global Priority Areas and Species Affected by Linear Infrastructures," *International Journal of Primatology* 43 (2022): 57–73.

252   **"cause the [additional] loss"**: Alex Bager, Carlos E. Borghi, and Helio Secco, "The Influence of Economics, Politics, and Environment on Road Ecology in South America," in van der Ree, Smith, and Grillo, eds., *Handbook of Road Ecology*, 408.

252   **"The expansion of roads in Brazil"**: Clara Grilo, interview with the author, July 31, 2019.

255   **how to "correct" its soil with lime**: Jonathan Mingle, "The Slow Death of Ecology's Birthplace," *Undark*, December 16, 2016.

256   **"Human Death Caused by a Giant Anteater"**: Vidal Haddad Jr. et al., "Human Death Caused by a Giant Anteater (*Myrmecophaga tridactyla*) in Brazil," *Wilderness and Environmental Medicine* 25, no. 4 (2014): 446–449.

257   **More than 80 percent of anteaters**: Michael J. Noonan et al., "Roads as Ecological Traps for Giant Anteaters," *Animal Conservation* 25, no. 2 (2022): 182–194.

258   **habitat fragmentation may pose a graver threat**: Fernando A. S. Pinto, "Giant Anteater (*Myrmecophaga tridactyla*) Conservation in Brazil: Analysing the Relative Effects of Fragmentation and Mortality Due to Roads," *Biological Conservation* 228 (2018): 148–157.

259   **$27 billion in highway contracts**: Sue Branford, "Amazon at Risk: Brazil Plans Rapid Road and Rail Infrastructure Expansion," *Mongabay*, February 12, 2019.

259   **"We can't turn Amazonas state"**: Quoted in John Otis, "A Brazilian Road Project Cuts through the Amazon, Paving the Way to Vast Deforestation," *Weekend Edition Saturday*, National Public Radio, October 30, 2022.

260   **"the most corrupt highway in the world"**: "Peru's Interoceanic: The Most Corrupt Highway in the World," Upper Amazon Conservancy, July 2018, https://www.upperamazon.org/news/fz8vmhz9ytujotmqksrxoex5rljfcv-mnsk7.

260   **"lines of penetration"**: Edward J. Taaffe, Richard L. Morrill, and Peter R. Gould, "Transport Expansion in Underdeveloped Countries: A Comparative Analysis," *Geographical Review* 53, no. 4 (1963): 503–529.

261   **"surplus population"; "a land without people"; *Ocupar para não entregar***: Andrew Revkin, *The Burning Season* (Washington, D.C.: Island Press, [1990] 2004), 103, 112, 104.

261 **floundered in its "green hell":** Robert G. Hummerstone, "Cutting a Road through Brazil's 'Green Hell,'" *New York Times*, March 5, 1972.

261 **"illusion of endlessness":** Philip M. Fearnside, *Human Carrying Capacity of the Brazilian Rainforest* (New York: Columbia University Press, 1986), 6.

261 **whether they cause deforestation or follow in its wake:** Arild Angelsen and David Kaimowitz, "Rethinking the Causes of Deforestation: Lessons from Economic Models," *World Bank Research Observer* 14, no. 1 (1999): 73–98.

262 **"a spreading grid of open space":** Revkin, *Burning Season*, 101.

262 **the skeleton of a monstrous fish:** For a description of "fishbone" deforestation, see Francisco José Barbosa Oliveira de Filho and Jean Paul Metzger, "Thresholds in Landscape Structure for Three Common Deforestation Patterns in the Brazilian Amazon," *Landscape Ecology* 21, no. 7 (2006): 1061–1073.

262 **a fatal feedback loop:** William F. Laurance et al., "Ecosystem Decay of Amazonian Forest Fragments: A 22-Year Investigation," *Conservation Biology* 16, no. 3 (2002): 605–618.

262 **One thing antbirds don't do:** Susan G. Laurance, Philip C. Stouffer, and William F. Laurance, "Effects of Road Clearings on Movement Patterns of Understory Rainforest Birds in Central Amazonia," *Conservation Biology* 18, no. 4 (2004): 1099–1109.

262 **nearly 140 new bird species:** Cameron L. Rutt, "Avian Ecological Succession in the Amazon: A Long-Term Case Study Following Experimental Deforestation," *Ecology and Evolution* 9, no. 24 (2019): 13850–13861.

263 **spread five hundred miles along one Brazilian highway:** Nigel J. H. Smith, "House Sparrows (*Passer domesticus*) in the Amazon," *Condor* 75, no. 2 (1973): 242–243.

263 **hantavirus sprang from soybean farms:** Daniele B. A. Medeiros, "Circulation of Hantaviruses in the Influence Area of the Cuiabá-Santarém Highway," *Memórias do Instituto Oswaldo Cruz* 105, no. 5 (2010): 665–671.

263 **"landscape immunity":** Raina K. Plowright et al., "Land-Use Induced Spillover: A Call to Action to Safeguard Environmental, Animal, and Human Health," *Lancet* 5, no. 4 (2021): e237–e245.

263 **"If you want to prevent pandemics":** Gary Tabor, interview with the author, April 5, 2021.

263 **The Chinese counties that suffered:** Li-Qun Fang et al., "Geographical Spread of SARS in Mainland China," *Tropical Medicine & International Health* 14 (2020): 14–20.

263 **"the transformation of a thread of dirt":** Richard Preston, *The Hot Zone* (New York: Anchor, 1995), 383.

264 **"Once again anger invaded my thought":** Davi Kopenawa and Bruce Albert, *The Falling Sky: Words of a Yanomami Shaman* (Cambridge, MA: Belknap Press of Harvard University Press, 2013), 235.

264 **linguistic diversity is lowest:** J. R. Stepp et al., "Biocultural Diversity: Roads and Languages, Tropical South America," Biocultural Diversity Mapping Project, University of Florida, http://www.bioculturaldiversity.org/.

264 **"When the road is open":** Rafael Mendoza, interview with the author, March 2, 2020.

264 **Ninety-five percent of deforestation:** Christopher P. Barber et al., "Roads, Deforestation, and the Mitigating Effect of Protected Areas in the Amazon," *Biological Conservation* 177 (2014): 203–209.

265 **The territories of sixty-three Indigenous tribes:** Lucas Ferrante, Mércio Gomes, and Philip Martin Fearnside, "Amazonian Indigenous Peoples Are Threatened by Brazil's Highway BR-319," *Land Use Policy* 94, no. 2 (2020): 104548.

265 **"malicious restoration":** Lauren J. Moore et al., "On the Road without a Map: Why We Need an 'Ethic of Road Ecology,'" *Frontiers in Ecology and Evolution* 16 (2021): 774286.

265 **"Passages won't make any difference":** Fernanda Zimmerman, interview with the author, September 24, 2019.

265 **edge habitats make up 70 percent:** Nick M. Haddad, "Habitat Fragmentation and Its Lasting Impact on Earth's Ecosystems," *Science Advances* 1, no. 2 (2015).

265 **"The best thing you could do for the Amazon":** Quoted in William Laurance, "Roads Are Ruining the Rainforests," *New Scientist*, August 26, 2009.

268 **invested $66 billion in Brazil:** Chris Devonshire-Ellis, "Brazil: South America's Largest Recipient of BRI Infrastructure Financing and Projects," *New Silk Road Briefing*, November 8, 2021.

268 **Some 80 percent of Brazil's soy exports:** Sue Branford and Mauricio Torres, "How Chinese Interests—and Money—Have Revived Brazil's Ambitious Amazon Rail Network," *Pacific Standard*, December 28, 2018.

268 **"vaguely visible hand":** Ben Mauk, "Can China Turn the Middle of Nowhere into the Center of the World Economy?" *New York Times Magazine*, January 29, 2019.

269 **More than 250 threatened species:** Divya Narain et al., "Best-Practice Biodiversity Safeguards for Belt and Road Initiative's Financiers," *Nature Sustainability* 3 (2020): 650–657.

269 **"There's not much you can do":** Peter Kibobi, interview with the author, April 8, 2021.

269 **"new version of colonialism":** Quoted in Evan Osnos, "The Future of America's Contest with China," *New Yorker*, January 6, 2020.

269 **"The bottom line is that developing nations":** William Laurance, interview with the author, September 3, 2020.

269 **even serve as a "magnet":** William F. Laurance, "A Global Strategy for Road Building," *Nature* 513, no. 7521 (2014): 229.

269 **planned rather than spontaneous:** See, for example, William Laurance, "If You Can't Build Well, Then Build Nothing at All," *Nature* 563 (2018): 295.

270 **funded by Chinese state banks:** Narain et al., "Best-Practice Biodiversity Safeguards."

270 **"appear to have little influence":** Rodney van der Ree, Daniel J. Smith, and Claro Grilo, "The Ecological Effects of Linear Infrastructure and Traffic: Challenges and Opportunities of Rapid Global Growth," in van der Ree, Smith, and Grilo, eds., *Handbook of Road Ecology*, 7.

272 **animal collisions continue apace:** See, for example, Corinna Riginos et al., "Reduced

Speed Limit Is Ineffective for Mitigating the Effects of Roads on Ungulates," *Conservation Science and Practice* 4, no. 3 (2022): e618.

272 **"The virtue of slowness":** Gary Kroll, "Snarge," *Aeon*, March 28, 2018.

13: REPARATIONS

274 **"[The engineer] does not hesitate to lay waste":** Lewis Mumford quoted in Joseph F. C. DiMento and Cliff Ellis, *Changing Lanes: Visions and Histories of Urban Freeways* (Cambridge, MA: MIT Press, 2013), 2.

274 **some 3,600 people around the world:** Road crashes are responsible for around 1.3 million deaths annually. See Etienne Krug, "It's Time to End Deaths on Our Roads," World Health Organization, June 28, 2022, https://www.who.int/news-room/commentaries/detail/it-s-time-to-end-deaths-on-our-roads.

274 **prone to committing domestic violence:** Louise-Philippe Beland and Daniel A. Brent, "Traffic and Crime," *Journal of Public Economics* 160 (2018): 96–116.

275 **"infrastructure species":** Jedediah Britton-Purdy, "The World We've Built," *Dissent Magazine*, July 3, 2018.

275 **40 percent more likely:** Centers for Disease Control and Prevention, "CDC Health Disparities and Inequalities Report," *Morbidity and Mortality Weekly Report* 62, no. 3 (2013).

275 **higher exposure to airborne carcinogens:** Benjamin J. Apelberg, Timothy J. Buckley, and Ronald H. White, "Socioeconomic and Racial Disparities in Cancer Risk from Air Toxics in Maryland," *Environmental Health Perspectives* 113, no. 6 (2005): 693–699.

275 **three times more likely to die of asthma:** "Asthma Prevention and Management in Bronx, New York and New York State at Large," prepared for the office of New York State Senator Jeffrey D. Klein, July 2011, https://legacy-assets.eenews.net/open_files/assets/2020/04/15/document_cw_03.pdf.

275 **"the same ethical terrain":** David Roberts, "Air Pollution Is Much Worse Than We Thought," *Vox*, August 12, 2020.

275 **"the elimination of unsightly and unsanitary districts":** Richard Rothstein, *The Color of Law* (New York: Liveright, 2017), 127.

275 **displace "undesirable slum areas":** Quoted in Megan Kimble, "If They Can Tear Down This Highway in Texas . . . Yes, Texas!" *Nation*, July 2021.

276 **the Harlem of the South:** For freeways' impacts on communities of color, see Eric Avila, *The Folklore of the Freeway* (Minneapolis: University of Minnesota Press, 2014).

276 **"the boundary between the white and Negro communities":** Kevin M. Kruse, "What Does a Traffic Jam in Atlanta Have to Do with Segregation? Quite a Lot," *New York Times Magazine*, August 14, 2019.

276 **"foremost segregation leader":** Rebecca Retzlaff, "Interstate Highways and the Civil Rights Movement: The Case of I-85 and the Oak Park Neighborhood in Montgomery, Alabama," *Journal of Urban Affairs* 41, no. 7 (2019): 934.

276 **"You hear machines tearing down"**: John A. Williams, "Portrait of a City: Syracuse, the Old Home Town," *Courier* 28, no. 1 (1993): 73.

276 **"boondoggle of concrete and steel"**: Quoted in "Revisiting and Revisualizing Syracuse's 15th Ward," Visualizing 81 (website), S. I. Newhouse School of Public Communications, Syracuse University, 2021, https://visualizing81.thenewshouse.com.

277 **"great central depot"**: Kim M. Williamson, "Explore the Underground Railroad's 'Great Central Depot,'" *National Geographic*, February 27, 2019.

277 **the Home Owners' Loan Corporation**: For a history of redlining and urban freeways, see Rothstein, *Color of Law*, 127–131.

278 **"population almost entirely Negro"**: Quoted in Dick Case, "Roosevelt's 'Rainbow' Held No Pot of Gold," *Syracuse Herald American*, January 9, 2000.

278 **"Literally thousands of roaches"**: Walter Carroll, "Block by Block Cleanup Proposed in Slum Area," *Syracuse Post-Standard*, January 22, 1954.

278 **the highway's path was settled**: David Haas, "I-81 Highway Robbery: The Razing of Syracuse's 15th Ward," *Syracuse New Times*, December 12, 2018.

278 **"direct intervention of our physical selves"**: Quoted in "Revisiting and Revisualizing Syracuse's 15th Ward."

278 **"cement octopus"**: The phrase "cement octopus" was popularized by Malvina Reynolds in her song of the same name (Schroder Music Company, 1964). The octopus "gets red tape to eat, gasoline taxes to drink."

279 **"It would be all right with me"**: Quoted in Joseph DiMento, "Stent (or Dagger?) in the Heart of Town: Urban Freeways in Syracuse, 1944–1967," *Journal of Planning History* 8, no. 2 (2009): 151.

280 **"Planing over the land at tree-top level"**: Lawrence Halprin quoted in DiMento and Ellis, *Changing Lanes*, 113.

282 **"the Champs-Elysees of Central New York"**: Quoted in Jeff Kramer, "81 Redesign: Road to Revitalization, or Gentrification?" *South Side Stand*, May 8, 2019.

282 **"learn from our past"**: Quoted in "Secretary Buttigieg Visits I-81 in Syracuse and Pushes for Passage of American Jobs Plan," *CNY Central*, June 28, 2021.

283 **birth up to eighteen acres**: Lanessa Owens-Chaplin, Johanna Miller, and Simon McCormack, "Building a Better Future: The Structural Racism Built into I-81, and How to Tear It Down," New York Civil Liberties Union, December 2, 2020, 29, https://www.nyclu.org/en/publications/building-better-future.

283 **"Negro Removal 2.0"**: Ken Jackson, "Negro Removal 2.0," *Urban CNY*, March 3, 2019.

283 **"The pavement started to move"**: Quoted in "200 Feared Dead in Freeway Collapse," *Petaluma Argus-Courier* via Associated Press, October 18, 1989.

284 **"We now have sunshine"**: Quoted in Kristin Bender, "Mandela Parkway Unveiled," *East Bay Times*, July 13, 2005.

284 **thirty prospective freeway teardowns**: Ben Crowther, "Freeways without Futures," Congress for the New Urbanism (website), 2021, https://www.cnu.org/highways-boulevards/freeways-without-futures/2021.

284 **fell by nearly 30 percent:** For this and preceding air-quality improvements, see Regan F. Patterson and Robert A. Harley, "Effects of Freeway Rerouting and Boulevard Replacement on Air Pollution Exposure and Neighborhood Attributes," *International Journal of Environmental Research and Public Health* 16, no. 21 (2019): 4072.

285 **training and apprenticeship programs; give it to a land trust:** Owens-Chaplin, Miller, and McCormack, "Building a Better Future," 29–31.

285 **"immense wealth [had been]":** Quoted in Ta-Nehisi Coates, "The Case for Reparations," *Atlantic*, May 21, 2014.

286 **"generational loss of wealth accumulation":** Owens-Chaplin, Miller, and McCormack, "Building a Better Future," 10.

286 **$157 million in home equity:** "Restorative Rondo: Building Equity for All: Past Prosperity Study," prepared for Reconnect Rondo by the Yorth Group, July 2020, 6, https://docslib.org/doc/3330659/rondo-past-prosperity-study.

287 **"racism [was] physically built into some of our highways":** Quoted in April Ryan, "Buttigieg Says Racism Built into US Infrastructure Was a 'Conscious Choice,'" *Grio*, April 6, 2021.

287 **"To me . . . a road's a road":** Quoted in Renzo Downey, "'A Road's a Road': Ron DeSantis Sideswipes Pete Buttigieg for Addressing Racist Highway Design," Florida Politics (website), November 9, 2021, https://floridapolitics.com/archives/471820-desantis-buttigieg-racist-roads/.

288 **"can't carry the same amount of traffic":** "How Cars Killed Syracuse," InTheSalt.City (website), June 21, 2021, https://inthesalt.city/2021/06/21/howcarskilledsyracuse/.

288 **"All these people walking":** Robert Haley, interview with the author, September 23, 2021.

288 **"Damn if it helps the minority community":** Vernon Williams, interview with the author, September 23, 2021.

EPILOGUE: THE ANTHROPAUSE

291 **the Great Acceleration:** Will Steffen et al., "The Trajectory of the Anthropocene: The Great Acceleration," *Anthropocene Review* 2, no. 1 (2015).

292 **dubbed it the Anthropause:** Christian Rutz et al., "COVID-19 Lockdown Allows Researchers to Quantify the Effects of Human Activity on Wildlife," *Nature Ecology & Evolution* 4 (2020): 1156–1159.

292 **In Costa Rica ocelot roadkill:** Esther Pomareda-García, communication with the author, July 1, 2020.

292 **twice as many frogs:** Gregory LeClair et al., "Influence of the COVID-19 Pandemic on Amphibian Road Mortality," *Conservation Science and Practice* 3, no. 11 (2021).

292 **British hedgehog deaths fell:** Lauren Moore, "Reports of U.K. Roadkill Down Two-Thirds—But Will Hedgehogs Thrive after Lockdown?" *Conversation*, May 12, 2020.

292 **analyzed carcass cleanup statistics:** Fraser Shilling et al., "Special Report 4: Impact of COVID-19 Mitigation on Wildlife-Vehicle Conflict," University of California, Davis,

Road Ecology Center, June 24, 2020, https://roadecology.ucdavis.edu/resource-type/report.

292 **"This is the biggest conservation action"**: Fraser Shilling, interview with the author, June 29, 2020.

293 **"created a proverbial silent spring"**: Elizabeth P. Derryberry et al., "Singing in a Silent Spring: Birds Respond to a Half-Century Soundscape Reversion during the COVID-19 Shutdown," *Science* 370, no. 6516 (2020): 575–579.

293 **"sound of the city aching"**: Quoted in Quoctrung Bui and Emily Badger, "The Coronavirus Quieted City Noise: Listen to What's Left," *New York Times*, May 22, 2020.

294 **"behavioral lags"**: Joel O. Abraham and Matthew A. Mumma, "Elevated Wildlife-Vehicle Collision Rates during the COVID-19 Pandemic," *Nature Scientific Reports* 11, no. 20391 (2021).

294 **"transit death spiral"**: Pranshu Verma, "Public Transit Officials Fear Virus Could Send Systems into 'Death Spiral,'" *New York Times*, July 19, 2020.

294 **"It's an infinite desire"**: Jochen Jaeger, interview with the author, April 6, 2020.

294 **results of its investigation into how Eastern Collier Property Owners**: Jimmy Tobias, "Defanged," *Intercept*, January 24, 2021.

295 **that framing deflects culpability**: David Zipper, "The Deadly Myth That Human Error Causes Most Car Crashes," *Atlantic*, November 26, 2021.

295 **"a pressure packed alliance"**: Quoted in Oliver A. Hauck, "The Vieux Carre Expressway," *Tulane Environmental Law Journal* 30, no. 1 (2016): 22.

295 **"irredeemably violent"**: Ross Andersen, "What the Crow Knows," *Atlantic*, March 2019.

295 **two billion motorized vehicles**: William Laurance, "Curbing an Onslaught of Two Billion Cars," *bioGraphic*, June 14, 2016.

296 **humanity's Great Work**: Thomas Berry, *The Great Work* (New York: Belltower, 1999), 7.

296 **"A thing is right"**: Leopold, *Sand County Almanac*, 224.

296 **an analogous *road* ethic**: For a much fuller exploration of what a road-ecology ethic might entail, see Moore et al., "On the Road without a Map."

297 **"decimal dust"**: Cramer, interview with the author.

297 **"We can't treat every mile"**: Rob Ament, interview with the author, November 8, 2021.

297 **"innovative technologies"**: Infrastructure Investment and Jobs Act, Sec. 11123, Wildlife Crossing Safety, signed into law November 15, 2021.

297 **reduce roadkill by about 50 percent**: Rytwinski, "How Effective Is Road Mitigation?"

297 **fiber-reinforced polymers**: For more on fiber-reinforced polymer crossings, see Kylie Mohr, "Wildlife Crossing Innovation," *Western Confluence*, September 27, 2021.

INDEX